Visual Prosthetics

Gislin Dagnelie
Editor

Visual Prosthetics

Physiology, Bioengineering, Rehabilitation

 Springer

Editor
Gislin Dagnelie
Lions Vision Research & Rehabilitation Center
Johns Hopkins University School of Medicine
550 N. Broadway, 6th floor
Baltimore, MD 21205-2020
USA
gdagnelie@jhmi.edu

ISBN 978-1-4899-8746-4   ISBN 978-1-4419-0754-7 (eBook)
DOI 10.1007/978-1-4419-0754-7
Springer New York Dordrecht Heidelberg London

Springer is part of Springer Science+Business Media (www.springer.com)

Preface

Visual Prosthetics as a Multidisciplinary Challenge

This is a book about the quest to realize a dream: the dream of restoring sight to the blind. A dream that may have been with humanity much longer than the idea that disabilities can be treated through technology – which itself is probably a very old idea. Long ago, when blindness was still considered a curse from the gods, some-one must have had the inspiration of building a wooden leg to replace one that had been crushed in a natural calamity or in battle. Many centuries lie between the concept of creating such a crude prosthesis to treat disability and today's endeavors to replace increasingly complex bodily functions, but the wish to restore useful function and the researchers' creative spirit remain the same.

Around 1980, the developers of the cochlear implant were performing the first modest clinical trials of a technology to make the deaf hear again, or even hear for the first time. From those humble first attempts sprang a field that has become a model for modern neuroprosthetics, with tens of thousands of cochlear implants used successfully around the world. The development of the cochlear prosthesis illustrates the importance of bringing together professionals from a wide range of disciplines, from basic biology and engineering to rehabilitation, to create a functional substitute for a human sensory organ.

In 1995, the editor of *IEEE Spectrum* magazine determined that artificial vision might be the next technological frontier, and that it should be the topic of a special issue. He invited a half dozen vision researchers to contribute articles about their expectations in two areas, visual prosthetics and machine vision, combined under the title "Towards an Artificial Eye." He instructed the authors not to feel con-strained by existing technology, but rather to envision the steps that would be required to replace natural vision. Most of the ideas presented in that May 1996 issue have not yet been realized, especially those for prosthetic vision. Machine vision has made larger strides, which just goes to show that biology is more stub-born than technology - but also more resourceful, as machine vision researchers realize on a daily basis: Segmenting and recognition tasks that our visual system performs effortlessly can pose formidable problems for a computer-based image analysis system. Yet, encouragingly, some visual prosthesis designs predicted in that 1996 magazine are now being tested in clinical trials.

This is an exciting time for the field of visual prosthetics. Obviously, it is exciting for the hope it brings that vision can be restored. It is exciting for its challenge to researchers, technicians, clinicians, rehabilitation workers, and people in many other fields to commit their talents to the solution of a problem with so many dimensions. It is exciting for the experimenters when, seemingly against all odds, a blind study participant with a few dozen electrodes on the retina recognizes an object or letter "E" and finds a path around traffic cones in the lab without a cane or guidance. It is exciting for the participants in these trials, who feel they can play an active role in realizing the dream. It is exciting for their loved ones and the public at large, for whom the developments can't come quickly enough. And it is, unfortunately, too exciting for some media types who can't stop themselves from running ahead of the facts.

This is also a field of setbacks, as when the new electrode coating that was supposed to withstand conditions inside the body for 20 years starts peeling off during its initial high-temperature soak test; of unpleasant surprises, as when the simple idea of putting together many small phosphenes to create an image runs up against the reality that phosphenes overlap and blur the image beyond recognition; and of patience put to the test, as when investors and the public do not get the miracle cure they may have been expecting.

But mostly this is a field of great dedication by hundreds of researchers in dozens of labs in countries on four continents; of amazing tenacity by study participants learning to make sense of a way of seeing that is so different from the vision they lost; and of true collegial spirit among all who share the dream, despite the realities of commercial interest. This collegial spirit was evident even in the days of the *IEEE Spectrum* issue: Throughout the 1990s, the National Institute of Neurological Disorders and Stroke sponsored an annual neural prosthesis workshop that was attended by all researchers competing for the scarce development funds then available for neuroprosthetics. Although the competition could be fierce, the annual workshop attendees formed a community that collectively solved stubborn problems of interfacing technology and biology, and attracted many new and talented researchers to the field. Looking back, I feel that these workshops had a limitation: They were, by the nature of the research contracts given out, strongly geared towards technology, and less towards integration with physiology or rehabilitation. This was inherent in NINDS's mission to foster development of devices with broad application, but non-engineers were less likely to attend these highly technical gatherings.

In the year 2000, Dr. Philip Hesburg at the Detroit Institute of Ophthalmology had the inspiration to foster a new collaboration among visual prosthesis researchers, clinicians, and workers in low vision rehabilitation by creating and sponsoring a series of biennial meetings that he calls "The Eye and the Chip." Successful beyond Dr. Hesburg's expectations, these meetings have become the premier gathering place for researchers from all parts of the world and from very different backgrounds. Invited speakers are scientists who are advancing the field, yet the scale and atmosphere allow all researchers, patients, and the media to come and be updated about progress over the past 2 years. More perhaps than at other scientific meetings, where investigators tend to gather within disciplines, participants at The Eye and the Chip are challenged to be open-minded, learn about and critique each other's work, and return home with fresh ideas for interdisciplinary approaches. The interdisciplinary character of this book reflects that same spirit.

This book is also a reality check, an assessment of where we stand in 2010, almost 50 years after G.S. Brindley put the first revolutionary electrode assemblies under a blind patient's skull, yet in a field that is still very young. And this book is an introduction for people outside the field who may want to join the quest, or just be better informed. The book is unusual in being aimed at a readership as diverse as the disciplines contributing to the field: basic scientists, tissue and biomedical engineers, clinical researchers, and rehabilitation specialists.

Most of all, this book is a tribute to the visionaries, the inventors, the creators of devices, the biomedical engineers, the surgeons and medical staff, the research psychophysicists, the occupational therapists, and the patient pioneers and their loved ones. In the chapters that follow, a few dozen workers in the field present their work and that of many colleagues. Each of their accounts conveys a passion for this multidisciplinary journey of discovery, a sense of urgency, a precise and meticulous effort to get it right and to learn – from the damaged visual system and from study participants – how to further improve the technology.

If the reader comes away from this book with a sense of the breadth of the enterprise, the hope for solutions that will truly help blind individuals, and the excitement shared by so many working in the field, then it has accomplished much of what the authors set out to do. If it allows practitioners in one discipline participating in this development to get a better appreciation for what their colleagues in other disciplines are trying to accomplish, then the authors have clearly hit the right notes. And if it inspires enthusiastic young minds to join the quest, and to help turn the visual prosthesis into the next cochlear implant, then we will truly have succeeded.

Baltimore, MD   Gislin Dagnelie
September 2010

Acknowledgments

This book reflects a group effort. Each contributor embraced the concept of a book that would span many disciplines, and reaching a consensus about what should be covered, and by which authors, proved surprisingly easy. I thank the authors for making time in their busy schedules to share their knowledge and create this overview.

I appreciate the encouragement of my colleagues at the Lions Vision Research and Rehabilitation Center of the Wilmer Eye Institute at Johns Hopkins, who encouraged me to take on the challenge of creating this book and who helped in large and small ways to bring it to completion. I am deeply grateful to Maryam Khan, M.D., who helped me turn a stack of diverse manuscripts into polished chapters that not only met the publisher's technical standards but are a pleasure to read. But most of all, I am grateful to the study participants who give meaning to our research, and who are an ongoing source of inspiration. This book is dedicated to them.

Baltimore, MD   Gislin Dagnelie

Contents

Contributors

Michael P. Barry
Lions Vision Research & Rehabilitation Center,
Johns Hopkins University School of Medicine, Baltimore, MD 21205-2020, USA
mbarry11@jhu.edu

Ava K. Bittner
Lions Vision Research & Rehabilitation Center, Wilmer Eye Institute,
Johns Hopkins University School of Medicine, Baltimore, MD 21205, USA
abittne1@jhmi.edu

Alex Butterwick
Department of Applied Physics, Stanford University, Stanford, CA, USA
abutterwick@gmail.com

Carlos J. Cela
Department of Electrical and Computer Engineering, University of Utah,
50 S. Central Campus Drive, Room 3280, Salt Lake City, UT 84112-9206, USA
carlos.cela@utah.edu

Gislin Dagnelie
Lions Vision Research & Rehabilitation Center,
Johns Hopkins University School of Medicine, Baltimore, MD, 21205-2020 USA
gdagnelie@jhmi.edu

James Deremeik
Johns Hopkins University, Baltimore, MD, USA
jderemeik@jhmi.edu

Barbara R. Evans
Oak Ridge National Laboratory, Oak Ridge, TN, USA
evansb@ornl.gov

Eduardo Fernández
Instituto de Bioingeniería, Universidad Miguel Hernández,
Avda de la Universidad s/n, 03202 Elche (Alicante), Spain
e.fernandez@umh.es

Ione Fine
University of Washington, Seattle, WA, USA
ionefine@u.washington.edu

Paul G. Finlayson
Departments of Otolaryngology and Ophthalmology,
Wayne State University, 550 E. Canfield
pfinlays@med.wayne.edu

Shelley I. Fried
VA Boston Healthcare System, 150 South Huntington Avenue,
Boston, MA 02130, USA
and
Massachusetts General Hospital & Harvard Medical School,
429 Their, 50 Blossom Street, Boston, MA 02114, USA
fried.shelley@mgh.harvard.edu

Duane R. Geruschat
Lions Vision Research & Rehabilitation Center, Wilmer Eye Institute, Johns
Hopkins University School of Medicine, 550 N, Baltimore, MD 21205, USA
dgeruschat@jhmi.edu

Elias Greenbaum
Oak Ridge National Laboratory, Oak Ridge, TN 37831, USA
greenbaum@ornl.gov

Luke E. Hallum
Graduate School of Biomedical Engineering, University of New South Wales,
ANZAC Parade, Sydney 2052, Australia
and
Center for Neural Science, New York University, New York, NY 10003, USA
hallum@cns.nyu.edu

Alan Horsager
Eos Neuroscience, Inc., 2100 3rd Street, 3rd floor, Los Angeles,
CA 90057, USA
and
Department of Ophthalmology, University of Southern California,
Los Angeles, CA 90089, USA
horsager@usc.edu

Philip Huie
Department of Ophthalmology, Stanford University, 450 Serra Mall,
Stanford, CA, 94305, USA
philhuie@stanford.edu

Raymond Iezzi
Department of Ophthalmology, Mayo Clinic, 200 First Street, SW,
Rochester, MN 55905, USA
iezzi.raymond@mayo.edu

Ralph J. Jensen
VA Boston Healthcare System, Boston, MA, USA
Ralph.Jensen@va.gov

Bryan W. Jones
Moran Eye Center, University of Utah, 65 Mario Capecchi Drive,
Salt Lake City, UT 84132, USA
bryan.jones@m.cc.utah.edu

Gianluca Lazzi
Department of Electrical and Computer Engineering,
University of Utah, Salt Lake City, UT, USA
lazzi@utah.edu

James Loudin
Department of Applied Physics, Stanford University, 450 Serra Mall,
Stanford, CA 94305, USA
loudin@stanford.edu

Nigel H. Lovell
University of New South Wales, Sydney, Australia
N.Lovell@unsw.edu.au

Stephen L. Macknik
Barrow Neurological Institute, Phoenix, AZ, USA
macknik@neuralcorrelate.com

Susana Martinez-Conde
Barrow Neurological Institute, 350 W. Thomas Road, Phoenix, AZ 85013, USA
smart@neuralcorrelate.com

Robert E. Marc
Moran Eye Center, University of Utah, Salt Lake City, UT, USA
robert.marc@hsc.utah.edu

Lotfi B. Merabet
Harvard Medical School, Boston, MA, USA
Lotfi_Merabet@meei.harvard.edu

Daniel Palanker
Department of Applied Physics, Stanford University, Stanford, CA, USA
palanker@stanford.edu

Aditi Ray
Department of Biomedical Engineering, 1042 Downey Way,
Denney Research Building (DRB) 140, Los Angeles, CA 90089, USA
Aditi.Ray@AlconLabs.com

Gernot Roessler
Department of Ophthalmology, RWTH Aachen University, Aachen, Germany
groessler@ukaachen.de

Edward M. Schmidt
National Institutes of Health (retired)
emschmidt@atlanticbb.net

Marilyn E. Schneck
Rehabilitation Engineering and Research Center, The Smith-Kettlewell Eye
Research Institute, 2318 Fillmore Street, San Francisco, CA 94115, USA
and
Vision Sciences Program School of Optometry-2020, University of California at
Berkeley, Berkeley, CA 94720-2020, USA
mes@ski.org

Nishant R. Srivastava
Department of Biomedical Engineering, Pritzker Institute of Biomedical Science
and Engineering, Illinois Institute of Technology, 3255 S. Dearborn, WH 314,
Chicago, IL 60616, USA
srivnis@gmail.com

H. Christiaan Stronks
Lions Vision Research and Rehabilitation Center, Wilmer Eye Institute, Johns
Hopkins University School of Medicine, 550 N. Broadway, 6th floor, Baltimore,
MD 21205, USA
hstronk1@jhmi.edu

Janet S. Sunness
Greater Baltimore Medical Center, Baltimore, MD, USA
jsunness@gbmc.org

Xoana G. Troncoso
California Institute of Technology, Pasadena, CA, USA
x.troncoso@neuralcorrelate.com

Philip R. Troyk
Department of Biomedical Engineering,
Pritzker Institute of Biomedical Science and Engineering, Illinois Institute
of Technology, 3255 S. Dearborn, WH 314, Chicago, IL 60616, USA
troyk@iit.edu

Peter Walter
Department of Ophthalmology, RWTH Aachen University, Pauwelsstr. 30,
52074 Aachen, Germany
pwalter@ukaachen.de

Carl B. Watt
Moran Eye Center, University of Utah, Salt Lake City, UT, USA
carl.watt@hsc.utah.edu

James D. Weiland
Department of Biomedical Engineering, University of Southern California,
Los Angeles, CA, USA
jweiland@doheny.org

Part I
Structure and Function
of the Visual System

Chapter 1
The Human Visual System: An Engineering Perspective

Gislin Dagnelie

Abstract This chapter provides a brief introduction to the architecture and function of the healthy visual system. Particular emphasis is placed on the diverse capabilities of the visual system that visual prosthesis researchers may want to emulate, to provide the reader with a realistic sense of the daunting challenges facing workers in this field.

Abbreviations

CCD Charge-coupled device
COS Cone outer segment
HL Henle fiber layer
INL Inner nuclear layer
LGN Lateral geniculate nucleus
NOT Nucleus of the optic tract
ONL Outer nuclear layer
RPE Retinal pigment epithelium
SC Superior colliculus
TN Terminal nucleus
V1 Primary visual cortex, striate cortex
V2 Secondary visual cortex, peristriate/extrastriate cortex

G. Dagnelie (✉)
Lions Vision Research & Rehabilitation Center, Johns Hopkins University School of Medicine,
550 N. Broadway, 6th floor, Baltimore, MD 21205-2020, USA
e-mail: gdagnelie@jhmi.edu

G. Dagnelie (ed.), *Visual Prosthetics: Physiology, Bioengineering, Rehabilitation*,
DOI 10.1007/978-1-4419-0754-7_1, © Springer Science+Business Media, LLC 2011

1.1   The Visual System as an Engineering Compromise

The purpose of this chapter is to outline the architecture and properties of the human visual system, but only to the extent required for a better understanding of its role as a substrate for visual prostheses. By sketching the properties of the healthy human visual system, we intend to provide the reader with an appreciation of the challenges one encounters in trying to reconstruct vision to the blind, even to a very modest level. Readers interested in more detailed or specific information regarding the visual system in health and disease are referred to some of the many excellent reviews in this area [12, 21, 22, 24, 49] and specifically to Chaps. 2–5 in this volume.

Evolution of the vertebrate visual system over several hundred million years has provided the human eye and higher visual processing centers with ingenious compromises to allow sharp central vision, a wide field of view, color perception, and an enormous range of light-to-dark adaptation. Note the following benchmarks, unparalleled by any single man-made system:

- The optic nerve, connecting the eye to the visual centers of the midbrain, has only approximately 1.2 million fibers [37] to represent the entire visual field (over $140°$ horizontally and $120°$ vertically, or roughly 3.6×10^7 arcmin^2), in full color; a digital color camera with similar output bandwidth would provide 333,000 pixels, i.e., about 630 pixels across the field or 13.3 arcmin resolution. Yet the human eye achieves 1 arcmin resolution in the center of the visual field by combining variable cone photoreceptor spacing – from 0.4 arcmin (i.e., 1/150th of a degree, or 0.0067 μm; foveola) to 3 arcmin (far periphery)[1, 2, 17, 38] – with variable post-receptoral convergence – from (on average) three ganglion cells per cone in the fovea to one ganglion cell per 6 cones in the far periphery [16, 56].
- The three color filters used in digital cameras have narrower bandwidths and wider color separation than the three human cone types. Yet the post-receptoral interactions in human vision allow discrimination over a wider range of color space than can be physically created with common light sources and pigments (pp. 306 ff. in [66]).
- Both traditional cameras and the human eye employ mechanical apertures to adjust to a limited range of light levels (over 100 to 1 in cameras, about 15 to 1 for the human pupil). This, however, represents only a fraction of the dark adaptation range required by changes in natural lighting conditions. Low noise properties of CCD chips and a variety of automatic gain control mechanisms allow modern cameras to function over a brightness range from less than 1 to over 100,000 lux. Rod photoreceptors in the human eye, however, extend the downward range by at least a factor of 1,000 [38], while cone dynamics extend the upward range by at least a factor of 10. The rod system transmits its information to the brain using the same optic nerve fibers used by the cones.
- Compared to man-made detection and shape-recognition systems, the human visual system works quickly and with great precision: Attentional shift mechanisms

linked to movement and change in our peripheral vision lead to rapid redirection of gaze, in order to perceive detail in this novel stimulus [48]. Depth information is acquired monocularly by relative movement and size of objects in foreground and background [32, 47], while cooperative imaging by the two eyes, with disparities as small as a few arcsec (i.e., a small fraction of the width of a foveal cone), provides detailed depth and three dimensional shape information for nearby objects [63].

- Continuous information updating through rapid involuntary eye movements (microsaccades) is built into the human visual system, which can retain an accurate image only through such frequent "refreshment:" The perception of an image stabilized on the retina would fade after a few seconds [4, 27]. Man-made image acquisition systems may not spontaneously lose image information over time, but they require a recording medium if information is to be retained.

Probably the most remarkable design accomplishment in the visual system is the fovea, or yellow spot, in the center of the retina: It combines the quality of the eye's optics close to the main axis, the density of photoreceptor cells in the central retina, the outward displacement of secondary neural elements and blood vessels from the retinal center [1, 68], and the ingenuity of retinal and cortical connectivity and processing to achieve the highest possible image resolution within the restrictions of limited anatomic resources and physiologic bandwidth set by a biological system.

One of the greatest challenges in designing a visual prosthesis is, therefore, to reproduce the principal properties lost by eye disease or abnormal development, while using the capabilities provided by the remaining visual system to the greatest possible advantage. In order to gain a perspective of what visual prosthetics can and cannot accomplish for their recipients, it is important to understand how normal visual function depends on the anatomy and physiology of visual system component: the eye's optics, the retina, the pathways leading from retina to visual cortex, and the cortical areas involved in visual perception. A limited understanding of the operation and role of eye movements and binocular vision is also required. Some of these topics are briefly covered here; more detail can be found in Chap. 2.

1.2 An Overview of Human Visual System Architecture

1.2.1 Architecture and Basic Function of the Eye

The structure and function of the eye correspond to the properties of visible light: Its optics form an image of the outside world on a photosensitive layer of cells, and the spectral properties of these cells construct multiple representations in different color bands. Specifically, the cornea, iris and crystalline lens (see Fig. 1.1), aided if necessary by corrective optics such as spectacles or contact lenses, cooperate to form a focused image on the back wall of the eye. The retina, a thin film covering the posterior half of the interior eye wall, contains multiple cell populations capturing

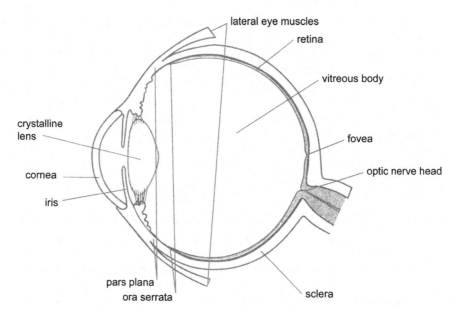

Fig. 1.1 Cross section through the human eye, showing the principal structures referred to in the text. Reprinted from [19], with permission

the image, pre-processing it for efficient information compression, and encoding it for transmission via the optic nerve to the visual centers in the brain.

Figure 1.1 shows a horizontal cross-section of the adult human eye, including the optical elements, the retina, and the optic nerve. Light entering the eye is refracted by the cornea and the crystalline lens, and aperture-limited by the iris. The lens is suspended in a ring of thin fibers (zonula). The ciliary muscle behind the iris through the connecting zonula, can adjust the refractive power of the lens to maintain a sharp image when objects are brought closer to the eye; a membrane called the capsular bag surrounds the lens and separates the anterior and posterior chambers. Light passes unhindered through the watery fluid (aqueous humor) in the anterior chamber between cornea and lens, through the gel-like fluid (vitreous body) in the posterior chamber, and through the inner retinal layers, to reach the photoreceptors – the light-sensitive cells that convert it into electrical and chemical signals, initiating the process of vision.

An intricate system of four straight and two oblique extraocular muscles allows the eye to be rapidly directed towards a visual target without requiring a head movement. More importantly, these muscles also allow for vergence (directing the two eyes to a common point at varying distances) and limited cyclo-rotation to counteract small rotations of the head or the scene and maintain a stable view of the world.

The photoreceptors are situated in the deepest retinal layer and they, along with other cells in the outer retinal layers, receive nutrients and oxygen via a network of small capillaries under the retina, the choroid plexus. The inner retinal layers have their own blood supply, which is fed through blood vessels in the optic nerve head; the arteries and veins of the inner retinal blood supply form two semi-circular patterns (the lower of these so-called arcades can be seen in Fig. 5.3, Chap. 5) around the central retina (or macula), which therefore has only narrow capillaries to limit interference with light projected onto the photoreceptors. The center-most portion of the macula (the fovea) does not contain any blood vessels and is called the foveal avascular zone. No inner retinal blood supply is required in this area, due to the outward displacement of all inner retinal cells, away from the center [1, 68]. Note the slight indentation of the retina at this so-called foveal pit, where the ability to capture details in the image is greatest. Also note the cupping of the retinal surface at the optic nerve (the optic nerve head; also called physiological blind spot, as it contains no photoreceptors), allowing for optic nerve fibers – actually the axons of retinal ganglion cells – to converge and enter the supporting structure of the optic nerve.

To gain surgical access to the interior of the eye, one can make an incision in or near the cornea to enter the anterior segment, or cut the sclera (white outer wall of the eye) through the so-called pars plana, i.e., posterior to the attachment of the lens capsule, but anterior to the ora serrata, the forward edge of the retina. Retinal surgeons routinely use this latter route of access, and both the inner and outer retinal layers can be reached this way, albeit that reaching the outer retina requires an incision through the full thickness of the retina and the creation of an artificial detachment of the retina from the underlying retinal pigment epithelium (RPE) layer. In a healthy eye, re-attachment occurs naturally by resorption of subretinal fluid through the RPE. Recently, surgeons have also gained access to the outer retina by entering through the sclera behind the equator [71].

The retina forms a layered structure against the back wall of the eye, with photoreceptors (rods and cones) capturing the light; bipolar and ganglion cells passing the visual signal on towards the optic nerve, and horizontal and amacrine cells providing lateral interactions among cells in neighboring locations. Chapter 2 provides greater detail regarding the different cell types in each retinal layer and their functions; here we will limit ourselves to the major structures that allow visual function to occur.

In the normal retina, a highly structured arrangement of cells is seen in each layer. Under the retina, a layer of retinal pigment epithelium (RPE) cells fulfills the roles necessary to sustain the metabolism of the photoreceptors: The metabolic level of the photoreceptor outer segments is among the highest in the human body [36]. RPE cells supply nutrients and oxygen, regenerate phototransduction products, and digest debris shed by the photoreceptors [10].

Photoreceptors, the cells capturing the light, come in two main classes: rods, whose high internal gain allows vision at very low light levels [67], and cones, in short, medium, and long wavelength-sensitive types to allow color perception [25]. In both classes of cells the actual light capture and conversion takes place in the outer segment – indicated for the foveal cones in Fig. 1.2 by the abbreviation "COS" while the cell's inner segment, situated in the outer nuclear layer (ONL), provides the transduction to secondary neurons and regulates cell function.

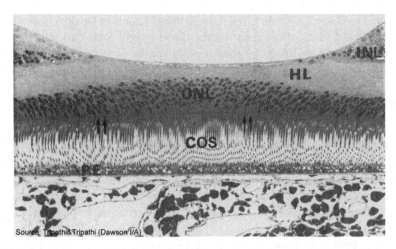

Fig. 1.2 Cross section through the human fovea, showing the dense packing of elongated cone outer segments and the absence of the inner retinal layers across the "foveal pit." *COS* cone outer segments; *ONL* outer nuclear laye; *HL* Henle fiber layer; *INL* inner nuclear layer. The Henle fibers connect foveal cones with the outwardly displaced bipolar cells in the INL. Reprinted from [57], with permission

The distribution and packing of rods and cones varies dramatically across the retina. In the foveola, only medium and long wavelength-sensitive cones are found. In the surrounding foveal area, where the width of individual cones increases, and their packing density decreases accordingly, short wavelength-sensitive cones are also found, while rods are found only beyond the fovea. Figure 1.3, created 75 years ago on the basis of anatomical studies of donor retinas, still provides a fair representation of the density distribution of rods and cones along a horizontal line through the retina of a left eye. One may note that rod densities are highest around 20° eccentricity. Cones are distributed throughout the entire retina, in roughly constant density beyond the central macula. Due to the decreasing convergence from photoreceptors to bipolar and ganglion cells in the inner retina, the visual acuity of both day and night vision gradually diminishes towards the periphery.

1.2.2 Layout of the Retino-Cortical Pathway

The connection between the eye and the central nervous system is formed by the fibers of the optic nerve. As noted above, these fibers, whose diameter is on the order of 1 μm, are the axons of retinal ganglion cells. Inside the eye, the fibers run along the inner retinal surface towards the optic nerve head in a characteristic pattern, such that fibers of the upper and lower retinal halves remain separated, and fibers close to the horizontal meridian, but far from the nerve head, arc away from this line to allow room for fibers originating closer to the nerve head. This orderly

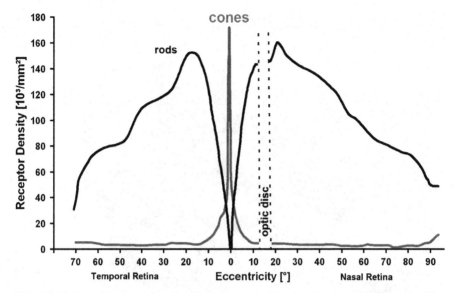

Fig. 1.3 Horizontal cross section through the human retina, showing the rod and cone packing densities in the normal human retina. Note the very narrow area of high cone density, the highest rod density near 10° eccentricity, and the absence of photoreceptors in the physiological blind spot. Originally in [45]; this version from [39], with permission

arrangement causes the fibers from the foveal area (which form 15 to 20% of all nerve fibers) to be located in the temporal quadrant of the optic nerve, at least for the anterior portion of its trajectory [27].

Once the axons enter the optic nerve, each fiber is encapsulated by a myelin sheath, formed by a class of cells called astrocytes; this sheath decreases the membrane conductance of the axons, increasing the conduction velocity and the length over which impulses can be conducted without severe attenuation [51]. Only at the so-called Ranvier nodes is the myelin sheath interrupted, allowing the impulses to be reinforced by virtue of the ion-gating properties of the local membrane.

A cross-section through the human visual pathways can be seen in Fig. 1.4. One may note that the predominant pathway leads from the eye to the lateral geniculate nucleus (LGN) of the thalamus, and from there to the occipital part of the cortex, while smaller numbers of fibers branch off to a tectal area, the superior colliculus, and to a number of pre-tectal nuclei. We will briefly discuss these subcortical pathways below.

Note also that the LGN and cortical areas exist in duplicate in the two halves of the brain. Each deals with one half of the visual world: The optic nerves from the two eyes meet in a structure called the optic chiasm, where fibers from the two nasal retinas cross over to combine with those from the temporal retina of the fellow eye; consequently each LGN and cortical hemisphere receive visual information from two corresponding retinal halves on their own side, and thus from the contralateral half of the visual field.

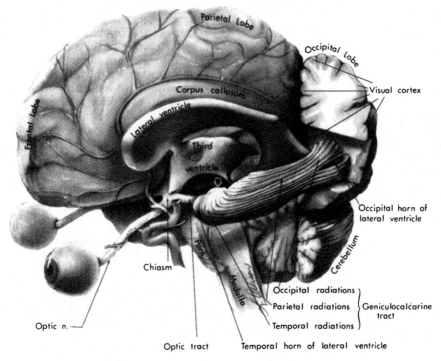

Fig. 1.4 Structure and location of the human primary visual pathways, in relation to other major brain structures. The left cerebral hemisphere, with the exception of the occipital cortex, has been removed; the left LGN is hidden by the optic radiations (*arrow*). Reprinted from [19], with permission

The LGN has a layered structure, with pairs of layers receiving axons of different ganglion cell types, and each layer in a pair receiving signals from one eye. Interactions between layers in the form of overall suppression when the retina in the fellow eye is stimulated layers have been demonstrated [53], but localized interactions across layers do not occur; this indicates that binocular processing required for stereopsis does not take place until the level of the visual cortex. The gateway function of the LGN, which in other mammals such as the cat appears to play a crucial role in adaptation and attention, and through which signals from the two eyes can mutually inhibit each other [29], is thought to be less prominent in primates, including humans. Yet anatomical feedback connections from a number of subcortical nuclei onto the LGN are as extensive in monkey as in cat [8], and gating functions related to circadian rhythms and other systemic conditions are therefore plausible in primates as well.

Forward pathways from the LGN lead to the primary visual cortex (V1, also called striate cortex; these fibers form the optic radiation), but also to higher visual cortical areas and to subcortical areas such as the superior colliculus (SC). The role of the extrastriate cortical pathways is still a topic of speculation and investigation;

from clinical cases it is evident, however, that patients with lesions to the striate cortex acquired after childhood retain little or no useful vision [11]. The roles of the tectal pathways, including mutual connections between cortical areas, the SC and the pulvinar, are also subject to active research. It has been found, for example, that cortical connections with midbrain areas are essential for maintaining and shifting attention, rather than for processing detailed visual information [35].

The visual cortex occupies the occipital and parts of the parietal and temporal lobes of the cerebral cortex. Like the entire cortex, it forms a highly folded structure, with a thickness of approximately 1.5 mm. It is surrounded by the cerebrospinal fluid, several layers of meninges – pia mater, arachnoid, and dura –, and the skull. Especially the skull forms an important barrier to any attempt at functional electrical stimulation of cortical cells.

Like the retina, the visual cortex is a layered structure, in which different cell groups perform different tasks. Along its two-dimensional surface, one finds an orderly mapped representation of the outside world. Contrary to the retina, however, the cortex consists of multiple areas, hierarchically organized, each of which performs a partial processing task in the analysis of the scene around us. At the present time, over 30 visual cortical areas per hemisphere are recognized in monkey, and a similar number of distinct areas is thought to exist in humans [61].

The first cortical representation, in the striate cortex, is shown schematically in Fig. 1.5. It presents a straightforward map of the visual world, but contains four major transformations:

The projection from the LGN (and thus retinal ganglion cells) onto V1 input cells has approximately constant density, which means that the central visual field is highly over-represented in the visual cortex: Roughly 20% of V1 represents the retinal fovea, and thus the central 1–2° of the visual field, with rapid drop-off of the density towards the periphery. This inhomogeneous map is conveniently expressed by the cortical magnification factor, $M(')$, i.e., the number of mm of cortex devoted to 1° of retina, as a function of eccentricity [20, 34].

The folding of the human cortex, prompted by the evolutionary expansion of higher (cognitive) processing, has resulted in an arrangement where most of the peripheral visual field is represented in portions of V1 that are buried in the medial walls and sulci of the cerebral hemispheres. Only the foveal representation, situated along the border of V1 and the adjacent area V2, is exposed at the surface of the occipital cortex, approximately 1 cm above the inion, a protruding portion near the bottom of the skull bone on the back of the head. More peripheral visual field areas are represented along the medial walls of the cerebral hemispheres, and in deep sulci embedded within these areas.

The complex representation of the visual field onto area V1, combined with the lack of accessibility due to cortical folding, greatly reduces our ability to investigate and stimulate the peripheral visual field. Area V2 and several higher areas form the exposed portion of the occipito-parietal cortex, and would seem to provide better opportunities for peripheral field stimulation. However, while these areas may appear to be more easily accessible than V1, they have an equally dense pattern of sulci, which greatly vary between individuals; moreover, receptive field properties and visual field maps become increasingly complex in higher cortical areas [41].

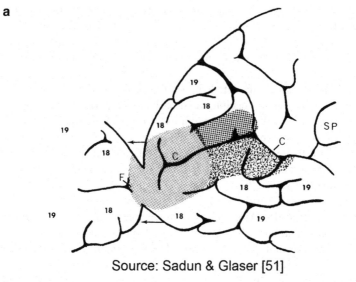

Source: Sadun & Glaser [51]

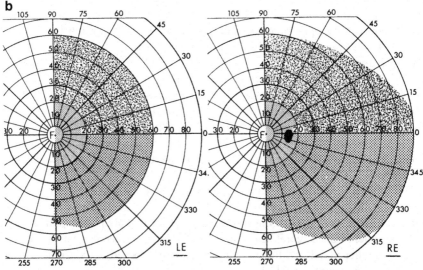

Fig. 1.5 V1 projection of the visual field. The medial wall and part of the occipital surface of the left cerebral hemisphere (**a**) and the corresponding visual fields for the two eyes (**b**) are shown. Note that the projection of the fovea (F) and a narrow surrounding hemicircle of the visual field project onto the occipital cortex, with the projection of the vertical meridian adjacent to area V2, whereas more peripheral areas – including most of the macula – projects to the medial wall of the cortex, with much of the projection buried in the calcarine fissure (C). Also note that the left hemisphere receives information from the right visual hemifield, that the superior visual field projects to the inferior part of V1 – i.e., gross localization is preserved from retina to V1, and that corresponding retinal locations in the two eyes project to the same cortical location. No matching locations exist for the far nasal segment of the right retina (60–90°), as the bridge of the nose blocks the corresponding area in the left eye. Reprinted from [19], with permission

1.2.3   Layout of the Subcortical Pathways

In addition to the visual pathway to LGN and striate cortex, which receives the great majority of retinal ganglion cell axons, there also are subcortical pathways, formed by optic nerve fibers projecting to pretectal nuclei and to the pulvinar. In primates, the projections to the pregeniculate nucleus and pulvinar are thought to be of minor importance, and may be thought of as anatomical remnants: In lower mammals, ablation of striate cortex at birth allows these projections to greatly increase in density, leading to the development of crude functional vision, but similar experiments in newborn monkeys show neither the proliferation of projections nor appreciable acquisition of visual function [14, 15]. Other projections, however, in particular those to the pretectal nucleus of the optic tract (NOT) and the terminal nuclei (TN) of the accessory optic system, have been demonstrated to play an important role in the rapid control of eye position through vestibulo-ocular reflex, saccades, and sustained fixation [33, 43].

Detailed studies of anatomy and physiology of the primate eye movement system over the last several decades in awake, trained animal models, have shown that the NOT receives information on "retinal slip," i.e., generalized displacement of the retinal image [58]. This retinal slip signal is encoded as a velocity signal, and serves as input to the neural integrator in the nucleus prepositus hypoglossi [50]. Pathways between the NOT and primary visual cortex (as well as multiple similar projections between cortical and subcortical structures) are also known to exist, and have been shown to compensate in part for lesions to the NOT or its retinal input [23, 31].

1.3   An Overview of Human Visual Function

The anatomy and physiology of the visual system presented above can help us understand many of the properties of normal vision, and some of the vision defects experienced by patients with blinding eye diseases. We will briefly discuss the aspects most pertinent in understanding the requirements for neural visual prostheses.

1.3.1   Roles of Central (Foveal) Vision

Central visual function is more than just the utilization of the denser packing of photoreceptors in the central retinal area and the higher density of ganglion cells per photoreceptor in this area. These properties of the retina would account for basic properties such as good two-point resolution, but they would not explain why foveal vision is superior to peripheral vision in many other ways. The following major areas of foveal specialization should be considered.

Spatial integration tasks, e.g., hyperacuity. Normally sighted observers have the ability to resolve small deviations in alignment of parallel or abutting lines, small angular differences and displacements, all on a scale well below 1 arcmin, the spacing of foveal photoreceptors. Such "hyperacuities" apparently rest on the ability of foveal projections in cortical areas to combine the precise positional coding of earlier stages in the visual system over increasing distances, using feedback and tuning mechanisms that have been honed by years of experience. The notions of learning and tuning are supported by the lack of hyperacuity in subjects with inherited abnormalities of foveal development and eye movements [64] or with developmental deficits [9], and by the gradual acquisition of hyperacuity performance throughout childhood [69].

Stereopsis. Combination of the signals from corresponding locations in the two retinas, in a highly systematic fashion, is required for perception of depth in stationary three-dimensional scenes. This function takes place at and beyond the V1 cortical level [55]. Fusion of the two retinal images on an object of interest defines a curved plane, the horopter, formed by the collection of points being imaged at exactly corresponding locations on the retinas of the two eyes. Finely-tuned disparity neurons detect left-right eye correspondence of retinal locations for points slightly in front of (crossed disparity) or beyond (uncrossed disparity) the horopter, with resolution on the order of arc seconds, similar to that seen in hyperacuity task performance.

Complex pattern recognition and discrimination tasks, e.g., face recognition. Beyond the ability to make precise visual judgments enabled by the high resolution of foveal vision, normally-sighted observers acquire great skill at memorizing, recognizing, and discriminating among patterns, varying from feature discrimination in the natural environment, such as recognizing human faces, to the processing of complex man-made forms and objects, such as reading text or maps. These capabilities require both high-level visual processing skills and cognitive brain functions such as leaning and memory. It is not necessarily true that these specialized skills cannot be acquired in peripheral vision: Certainly, a person with a central scotoma (blind area) due to macular degeneration can read, if given text with enough magnification and contrast [26, 40]. Nonetheless, these skills appear to depend critically on specialization during early phases of development, and functions such as reading, that once were linked to foveal visual function, can only partially, and with great effort, be taken over by extrafoveal vision, as if the task of vision itself has to be re-learned [28]. On the other hand, children with poorly developed foveal vision, such as those with albinism or aniridia, can learn to proficiently read and recognize patterns or faces, if given adequate magnification, and the same intensive exposure as their normally-sighted peers [30].

Visuomotor integration tasks, e.g., handwriting. These tasks are very similar to the pattern recognition tasks described above, in that they require complex visual processing and memory functions, but moreover they require integration with proprioceptive and motor functions distributed across many different brain areas. Some of these tasks may depend less critically on foveal function, but inasmuch as they are based on skills learned during early development, their execution often proves difficult when foveal vision becomes impaired later in life [44].

Note that all the skills referred to as specific for foveal vision involve the ability to see fine detail, combined with extensive learning throughout the critical period of development.

1.3.2 Roles of Peripheral Vision

The role of peripheral vision in performing daily activities is often underappreciated. Most human-designed visual tasks rely on the perception of detailed shapes, but there are important exceptions: Noticing traffic off to the side while driving and keeping track of fellow and opponent players during team sports require continuous processing of events and objects throughout the visual field. Similarly, observing wildlife and other outdoor activities require the use of our entire visual field. In almost all cases, these visual tasks require us to perceive motion and other changes, and it should come as no surprise that the evolution of the vertebrate visual system has favored use of the periphery for precisely these functions [6]. On the other hand, our attention tends to be focused on objects and events in central vision, whereas school children with severely impaired central vision appear to use their peripheral vision much more efficiently than normally-sighted individuals [65].

One of the surprising aspects of peripheral vision is how much of it can be lost before a person becomes aware of the change. Thus disorders such as glaucoma and retinitis pigmentosa may go undetected well beyond the point where irreversible damage to cells in the peripheral retina has occurred [13, 42].

1.3.3 Roles of Dark-Adapted Vision

As was mentioned above, cones are unable to function effectively at light levels below 0.003 cd/m². At these low illumination levels, rod photoreceptors continue to be effective, by virtue of the high gain, multi-stage phototransduction cascade in the rod outer segment. On the other hand, at intensities above 3 cd/m² rod function is actively suppressed. Rods are not distributed evenly throughout the retina: The center of the retina, with a diameter of approximately 5°, forms a rod-free zone, and the highest density of secondary retinal cells receiving rod signals is situated between at eccentricities between 5 and 10°, as evidenced by the common experience that a dim object at night is best observed by intentionally looking slightly away from the object.

Dark-adapted vision differs from daytime vision in two important respects, both related to the need for maximum sensitivity, i.e., the detection of a very small stimulus signal in the ongoing background of visual noise (spontaneous activity of retinal cells). Ganglion cells in the dark-adapted retina integrate signals from a much larger number of photoreceptors than in the light-adapted retina [60], and the time course over which this integration takes place is significantly extended [7].

For this reason it is not possible to see small or rapidly moving objects at very low light levels that allow only rod vision.

In designing image capture systems for prosthetic vision, researchers may want to borrow some of the principles employed by the retina. Increased integration times are commonly employed in CCD arrays and other electronic camera detectors, but integration across space is rarely employed for real-time image acquisition, as it runs counter to the designers' wish to maximize spatial detail. Given the limited spatial resolution of early visual prostheses, however, such loss of spatial detail at the input should be of no consequence to the image perceived by the prosthesis wearer, and could be employed to achieve maximum sensitivity.

1.3.4 A Few Remarks Regarding Visual Development

Throughout this chapter we have seen that the principal pathway on which functional vision depends is that from the retina through the LGN to cortical area V1. Even the mechanisms of involuntary eye movement control (maintaining off-center gaze, microsaccades), which are served by pretectal pathways, can be compensated for – as appears from primate experiments –, presumably by virtue of cortico-subcortical connections like those from V1 to the NOT and accessory optic nuclei. Hence, if visual impairment or blindness is caused by a disorder at the level of the eyes, optic nerve, or primary visual cortex, in a person whose visual function had followed its normal development earlier in life, one can assume that all cortical processing mechanisms and functions are intact, and may be successfully restored if adequate input signals are provided. If, on the other hand, normal visual development did not occur, as in the case of a congenital deficit of the retina or optic nerve, then a visual prosthesis implanted at a later age is unlikely to enable functional vision, similarly to the lack of functional vision documented in adult corneal transplant recipients who had congenital corneal opacities or corneal trauma in infancy [59]. Just like the cochlear prosthesis, however [54, 62], the visual prosthesis may provide opportunities for partial development of functional vision in children, provided the implantation takes place at a very young age, presumably in the first or second year of life. Obviously, this will require the technology to have been proven safe and effective in adults.

1.4 Prospects for Prosthetic Vision Restoration

On the basis of the architectural and functional layout of the human visual system described above, it should be clear that vision restoration through stimulation of intact structures along the retino-cortical pathway is feasible, in principle. Given the transformation of visual information that occurs at every stage along the visual pathway, the prudent approach in visual prosthetics would seem to be to implant as

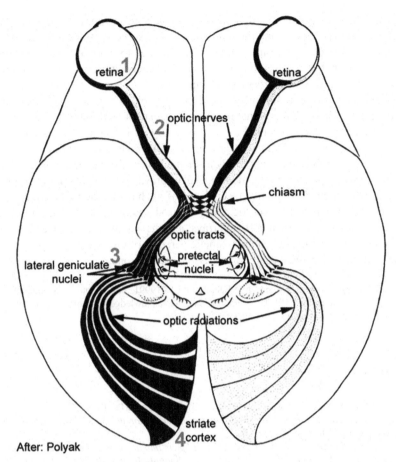

Fig. 1.6 In this cross-section of the human retina-cortical pathway, seen from below, the numbers 1 through 4 indicate locations currently being considered for visual prosthesis implantation. Adapted from [46]

distally – that is, as early along the pathway – as feasible: If the photoreceptors are non-functional, then the bipolar or ganglion cells in the retina would be the best target for stimulation; if the retina is detached so a retinal prosthesis cannot be placed reliably, then an optic nerve implant would be indicated; if the retinal ganglion cells, and thus the optic nerve fibers, are damaged by glaucoma, then an implant in the LGN or primary visual cortex may be in order; etc. Figure 1.6 illustrates the four locations that seem best suited for the placement of visual prostheses.

In all these examples the assumption is that the visual pathway proximal to the lesion – that is, towards the brain – is intact, but their success will depend on the extent of secondary degeneration that may have occurred further along the visual pathway. Certainly it is known that many ganglion cells are lost after an extended period of outer retinal degeneration, but a substantial percentage survives, more than enough to carry the small number of signals from today's implants [52]. Moreover, the more central and the more invasive the surgery, the greater the risk of systemic and irreversible complications. Thus the idea to implant on the proximal side of the lesion, but as close to it as can be done safely, appears to be the implicit practice among most visual prosthesis groups.

The extent to which prosthesis recipients will be able to regain useful vision, and the duration required for functional rehabilitation, cannot be predicted until a larger number of patients has received a greater variety of implants than is currently the case; recent reports from two groups regarding letter recognition [18, 70], wayfinding [3], and maze tracing [5] by a small number of retinal implant recipients are encouraging indicators that a modest level of prosthetic vision is possible. From simulations in sighted volunteers (see Chap. 16) we have learned that seeing with pixelized vision is possible; yet the small electrode numbers in retinal arrays, the irregularity of phosphenes in cortical arrays, and the apparent differences between simulations with distinct dots and prosthetic percepts of broadly overlapping phosphenes will make the rehabilitation process an arduous one.

Acknowledgment Supported in part by PHS grant # EY019991. This chapter is an adaptation of parts of an earlier chapter. [19] The author wishes to acknowledge the contributions of Eyal Margalit, M.D., who co-authored that chapter.

References

1. Ahnelt PK (1998), *The photoreceptor mosaic.* Eye (Lond), **12**(Pt 3b): p. 531–40.
2. Ahnelt PK, Kolb H (2000), *The mammalian photoreceptor mosaic-adaptive design.* Prog Retin Eye Res, **19**(6): p. 711–77.
3. Ahuja AK, Dorn JD, Caspi A, et al. (2010). *Subjects Implanted With the Argus™ II Retinal Prosthesis Are Able to Improve Performance in a Spatial-Motor Task.* in ARVO Annual Mtg.
4. Arend LE, Timberlake GT (1986), *What is psychophysically perfect image stabilization? Do perfectly stabilized images always disappear?* J Opt Soc Am A, **3**(2): p. 235–41.
5. Barry MP, Dagnelie G, Group AIS (2010), *Use of the Argus™ II Retinal Prosthesis to Improve Visual Guidance of Fine Hand Motion.* in ARVO Annual Mtg.
6. Battaglini PP, Galletti C, Fattori P (1993), *Functional properties of neurons in area V1 of awake macaque monkeys: peripheral versus central visual field representation.* Arch Ital Biol, **131**(4): p. 303–15.
7. Baylor DA (1987), *Photoreceptor signals and vision. Proctor lecture.* Invest Ophthalmol Vis Sci, **28**(1): p. 34–49.
8. Bickford ME, Ramcharan E, Godwin DW, et al. (2000), *Neurotransmitters contained in the subcortical extraretinal inputs to the monkey lateral geniculate nucleus.* J Comp Neurol, **424**(4): p. 701–17.
9. Birch EE, Swanson WH (2000), *Hyperacuity deficits in anisometropic and strabismic amblyopes with known ages of onset.* Vision Res, **40**(9): p. 1035–40.
10. Bok D (1993), *The retinal pigment epithelium: a versatile partner in vision.* J Cell Sci Suppl, **17**: p. 189–95.
11. Brindley GS, Gautier-Smith PC, Lewin W (1969), *Cortical blindness and the functions of the non-geniculate fibres of the optic tracts.* J Neurol Neurosurg Psychiatry, **32**(4): p. 259–64.
12. Buser P, Imbert M (1992), *Vision.* Cambridge, MA: MIT Press.

13. Coleman AL (1999), *Glaucoma*. Lancet, **354**(9192): p. 1803–10.
14. Cowey A, Johnson H, Stoerig P (2001), *The retinal projection to the pregeniculate nucleus in normal and destriate monkeys*. Eur J Neurosci, **13**(2): p. 279–90.
15. Cowey A, Stoerig P, Bannister M (1994), *Retinal ganglion cells labelled from the pulvinar nucleus in macaque monkeys*. Neuroscience, **61**(3): p. 691–705.
16. Curcio CA, Allen KA (1990), *Topography of ganglion cells in human retina*. J Comp Neurol, **300**(1): p. 5–25.
17. Curcio CA, Sloan KR (1992), *Packing geometry of human cone photoreceptors: variation with eccentricity and evidence for local anisotropy*. Vis Neurosci, **9**(2): p. 169–80.
18. da Cruz L, Coley B, Christopher P, et al. (2010). *Patients Blinded by Outer Retinal Dystrophies Are Able to Identify Letters Using the Argus™ II Retinal Prosthesis System*. in *ARVO Annual Mtg*.
19. Dagnelie G, Margalit E (2004), *The visual system as a neuroprosthesis substrate: Anatomy, physiology, function*, in *Neuroprosthetics, theory and practice*, Horch KW, Dillon G, Editors. World Scientific Press: Singapore. p. 235–59.
20. Daniel PM, Whitteridge D (1961), *The representation of the visual field on the cerebral cortex in monkeys*. J Physiol, **159**: p. 203–21.
21. Daroff R, Neetens A (1990), *Neurological organization of ocular movement*. Berkeley, CA: Kugler.
22. Daw NW (1995), *Visual Development*. New York, NY: Plenum Press.
23. Distler C, Hoffmann KP (2001), *Cortical input to the nucleus of the optic tract and dorsal terminal nucleus (NOT-DTN) in macaques: a retrograde tracing study*. Cereb Cortex, **11**(6): p. 572–80.
24. Dowling JE (1987), *The retina: an approachable part of the brain*. Cambridge, MA: Belknap Press.
25. Eckmiller M (1997), *Morphogenesis and renewal of cone outer segments*. Progr Ret Eye Res, **16**: p. 401–41.
26. Fine EM, Rubin GS (1999), *Reading with central field loss: number of letters masked is more important than the size of the mask in degrees*. Vision Res, **39**(4): p. 747–56.
27. Fitzgibbon T, Taylor SF (1996), *Retinotopy of the human retinal nerve fibre layer and optic nerve head*. J Comp Neurol, **375**(2): p. 238–51.
28. Frennesson C, Jakobsson P, Nilsson UL (1995), *A computer and video display based system for training eccentric viewing in macular degeneration with an absolute central scotoma*. Doc Ophthalmol, **91**(1): p. 9–16.
29. Freund JH (1973), *Neuronal mechanisms of the lateral geniculate body*, in *Handbook of sensory physiology*, Jung R, Editor. Springer: Berlin. p. 177–246.
30. Fulcher T, O'Keefe M, Bowell R, et al. (1995), *Intellectual and educational attainment in albinism*. J Pediatr Ophthalmol Strabismus, **32**(6): p. 368–72.
31. Giolli RA, Blanks RH, Lui F (2006), *The accessory optic system: basic organization with an update on connectivity, neurochemistry, and function*. Prog Brain Res, **151**: p. 407–40.
32. Gray R, Regan D (1996), *Cyclopean motion perception produced by oscillations of size, disparity and location*. Vision Res, **36**(5): p. 655–65.
33. Hoffmann KP (1996), *Comparative neurobiology of the optokinetic reflex in mammals*. Rev Bras Biol, **56S1**(2): p. 303–14.
34. Horton JC, Hoyt WF (1991), *The representation of the visual field in human striate cortex. A revision of the classic Holmes map*. Arch Ophthalmol, **109**(6): p. 816–24.
35. Inoue Y, Takemura A, Kawano K, Mustari MJ (2000), *Role of the pretectal nucleus of the optic tract in short-latency ocular following responses in monkeys*. Exp Brain Res, **131**(3): p. 269–81.
36. Jindrova H (1998), *Vertebrate phototransduction: activation, recovery, and adaptation*. Physiol Res, **47**(3): p. 155–68.
37. Jonas JB, Schmidt AM, Muller-Bergh JA, et al. (1992), *Human optic nerve fiber count and optic disc size*. Invest Ophthalmol Vis Sci, **33**(6): p. 2012–8.

38. Kimble TD, Williams RW (2000), *Structure of the cone photoreceptor mosaic in the retinal periphery of adult humans: analysis as a function of age, sex, and hemifield.* Anat Embryol (Berl), **201**(4): p. 305–16.
39. Kolb H, Fernandez E, Nelson R (2009), *Facts and figures concerning the Human Retina*, in *WebVision – The Organization of the Retina and the Visual System*, Jones BW, Editor. http://webvision.med.utah.edu/Facts.html: Salt Lake City, UT.
40. Legge GE, Rubin GS, Pelli DG, Schleske MM (1985), *Psychophysics of reading-II. Low vision.* Vision Res, **25**(2): p. 253–65.
41. Livingstone MS, Pack CC, Born RT (2001), *Two-dimensional substructure of MT receptive fields.* Neuron, **30**(3): p. 781–93.
42. Massof RW, Dagnelie G, Benzschawel T, et al. (1990), *First order dynamics of visual field loss in retinitis pigmentosa.* Clin Vision Sciences, **5**: p. 1–26.
43. Mustari MJ, Fuchs AF (1989), *Response properties of single units in the lateral terminal nucleus of the accessory optic system in the behaving primate.* J Neurophysiol, **61**(6): p. 1207–20.
44. O'Connell WF (1996), *Eccentric viewing*, in *Remediation and management of low vision*, Cole RG, Rosenthal BP, Editors. Mosby: St. Louis, MO. p. 27–57.
45. Osterberg G (1935), *Topography of the layer of rods and cones in the human retina.* Acta Ophthalmol Scand, **13**(S6): p. 11–103.
46. Polyak SL (1957), *The vertebrate visual system*, ed. Kluver H. Chicago, IL: Univ of Chicago Press.
47. Portfors-Yeomans CV, Regan D (1996), *Cyclopean discrimination thresholds for the direction and speed of motion in depth.* Vision Res, **36**(20): p. 3265–79.
48. Remington RW (1980), *Attention and saccadic eye movements.* J Exp Psychol Hum Percept Perform, **6**(4): p. 726–44.
49. Rizzo M, Barton JJS (2001), *Retrochiasmal visual pathways and higher cortical function*, in *Duane's clinical ophthalmology*, Tasman W, Jaeger EA, Editors. Lippincott Williams & Wilkins: Philadelphia, PA. p. Ch. 7.
50. Robinson DA (1968), *Eye movement control in primates. The oculomotor system contains specialized subsystems for acquiring and tracking visual targets.* Science, **161**(847): p. 1219–24.
51. Sadun AA, Glaser JS (2001), *Anatomy of the visual sensory system*, in *Duane's clinical ophthalmology*, Tasman W, Jaeger EA, Editors. Lippincott Williams & Wilkins: Philadelphia, PA. p. Ch. 4.
52. Santos A, Humayun MS, de Juan Jr. E, et al. (1997), *Preservation of the inner retina in retinitis pigmentosa.* Arch Ophthalmol, **115**: p. 511–5.
53. Schroeder CE, Tenke CE, Arezzo JC, Vaughan HG, Jr. (1990), *Binocularity in the lateral geniculate nucleus of the alert macaque.* Brain Res, **521**(1–2): p. 303–10.
54. Shannon RV, Zeng FG, Kamath V, et al. (1995), *Speech recognition with primarily temporal cues.* Science, **270**: p. 303–4.
55. Singer W (1990), *The formation of cooperative cell assemblies in the visual cortex.* J Exp Biol, **153**: p. 177–97.
56. Sjostrand J, Olsson V, Popovic Z, Conradi N (1999), *Quantitative estimations of foveal and extra-foveal retinal circuitry in humans.* Vision Res, **39**(18): p. 2987–98.
57. Tripathi RC, Tripathi BJ (1984), *Anatomy of the human eye, orbit & adnexa*, in *The Eye, Vol. 1A, Vegetative Physiology and Biochemistry*, Davson H, Editor. Academic Press: San Diego, CA. p. 1–268.
58. Tusa RJ (1990), *Saccadic eye movements. Supranuclear control*, in *Neurological organization of ocular movement*, Daroff R, Neetens A, Editors. Kugler: Berkeley, CA. p. 67–111.
59. Valvo A (1971), *Sight restoration after long-term blindness: the problems and behavior patterns of visual rehabilitation.* New York: American Foundation for the Blind.
60. van de Grind WA, Koenderink JJ, van Doorn AJ (2000), *Motion detection from photopic to low scotopic luminance levels.* Vision Res, **40**(2): p. 187–99.

61. Van Essen DC, Lewis JW, Drury HA, et al. (2001), *Mapping visual cortex in monkeys and humans using surface-based atlases.* Vision Res, **41**(10–11): p. 1359–78.
62. Waltzman SB, Cohen NL, Gomolin RH, et al. (1994), *Long-term results of early cochlear implantation in congenitally and prelingually deafened children.* Am J Otol, **15** (Suppl 2): p. 9–13.
63. Westheimer G (1994), *The Ferrier Lecture, 1992. Seeing depth with two eyes: stereopsis.* Proc Biol Sci, **257**(1349): p. 205–14.
64. Wilson HR, Mets MB, Nagy SE, Kressel AB (1988), *Albino spatial vision as an instance of arrested visual development.* Vision Res, **28**(9): p. 979–90.
65. Wolffe M (1995), *Role of peripheral vision in terms of critical perception – its relevance to the visually impaired.* Ophthalmic Physiol Opt, **15**(5): p. 471–4.
66. Wyszecki G, Stiles WS (1982), *Color science (2nd ed.).* New York: Wiley.
67. Yau KW, Baylor DA (1989), *Cyclic GMP-activated conductance of retinal photoreceptor cells.* Ann Rev Neurosci, **12**: p. 289–327.
68. Yuodelis C, Hendrickson A (1985), *A qualitative and quantitative analysis of the human fovea during development.* Vision Res, **26**: p. 847–55.
69. Zanker J, Mohn G, Weber U, et al. (1992), *The development of vernier acuity in human infants.* Vision Res, **32**(8): p. 1557–64.
70. Zrenner E (2009). *Blind retinitis pigmentosa patients can read letters and recognize the direction of fine stripe patterns with subretinal electronic implants.* in *ARVO Annual Mtg.*
71. Zrenner E, Miliczek KD, Gabel VP, et al. (1997), *The development of subretinal microphoto-diodes for replacement of degenerated photoreceptors.* Ophthalmic Res, **29**(5): p. 269–80.

Chapter 2
Vision's First Steps: Anatomy, Physiology, and Perception in the Retina, Lateral Geniculate Nucleus, and Early Visual Cortical Areas

Xoana G. Troncoso, Stephen L. Macknik, and Susana Martinez-Conde

Abstract This chapter reviews the functional anatomical bases of visual perception in the retina, the lateral geniculate nucleus (LGN) in the visual thalamus, the primary visual cortex (area V1, also called the striate cortex, and Brodmann area 17), and the extrastriate visual cortical areas of the dorsal and ventral pathways.

The sections dedicated to the retina and LGN review the basic anatomical and laminar organization of these two areas, as well as their retinotopic organization and receptive field structure. We also describe the anatomical and functional differences among the magnocellular, parvocelullar and koniocellular pathways.

The section dedicated to area V1 reviews the functional maps in this area (retinotopic map, ocular dominance map, orientation selectivity map), as well as their anatomical relationship to each other. Special attention is given to the modular columnar organization of area V1, and to the various receptive field classes in V1 neurons.

The section dedicated to extrastriate cortical visual areas describes the "where" and "what" pathways in the dorsal and ventral visual streams, and their respective physiological functions.

The temporal dynamics of neurons throughout the visual pathway are critical to understanding visibility and neural information processing. We discuss the role of lateral inhibition circuits in processing spatiotemporal edges, corners, and in the temporal dynamics of vision.

We also discuss the effects of eye movements on visual physiology and perception in early visual areas. Our visual and oculomotor systems must achieve a very delicate balance: insufficient eye movements lead to adaptation and visual fading, whereas excessive motion of the eyes produces blurring and unstable vision during fixation. These issues are very important for neural prosthetics, in which electrodes are stabilized on the substrate.

S. Martinez-Conde (✉)
Barrow Neurological Institute, 350 W. Thomas Road, Phoenix, AZ 85013, USA
e-mail: smart@neuralcorrelate.com

G. Dagnelie (ed.), *Visual Prosthetics: Physiology, Bioengineering, Rehabilitation*,
DOI 10.1007/978-1-4419-0754-7_2, © Springer Science+Business Media, LLC 2011

Finally, another critical issue for neural prosthetics concerns the neural code for visual perception: How can the electrical activity of a neuron, or a neuronal population, encode and transmit visual information about an object? Here we will discuss how neurons of early visual areas may communicate information about the visible world to each other.

Abbreviations

area MST   Medial superior temporal area
area MT   Middle temporal visual area
area V1   Primary visual cortex
DOG   Difference of gaussians
GABA   Gamma-aminobutyric acid
LGN   Lateral geniculate nucleus

2.1 Introduction

The process of "seeing" is complex and not well understood. But we do know that individual neurons in the early visual system are tuned to stimuli with specific attributes (such as color, shape, brightness, position on the retina, etc.). The receptive field of a visual neuron is the area of the visual field (or its corresponding region on the retina) that when stimulated (by light or electrical impulses) can influence the response of the neuron (Fig. 2.1). Visual stimuli outside a neuron's receptive field

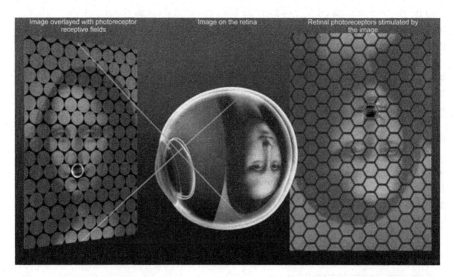

Fig. 2.1 Activation of retinal photoreceptors and their corresponding receptive fields during visual exploration. The eye focuses light that is reflected from the visual image onto the retina, upside down and backwards. Adjacent photoreceptors within the retina are activated by adjacent points of light from the painting. Figure by the Barrow Neurological Institute Illustrations Department

produce no effect on the neuron's responses. Understanding the precise receptive field structure of a given neuron is crucial to understanding and predicting its responses to specific stimuli. For instance, some early receptive fields have a spatial substructure (while others do not), and stimulating their different subregions results in increases or decreases in neural activity. A visual neural prosthesis should ultimately replace the visual processing represented by the receptive fields at a given (damaged or otherwise impaired) stage of the visual hierarchy. A close replication of the output of the replaced neurons will ensure that the healthy tissue farther along the visual pathway receives properly structured inputs.

2.2 Retina

2.2.1 Anatomy

Vision starts in the retina: it is here where photons are converted into electrical signals, to be then interpreted by the brain to construct our perception of the visual world.

The retina has the shape of a bowl (about 0.4 mm thick in adult humans). It is a well organized structure with three main layers (called the nuclear layers) of neuronal bodies. These main layers are separated by two other layers containing synapses made by axons and dendrites (called the plexiform layers). The basic retinal cell classes and their interconnections were revealed by Ramón y Cajal over a century ago [175] (Fig. 2.2).

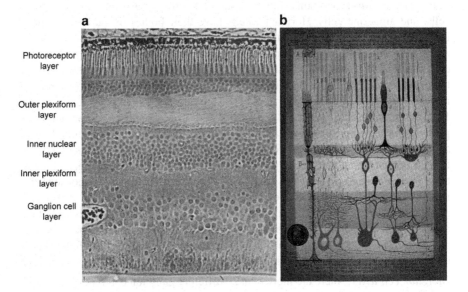

Fig. 2.2 Retinal layers. (**a**) Light micrograph of a vertical section of the human retina from [29]. (**b**) Cross-sectional microscopic drawing by Ramón y Cajal from [127, 176]

The functional anatomy of the retina is enormously rich and complicated. A short overview is provided here, to set a basis to understand the next few stages of the visual hierarchy.

The three nuclear layers are the photoreceptor layer (which lies on the back on the retina, farthest from the light coming in), the inner nuclear cell layer (in the middle) and the ganglion cell layer (nearest to the center of the eye).

Photoreceptor layer: light is transduced into electrical signals by photoreceptors: rods and cones. Cones are not sensitive to dim light, but under photopic conditions (bright light) they are responsible for fine detail and color vision. Rods are responsible for our vision under scotopic conditions (dim light), and saturate when the level of light is high. Rods and cones are distributed across the retina with very different profiles: in the fovea, where our fine vision is most detailed, cones are very densely packed (up to 160,000 cones/mm^2) but cone density drops rapidly as we move away from the fovea. Rods are absent from the fovea [190], but their density rises quickly to reach a peak at an eccentricity between 5 and 7 mm, beyond which they steadily decline in number [45, 46, 164]. Humans have one type of rod and three types of cones. The three types of cones, responsible for color vision, are called L (or red) cones, M (or green) cones, and S (or blue) cones, and they are most sensitive to different segments of the spectrum of light: L cones are most sensitive to long wavelengths (peak sensitivity at 564 nm), M cones are most sensitive to middle wavelengths (peak sensitivity at 533 nm) and S cones are most sensitive to short wavelengths (peak sensitivity at 437 nm) [32, 33, 131]. L, M, and S cones are distributed in the retina in a particular way: only 10% of the cones are S cones, and they are absent from the fovea. Although L cones and M cones are randomly intermixed, there are ~2 times more L cones than M cones [1, 41, 44, 152, 187].

Inner nuclear layer: contains three classes of neurons: horizontal cells, bipolar cells, and amacrine cells. Horizontal cells have their bodies in the inner nuclear layer and connect to photoreceptors (through chemical synapses) and other horizontal cells (through gap junctions) in the outer plexiform layer [223]. Horizontal cells receive input from photoreceptors, but they also give output to the same photoreceptors, providing lateral inhibition, which acts to enhance spatial differences in photoreceptor activation at the level of the bipolar cells [49, 222]. There are over 13 different types of bipolar cells [30, 105] and all of them have some dendritic processes in the outer plexiform layer, the soma in the inner nuclear layer and some axon terminals in the inner plexiform layer [66]. The dendritic processes of a bipolar cell receive input from one type of photoreceptor (either from cones or from rods, but never from both) [186]. Each bipolar cell then conveys its response to the inner plexiform layer, where it contacts both amacrine and ganglion cells [49]. Amacrine cells (over 30 different types), receive input from bipolar cells and other amacrine cells, and pass their messages onto bipolar cells, other amacrine cells, and ganglion cells [50, 128]. Different types of amacrine cells may have different functions in retinal processing, but their specific roles remain unknown for the most part.

Ganglion cell layer: there are more than 20 different ganglion cell types [105], and many of them are specialized on coding some particular aspect of the visual world such as sign-of-contrast and color [186]. Ganglion cells receive their input from amacrine and bipolar cells, and send their outputs to the brain in the form of action potentials through the optic nerve. These are the first cells in the visual pathway that produce action potentials (all-or-none) as their output; all the previous cell classes (photoreceptors, horizontal, bipolar and amacrine cells) release their neurotransmitters in response to graded potentials. Even though there are over 20 different types of ganglion cells, two of them account for almost 80% of the ganglion cell population [171]: the midget and the parasol ganglion cells, named by Polyak [173]. Near the fovea each midget ganglion cell receives direct input from only one midget bipolar cell [104, 106] and thus has a very small and compact receptive field (it collects input from a small number of cones). Parasol cells receive their direct input from diffuse bipolar cells, have larger dendritic fields, and thus receive input from many more cones [224]. The dendritic field size increases with eccentricity for both types of cells [48, 51, 224]. Away from the fovea, the increase in dendritic field size with retinal eccentricity is more or less matched by a decrease in spatial density, so the amount of retina covered is approximately constant over most of the retina [224].

2.2.2 Physiology and Receptive Fields

The receptive fields of ganglion cells in the retina are approximately circular and have functionally distinct central and peripheral regions (called center and surround); stimulation of these two regions produces opposite and antagonistic effects upon the activity of the ganglion cells. Ganglion cells respond optimally to differential illumination of the receptive field center and surround. Diffuse illumination of the whole receptive field produces only weak responses. There are two main types of center-surround receptive fields: on-center receptive fields respond best to light falling on the center, and darkness falling on the surround; off-center receptive fields respond best to darkness on the center and light on the surround (Fig. 2.3). The properties of center-surround receptive fields change during scotopic conditions: the size of the receptive field center usually increases, the surround strength diminishes and there is a longer latency for the response [16, 28, 65, 144, 156, 167].

Werblin and Dowling [225], and Kaneko [99] discovered that bipolar cells also have center-surround receptive fields.

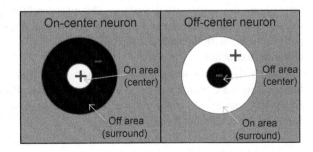

Fig. 2.3 Concentric receptive fields of retinal ganglion neurons

In the dark, photoreceptors are depolarized and continuously active [205], releasing glutamate to bipolar and horizontal cells. When light arrives and the photo pigments bleach within a photoreceptor, that photoreceptor hyperpolarizes, and the amount of glutamate released decreases in a graded manner, as a function of the number of photons [204]. All photoreceptors use the same neurotransmitter, glutamate, and so on-center and off-center bipolar cells acquire their preference by having one of two types of glutamate receptor [150]:

– On-center bipolar cells have metabotropic receptors that make the cell hyperpolarize when they receive glutamate [159, 199]. When light hits photoreceptors, they hyperpolarize and release less glutamate. This reduces the inhibition in the bipolar cells that therefore increase their activity. In the dark, photoreceptors depolarize and release more glutamate. Therefore the bipolar cells hyperpolarize.
– Off-center bipolar cells have ionotropic receptors that depolarize the cell when receiving glutamate [161, 200]. In this case, when light arrives to the retina, the photoreceptors hyperpolarize and release less glutamate. Consequently, the bipolar cells decrease their activity. In the dark, the photoreceptors depolarize and release more glutamate. As a consequence, the bipolar cells depolarize.

Both on- and off-bipolar cells make the same kind of contacts in the inner plexiform layer. All bipolar cells release glutamate as their neurotransmitter and all the ganglion cells have ionotropic receptors: therefore, ganglion cells that receive input from on-center bipolar cells are also on-center. Ganglion cells that receive input from off-center bipolar cells are off-center [186]. In 1978 Nelson et al. discovered that there is a clear anatomical difference between on- and off-bipolar cells: they synapse onto ganglion and amacrine cells within different sublayers within the inner plexiform layer. The off-center bipolar dendrites make synapses closer to the inner nuclear layer whereas the on-center bipolar dendrites terminate closer to the ganglion cell layer [47, 160].

As described earlier, there are two predominant types of ganglion cells: midget and parasol [173]. Both types of ganglion cells have center-surround receptive fields with similar spatial organization, but physiological studies have described several differences between them: parasol cells respond more transiently to light onset or offset than midget cells [82]; parasol cells have larger receptive fields centers than midget cells at the same eccentricity [55]; most midget cells have spectral selectivity and antagonism while most parasol cells do not [55, 57, 82]; parasol cells respond much more vigorously than midget cells to small changes in luminance contrast [102]. The anatomical and functional differences between midget and parasol cells lead to two different visual pathways that remain segregated throughout the early visual system. The parvocellular pathway starts with the midget cells and is very sensitive to color and spatial frequency. The magnocellular pathway starts with the parasol cells and is most sensitive to luminance contrast and temporal frequency.

Due to the center-surround organization of the ganglion cell receptive fields, these neurons are quite insensitive to changes in overall levels of luminance. They signal differences within their receptive fields by comparing the degree of illumination between the center and the surround.

2.3   LGN

All retinal ganglion cells send their axons to the brain via the optic nerve. The axons decussate at the optic chiasm, so the information from each nasal hemiretina is sent to the contralateral hemisphere. Retinal ganglion cells project to three major subcortical targets: the pretectum, the superior colliculus, and the lateral geniculate nucleus (LGN) of the thalamus. The LGN is the principal structure that sends visual information to the visual cortex, with input from 90% of the retinal ganglion cells. The LGN is laid out so that neighboring neurons are stimulated by adjacent regions in visual space. This property is called retinotopic organization.

In primates, the LGN contains six layers of cell bodies that can be classified in two groups according to their histological characteristics: the two bottom layers (ventral) contain large cell bodies and are called magnocellular layers; cells in the four upper layers (dorsal) are smaller and are called the parvocellular layers. The parvocellular layers receive their main inputs from the midget ganglion cells in the retina. The magnocellular layers receive their main inputs from parasol ganglion cells [42, 109, 170, 189, 192]. Between each of the magno and parvo layers lies a zone of very small cells: the koniocellular layers. Konio cells are functionally and neurochemically distinct from magno and parvo cells [87]. The finest caliber retinal axons, presumably originating from retinal ganglion cells that are morphologically distinct from those projecting to magno and parvo layers [109], innervate the koniocellular layers [42]. The koniocellular pathway starts with the small bistratified ganglion cells of the retina that are sensitive to blue (or S-cone) activation. The koniocellular layers interdigitate between the primary six layers of the LGN [86].

Each LGN receives input from both eyes, but the input from each eye is segregated to different monocular layers: layers 1, 3, and 6 get input from the contralateral eye, whereas layers 2, 4, and 5 get input from the ipsilateral eye [94].

Hubel and Wiesel discovered that LGN receptive fields have a similar center-surround configuration to retinal ganglion cells, however the suppressive strength of the surround is stronger than in retinal cells [90].

Virtually all parvocellular cells (99%) present linear spatial summation. That is, the response to two elements presented simultaneously to the receptive field equals the sum of the response to each of the elements presented separately. About 75% of magnocellular cells are also linear, the other 25% are not [101].

The LGN is often called a relay nucleus because it is the only structure between the retina and the cortex. However, LGN neurons are part of a complex circuit that involves ascending, descending and recurrent sets of neuronal connections [5, 194, 201]. The major source of descending input comes from neurons in layer 6 of V1. These feedback connections can be excitatory (through direct monosynaptic connections) or inhibitory (through inhibitory interneurons in the LGN or the reticular nucleus of the thalamus) [67, 83]. The functions of the corticothalamic pathway are still under discussion [5]. These connections could help to explain LGN neurons' extra-classical receptive field properties, such as the effects of suppressive field [5, 27, 38, 158]. It is generally agreed that these feedback connections act by modulating the responsiveness of the LGN neurons, and not by driving the actual responses [193]. It is possible that the major role of feedback in the visual system is to maintain top-down attention [124, 125].

2.4   V1

2.4.1   Anatomy

LGN neurons send their axons through the optic radiations to the back of the brain, where the primary visual cortex, area V1, is located. V1 is virtually the only target of primate LGN neurons [19, 35]. The magnocellular and parvocellular pathways that started in the retina remain largely separated.

V1, like most cortical areas, has six main layers [31]. Most of the LGN inputs arrive to layer 4, which is divided in four sublayers: sublayer 4Cα receives axons mostly from magnocellular neurons. Sublayer 4Cβ (and sublayer 4A to a lesser extent) receives axons mostly from parvocellular neurons. Layer 6 receives weak input from collaterals of the same LGN axons that provide strong input to layer 4C [22, 85, 94, 116]. Neurons from the koniocellular layers in the LGN send their axons to layer 1 and layers 2–3 [87, 112].

Layer 4Cα sends its output to 4B [37, 72, 117]. Axons from neurons in 4Cβ terminate in the deepest part of layer 3 [37, 72, 107]. Layers 2, 3, and 4B project mainly to other cortical regions [36] and also send axons to layer 5 [23]. Layer 5 projects back to layers 4B, 2, 3 [37] and to the superior colliculus [118]. Layer 6 projects to the LGN [73, 227] and also sends axons to several V1 layers [117, 227]. Many of the projection pyramidal cells in layers 2, 3, 4B, 5 and 6 have collaterals that connect locally. Layer 1 contains few cell bodies, but many axons and dendrites synapse there [119]. Figure 2.4 shows a schematic representation of the main connections.

In addition to the feedforward input coming from the LGN, V1 receives direct feedback from areas V2, V3, V4, V5 (or MT), MST, FEF, LIP and inferotemporal cortex [17, 169, 184, 196, 203, 213, 214]. The projections from these areas terminate in layers 1, 2, and 5 of V1, with occasional arbors in layer 3 [185, 197].

2.4.2   Physiology and Receptive Fields

In primates, the receptive fields of most V1 input neurons (layer 4C) have the same center-surround organization as the LGN neurons they receive direct input from [18, 21, 34, 113, 172]. Outside of layer 4C, the receptive field structure is very

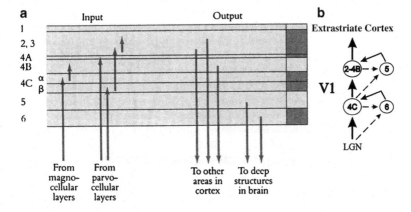

Fig. 2.4   Schematic representation of V1 inputs, outputs and vertical interconnections. (**a**) From [88]. (**b**) From [36]

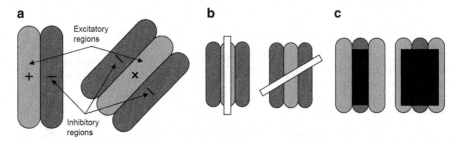

Fig. 2.5   (**a**) Schematic representation of simple cell receptive fields with different orientations and number of subregions. (**b**) Receptive field selective to vertical orientations. A vertical light bar over the excitatory region is the optimal stimulus (*left*). A non-vertical light bar (*right*) that partially falls on the inhibitory regions makes the cell fire less. (**c**) Cell stimulated with a bar of the preferred spatial frequency (*left*) and with a bar that is too wide and thus falls on the opposite contrast subregions

different and we can distinguish two main groups of cells according to their receptive field type: simple cells and complex cells [91].

Simple cells: Hubel and Wiesel first described the receptive fields of "simple cells" in area V1 [89]. The receptive fields of simple cells are organized in distinct elongated on and off antagonistic subregions, whose spatial arrangement determines the responses of the neuron to different stimuli. Simple cells are selective to the orientation and spatial frequency of the stimulus (Fig. 2.5). The response of simple neurons is reduced when there is a mismatch between the light and dark parts of the stimulus and the on- and off-regions of the receptive field. By testing the neuron's responses to different stimuli, it is possible to generate tuning curves for orientation and spatial frequency.

Hubel and Wiesel [91] proposed that each simple cell gets its input from an array of center-surround receptive fields of the same sign that have their centers arranged

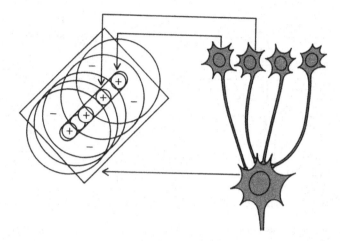

Fig. 2.6 Schematic representation of the feed forward excitatory model proposed by Hubel and Wiesel in 1962. From [91]

Fig. 2.7 A complex cell gives the same response to bars anywhere within the receptive field, and does not prefer either light or dark bars

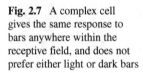

Position invariance

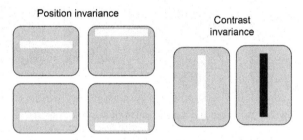

Contrast invariance

along a straight line on the retina. The synapses from the center-surround receptive fields to the simple cell are excitatory and this gives the simple receptive fields its elongated shape and orientation selectivity (Fig. 2.6). Recent studies have provided strong support for this model [9, 69–71, 180, 216].

Complex cells: complex cells in the primary visual cortex, discovered by Hubel and Wiesel, are selective to the orientation and spatial frequency of stimuli (like simple cells) but their receptive fields do not have distinctive on and off subregions [91]. Consequently, complex receptive fields are invariant to the spatial phase (position of the stimulus within the receptive field) and contrast polarity of the stimulus. When a single bar is presented within the receptive field, complex cells respond equally well regardless of the bar's position and contrast, as long as the bar has the preferred orientation and width (Fig. 2.7). When pairs of bars are presented simultaneously within the receptive field, complex cells exhibit nonlinearity in spatial summation [91]: the response to simultaneous presentation of two stimuli cannot be predicted from the sum of the responses to the two stimuli presented individually.

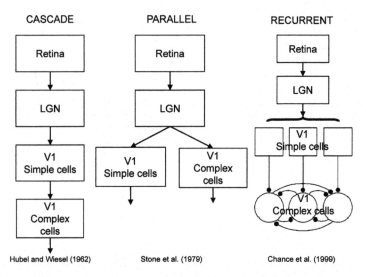

Fig. 2.8 Different hypothesis about the connectivity of complex cells. After [142]

This is a fundamental property of complex cells; simple cells are more or less linear [39, 91, 155, 182].

The circuits that gives rise to complex cells is not fully understood; there are several different hypotheses in the literature, some of which are shown in Fig. 2.8. The "cascade model" [91] suggests that simple cells and complex cells represent two successive stages in hierarchical processing: in a first stage, simple cells are created from the convergence of center-surround inputs that have receptive fields aligned in visual space. In the second stage, complex cells are then generated by the convergence of simple cells inputs with similar orientation preferences (Fig. 2.8, left). "Parallel models" [202] propose that simple cells and complex cells are both constructed from direct geniculate inputs. Simple cells are created from the convergence of linear LGN inputs, and complex cells from the convergence of non-linear LGN inputs (Fig. 2.8, middle). "Recurrent models" [40] use a combination of weak simple cell inputs and strong recurrent complex cell inputs to generate complex cell nonlinearities (Fig. 2.8, right). Martinez and Alonso [8, 142, 143] published evidence supporting the Hubel and Wiesel cascade model.

End-stopped cells: ordinary simple and complex cells show length summation: the longer the bar stimulus, the better the response, until the bar is as long as the receptive field; making the bar even longer has no further effect. For end-stopped cells, lengthening the bar improves the response up to some limit, but exceeding that limit in one or both directions results in a weaker response. The same stimulus orientation evokes maximal excitation on the activating region and maximal inhibition on the outlying areas. Hubel and Wiesel discovered and characterized end-stopped cells in cat areas 18 and 19 and initially called them hypercomplex cells [92]. Later Gilbert showed that some simple and complex cells in cat area 17 are also end-stopped [25, 78]. Several recent studies suggest that most primate V1 cells

Fig. 2.9 A *curved border* would be a good stimulus for the end-stopped cell represented in the diagram. From [88]

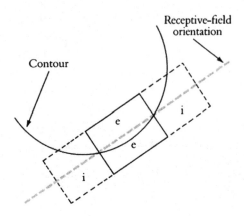

Contour

Receptive-field orientation

e

e

i

i

are somewhat end-stopped [97, 100, 103, 165, 188]. The receptive field structure of end-stopped cells makes them especially sensitive to corners, curvature and terminators [88, 92] (Fig. 2.9).

Columnar organization: A fundamental feature of cortical organization is the spatial grouping of neurons with similar properties. V1 is functionally organized in layers and cortical columns, which are roughly perpendicular to the layers. The concept of cortical columns was introduced by Mountcastle in the somatosensory system [153, 154, 174], although Lorente de Nó had envisaged their existence through his anatomical studies [114]. Hubel and Wiesel discovered columnar organization in area V1, first in the cat [91] and then in the primate [93, 95, 226]. They showed that V1 cells with similar properties are grouped into columns: as they advanced an electrode in an orthogonal penetration from the cortex surface, they found that the neurons recorded by the electrode had similar receptive field axis orientation, ocular dominance, and position in the visual field.

– Ocular dominance columns: the inputs from the two eyes are segregated in layer 4, where cortical neurons are driven monocularly. In any given column extending above and below layer 4, all the cortical neurons, even if driven by both eyes, share the same eye preference. Ocular dominance columns form an interdigitating pattern on the cortex [91, 93, 226]. Figure 2.10 shows an ocular dominance map obtained with intrinsic optical imaging: we can see distinct strips in a 1 cm^2 patch of cortex, activated by a stationary bar presented monocularly to the visual system of a rhesus monkey.

– Orientation columns: Hubel and Wiesel [91, 93, 95] found that, just as with eye dominance, orientation preference remains constant in orthogonal penetrations through the cortical surface: the cortex is subdivided into narrow regions of constant orientation, extending from the surface to the white matter but interrupted by layer 4C, where most cells have no orientation preference [18, 21, 34, 113, 172] (although some recent studies have found orientation selective cells in layer 4C [84, 181, 191]). In a tangential electrode penetration, the orientation

Fig. 2.10 A 1 cm^2 image from cortical area V1 in a primate. The *stripes* indicate an ocular dominance map created when visual stimuli are displayed to the right eye versus the left eye [121]

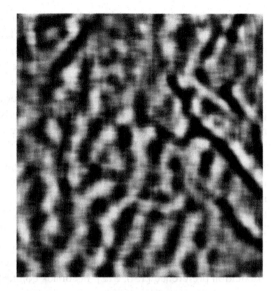

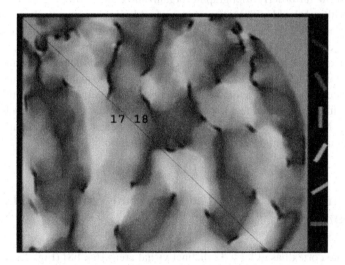

Fig. 2.11 An orientation map of the V1/V2 border from a cat (V1 and V2 are called area 17 and 18 in cats, by convention) obtained with intrinsic optical imaging. The legend on the right shows the relationship between the color of each pixel and orientation. The brightness of each pixel indicates the selectivity of each point in the map: *dark* indicates points in the map that are not particularly selective to any orientation, while *bright* points signify points in the map that are tuned specifically to a given orientation [127]

preference usually changes gradually. Figure 2.11 shows an orientation selectivity map in areas 17 and 18 of the cat visual cortex (equivalent to areas V1 and V2 in the primate) obtained with optical intrinsic signal: the image shows the preference of neurons to lines of different orientations, when presented to the retina.

Optical imaging studies have provided precise details about the columnar organization: orientation columns are arranged radially into pinwheel-like structures with orientation preference shifting gradually along contours circling the pinwheel center [20, 26]. Each pinwheel center tends to occur near the center of an ocular dominance patch [43, 115], and iso-orientation contours tend to cross ocular dominance boundaries at right angles [162]. Cortical columns where orientation preference changes smoothly or remains essentially constant are interspersed with regions containing orientation singularities where the orientation changes abruptly by up to 90° [24, 26, 54, 95, 211].

Horizontal and feedback connections: Many local connections in V1 have a wide lateral distribution, including long intralaminar connections spreading several millimeters [36]. Prominent horizontal connections are those originating from and terminating in layers 2–3 and 4B; these connections arise from neurons whose long-distance axon collaterals form periodic clusters [10, 37, 80, 81, 146, 183]. These clusters tend to preferentially link columns of neurons with similar response properties: in cats, ferrets, and monkeys they preferentially link columns with similar orientation preference [130, 212]. Feedback connections from extrastriate cortex to V1 also show an orderly topographic organization and terminate in a patch-like manner within V1 [11]. These two types of orderly connections (horizontal and feedback) may be involved in the generation of suppressive fields in V1 neurons, as well as other extra-classical receptive field modulations [11, 38, 39, 79, 110]. Intra cortical connections may be important to understand the neural computations carried out in V1. Zhaoping has proposed that V1 creates a saliency map using intra cortical mechanisms. This saliency map can be used to attract attention to a visual location without top-down factors, which may explain certain visual search properties [233]. Macknik and Martinez-Conde have proposed that the primary role of feedback may be the maintenance of top-down attention [124, 125].

2.5   Extrastriate Cortex: The Dorsal and Ventral Visual Pathways

The primate cortex has at least 32 distinct visual areas [64, 68] (Fig. 2.12).

In the first two stages of cortical processing (V1 and V2), the magnocellular and the parvocellular pathways are largely segregated: inputs from the LGN arrive to different sublayers in V1 according to their magno/parvo origin and projections from V1 layer 4C are also fairly separated in V1 and V2 as revealed by cytochrome oxidase staining [111, 113, 163, 219]. After V1 there are two main processing streams, associated with different visual capabilities [64, 144, 215]:

– The dorsal or parietal stream is tuned to moving stimuli (with similar properties to the magnocellular pathway). After V2 the information flows to MT, MST and other intermediate areas. MT neurons are selective to the direction of stimulus motion, speed and binocular disparity [4, 13, 229, 230]. The highest stages of this stream are clustered in the posterior parietal cortex. This stream is involved

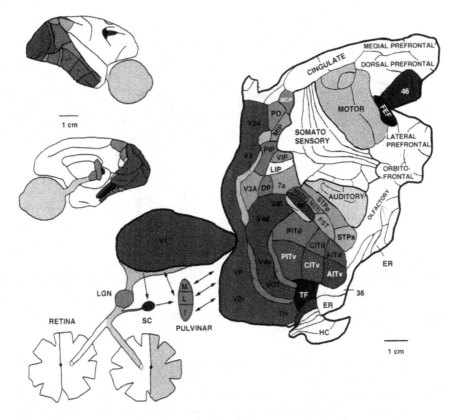

Fig. 2.12 Visual areas in the primate shown in a flattered brain. From [218]

in assessment of spatial relationships and it is often called the "Where" pathway.

– The ventral or temporal stream emphasizes form and color analysis (similar properties as the parvocellular pathway). After V2 the information flows to V4 and other intermediate areas; many V4 neurons are selective to stimulus color [231, 232], orientation, width, and length of bars [61], curvilinear and linear gratings [74, 75], and contour features like angles and curves [166]. The highest stages of this stream are clustered in the inferotemporal cortex. This stream is concerned with visual recognition of objects as it is often called the "What" pathway.

The transformations of the visual image that occur along each of these pathways do not appear to result in increased selectivity for basic parameters [145] such as direction or speed [4] in the dorsal pathway or wavelength [56] or orientation [62] in the ventral pathway. Rather than sharpening basic tuning curves, the transformation of information along each of the pathways appears to construct new, more complex response properties; both pathways may use similar computational strategies for processing information [145]. Also, retinotopic specificity decreases progressively in successive levels of each of the pathways: the average receptive field size in MT is 100 times

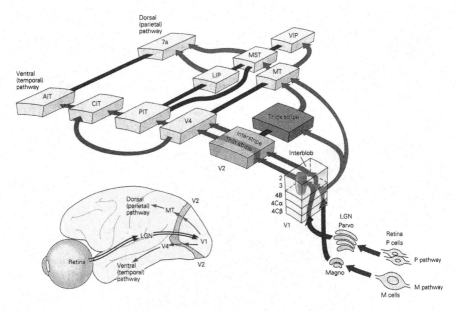

Fig. 2.13 Schematic of the two visual pathways in the primate, showing the main connections between the different areas. From [98]

larger than in V1 [76]. In MST receptive fields can cover a full quadrant of the visual field [63]. V4 receptive fields are about 30 times larger than V1 receptive fields [129, 220], and downstream in the ventral pathway they become over 100 times larger [60].

The hypothesis of two distinct streams of processing was initially formulated by Ungerleider and Mishkin [215]. Many different groups have provided anatomical, physiological, and behavioral support to this idea. In humans, clinical observations indicate that damage to the parietal cortex can affect visual perception of position, leaving object recognition unimpaired [52, 178, 234]. Temporal lobe lesions can produce specific deficits related to object recognition [53, 147, 148, 167]. Systematic lesion studies in primates have found a functional separation between the temporal and the parietal cortices [58, 151, 213].

While it is widely accepted that information is computed in these two largely parallel visual pathways (as shown in schematic on Fig. 2.13 taken from [98]), it is important to note that the separation between the two pathways is far from complete. There is anatomical and physiological evidence of substantial cross-talk between the two streams [68, 149, 218].

2.6   The Role of Spatiotemporal Edges in Early Vision

Information flows from one visual area to the next in the form of excitatory signals carried through glutamate synapses. Therefore, all inhibition between neurons, for instance to form receptive fields, is a function of local inhibitory circuits.

Local inhibition, which underlies center-surround receptive field organization, is enacted through the neurotransmitter GABA (gamma-aminobutyric acid). In 1965 Hartline and Ratliff delineated the far-reaching consequences of this simple arrangement, in terms of spatial and temporal visual processing [179]. They showed that the three components of a laterally inhibitory circuit:

1. Excitatory input and output: information arrives at a given visual area of the brain in the form of excitatory neural responses, and the information is sent to the next visual area(s) in the visual hierarchy as excitatory neural responses as well.
2. Lateral inhibition: occurs as a function of excitatory activation (thus inhibition follows excitation in time).
3. Self-inhibition: neurons that laterally inhibit their neighbors also inhibit themselves.

Figure 2.14 shows a plausible mammalian descriptive model of lateral inhibition, based on Hartline and Ratliff's original Limulus model [123]. The model predicts

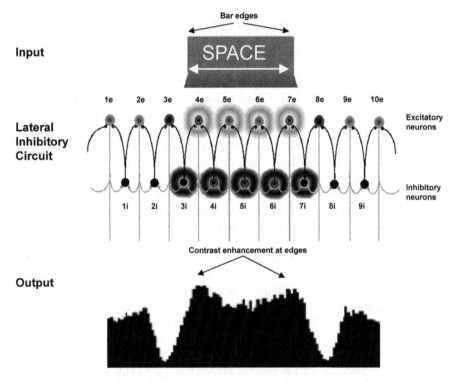

Fig. 2.14 A mammalian representation of the spatial lateral inhibition model originally proposed by Hartline and Ratliff. The excitatory neurons in the center of the upper row receive excitatory input from a visual stimulus. This excitation is transmitted laterally to the inhibitory neurons just outside the stimulus, and also within the area impinged upon by the stimulus. The inhibitory interactions between excited neurons at the edges of stimuli and their non-excited neighbors results in apparent contrast enhancement at the borders of the stimulus. Output of each of the excitatory neurons is represented in action potentials per unit time at the bottom [127]

Fig. 2.15 This Mach Band demonstration was originally designed by Chevreul in 1839. Notice how each vertical stripe appears to be lighter on the left than on the right. This illusory effect is due to contrast enhancement at the borders

that the strongest neural excitatory signals to a visual stimulus will occur just inside the stimulus' spatial borders. Neural inhibition, moreover, is strongest just outside of the borders. The spatial interiors of stimuli do not cause responses in visual neurons. It is hypothesized that the interiors of large spatial stimuli are visible through the illusory process of filling-in. One perceptual consequence of lateral inhibition is that stimuli to both sides of a luminance border are differentially enhanced in an illusory fashion (as in Fig. 2.15).

If we now examine two of the neurons in a lateral inhibitory network through time, one neuron being excitatory and the other inhibitory, we should expect the following specific temporal pattern of response (Fig. 2.16). Visual information enters a given visual area as excitatory input to specific neurons that are tuned to the specific visual stimulus being presented. The excited neurons then locally inhibit their neighbors, and also themselves, in a delayed inhibitory response that serves to bring suppress the initial transient onset-response. This state of excitatory-inhibitory equilibrium continues until such point that the excitatory input representing the stimulus is extinguished. After that point, the neurons briefly enter a state of suppression due to the fact that delayed inhibition is unopposed by excitation (a refractory period called the time-out), followed by a disinhibitory rebound, called an after-discharge. Just as neurons respond strongly to the spatial borders of stimuli due to lateral inhibition, so too do they respond strongly to the temporal borders (the stimulus onsets and terminations (also commonly called "stimulus offsets," although this term is linguistically incorrect)). The perceptual result of this is contrast enhancement at the temporal borders of stimuli. Lateral inhibition is thus responsible not only for the spatial layout of receptive fields, but also for their temporal response properties. The perceptual result of this process is that the perceived

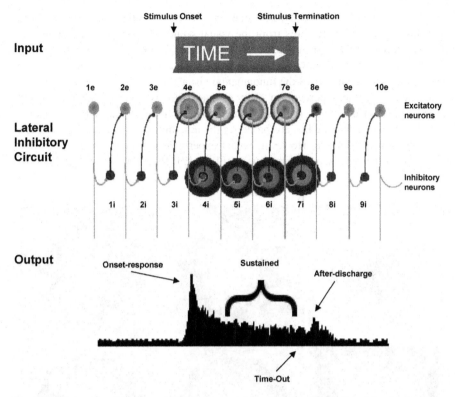

Fig. 2.16 One excitatory and one inhibitory neuron, followed through a period of time in which the stimulus is off (times 1, 2 and 3), on (times 4, 5, 6, and 7), and then off (times 8, 9 and 10) [127]

contrast of a stimulus is highest just after it turns on and then again after it turns off. Visual masking (the effect in which the visibility of a target stimulus is reduced by a masking stimulus that does not overlap the target in space or time) occurs perceptually when the neural responses to the target onset and/or termination are inhibited, suggesting that the onset-response and after-discharge are critical for the visibility of stimuli [120, 122, 126].

2.7 The Role of Corners in Early Vision

2.7.1 Overview

Our perception of the visual world is constructed, step-by-step, by neurons in different visual areas of the brain [59, 68, 91, 195]. While feedback certainly plays a role in the visual system [6, 7, 96, 124, 125, 133, 157], the visual system's overall tendency is towards a hierarchy, in which neurons in sequential levels extract more and more complicated features from the visual scene. These features include (but are not limited to) color, brightness, movement, shape, and depth.

In order to determine how visual perception is constructed in our brain, we need first to establish the nature of the fundamental visual features in a scene. Theories of shape and brightness perception have primarily focused on the detection and processing of visual edges. Early visual neurons are thought of as "edge detectors" [91, 132], and current studies are based on the assumption that edges are the most elementary visual feature. However, recent experiments show that corners can be more salient than edges, both perceptually (Fig. 2.17) [206, 208] and in the responses

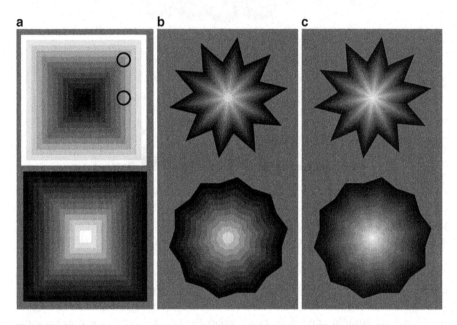

Fig. 2.17 Vasarely's nested squares and alternating brightness star illusions. (**a**) Nested squares illusion, based in Vasarely's "Arcturus" [221]. *Top*: The stimulus is made out of multiple concentric squares of increasing luminance (going from *black* in the center to *white* in the outside). The two circles indicate two regions that appear to have significantly different brightness. The area inside the upper circle has higher average luminance than the region inside the lower circle; however the region inside the upper circle appears perceptually darker. *Bottom*: Nested square stimulus, with a gradient of decreasing luminance (from the center to the outside). From [206]. (**b, c**) The Alternating Brightness Star illusion [134]. The stimulus is made of concentric stars of graded luminance. In the examples illustrated, the innermost star is white; the outermost star is black. The illusory corner-folds that radiate from the center appear as light or dark depending on the polarity of the corner angle; Corner Angle Brightness Reversal effect. Moreover, the illusory folds appear more salient with sharp corners (*top stars*), and less salient with shallow corners (*bottom stars*); Corner Angle Brightness Variation effect. However, all illusory folds are physically equal to each other in luminance. (**b**) The gradient from the center to the outside has ten luminance steps, and so the individual stars forming the polygonal constructs are easy to identify. (**c**) The gradient from the center to the outside has 100 luminance steps. From [208]

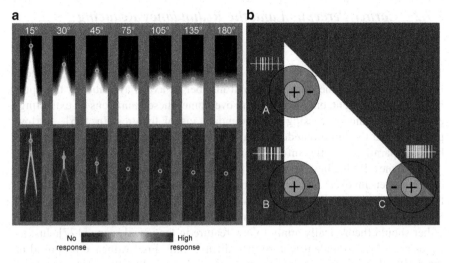

Fig. 2.18 Center-surround receptive field responses to corners of varying angles. (**a**) Computational simulations with a DOG filter. The filter parameters were chosen to match physiological center-surround receptive fields at the eccentricity used in the psychophysical experiments (3°). *Top*: Examples of corner-gradient stimuli analyzed in the simulations. The circles mark the point of 50% luminance. *Bottom*: Convolving the DOG filter with the stimuli in (A) simulates the output of an array of center-surround neurons. The circles indicate the responses of the model at the point of 50% luminance on the actual gradient. (**b**) Generalized model of corner processing. Three on-center receptive fields are respectively placed over one edge and two corners of a *white triangle*. The center of the receptive field over the edge (position A) is well stimulated by light, but most of the surround also falls in the light region, so the response of the neuron is partially inhibited. The center of the receptive field over the 90° corner (position B) is also stimulated by light and most of the surround falls in the dark area. This is a more optimal stimulus than in (A) and leads to a stronger neural response. The receptive field over the 45° corner (position C) receives even more optimal contrast between center and surround, leading to an even stronger response. The spiking responses depicted in the cartoon are hypothetical. From [206]

of neurons throughout the visual hierarchy, even in early stages (Fig. 2.18) [206, 210]. Combined results from human psychophysics experiments, human brain imaging, and computational modeling suggest that deflections or discontinuities in edges, such as corners, curvature, and terminating line endings, may be first processed by center-surround receptive fields [206, 208, 210]. These data suggest that corners may be a fundamental feature for shape and brightness perception.

This hypothesis in no way rules out a critical role for later cortical areas in more complex processing of corner angles. For instance, specific orientations of corner angles must be processed cortically, given that the first orientation-selective cells are cortical.

2.7.2   Corner Perception and the Redundancy-Reducing
Hypothesis

The information transmitted by our visual system is constrained by physical limitations, such as the relatively small number of axons available in the optic nerve. To some extent, our visual system overcomes these limitations by extracting, emphasizing, and processing non-redundant visual features. In 1961, Barlow proposed that the brain recodes visual data "so that their redundancy is reduced but comparatively little information is lost." This idea is known as the "Redundancy-Reducing Hypothesis" [14, 15]. The redundancy-reducing hypothesis has been invoked as an explanation for why neurons at the early levels of the visual system are suited to perform "edge-detection," or "contour-extraction." However, redundancy reduction is not necessarily constrained to edges, but rather should theoretically apply to any feature in the visual scene [177]. Just as edges are a less redundant feature than diffuse light, Fred Attneave proposed in the 1950s that "points of maximum curvature" (i.e., discontinuities in edges, such as curves, angles and corners – any point at which straight-lines are deflected) are even less redundant than edges themselves, and thus contain more information [12]. If points of high curvature are less redundant than points of low curvature, then sharp corners should also be less redundant than shallow corners. This hypothesis is consistent with experiments showing that sharp corners are perceptually more salient and generate stronger physiological responses than shallow corners [206, 208, 210].

2.8   Effects of Fixational Eye Movements in Early Visual
Physiology and Perception

2.8.1   Overview

As we read a page of text, our eyes rapidly flick from left to right in small hops, bringing each word sequentially into focus. When we look at a person's face, our eyes similarly dart here and there, resting momentarily on one eye, the other eye, mouth and other features. But these large eye movements, called saccades (Fig. 2.19a), are just a small part of the daily workout our eye muscles get. Our eyes *never* stop moving: even when they are apparently fixated on something, they still jump and jiggle imperceptibly in ways that turn out to be essential for seeing. The tiny eye motions that we produce whenever we fixate our gaze are called fixational eye movements (Fig. 2.19b) [139]. If these miniature motions are halted during fixation, all stationary objects simply fade from view.

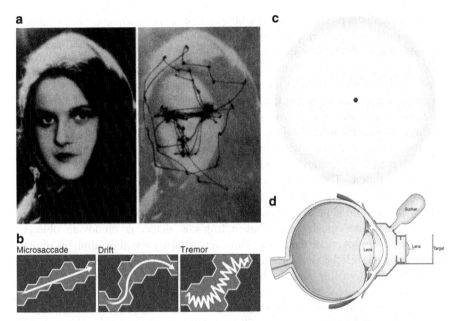

Fig. 2.19 Fixational eye movements and visual fading. (**a**) An observer views a picture (*left*) while eye positions are monitored (*right*). The eyes jump, seem to fixate or rest momentarily, producing a small dot on the trace, then jump to a new region of interest. The large jumps in eye position illustrated here are called saccades. However, even during fixation, or "rest" times, eyes are never still, but continuously produce fixational eye movements: drifts, tremor, and microsaccades. From [228]. (**b**) Cartoon representation of fixational eye movements in humans and primates. Microsaccades (straight and fast movements), drifts (curvy slow movements) and tremor (oscillations superimposed on drifts) transport the visual image across the retinal photoreceptor mosaic. From [135]. (**c**) Troxler fading. In 1804 Swiss philosopher Ignaz Paul Vital Troxler discovered that deliberately fixating on something causes surrounding stationary images to fade away. To elicit this experience, stare at the central dot while paying attention to the surrounding pale ring. The ring soon vanishes, and the central dot appears set against a while background. Move your eyes, and it pops back into view. Modified from [139]. (**d**) This drawing illustrates the suction cup technique, used by Yarbus [228] and others. This technique was very popular in early retinal stabilization studies for its simplicity, but it is now considered old-fashioned, and other, less invasive stabilization techniques are preferred. The target image is directly attached to the eyeball by means of a contact lens assembly. The target is viewed through a powerful lens. The assembly is firmly attached to the eye by a suction device. Modified from [139]

2.8.2   Neural Adaptation and Visual Fading

That the eyes move constantly has been known for centuries. In 1860 Hermann von Helmholtz pointed out that keeping one's eyes motionless was a difficult proposition and suggested that "wandering of the gaze" prevented the retina from becoming tired.

Animal nervous systems may have evolved to detect changes in the environment, because spotting differences promotes survival. Motion in the visual field may indicate that a predator is approaching or that prey is escaping. Such changes prompt visual neurons to respond with neural impulses. Unchanging objects do not generally pose a threat, so animal brains – and visual systems – did not evolve to notice them. Frogs are an extreme case, as they produce no spontaneous eye movements in the absence of head movements. For a resting frog, such lack of eye movements results in the visual fading of all stationary objects. Jerome Lettvin and colleagues stated that a frog "will starve to death surrounded by food if it is not moving." Thus a fly sitting still on the wall will be invisible to a resting frog, but once the fly is aloft, the frog will immediately detect it and capture it with its tongue.

Frogs cannot see unmoving objects because an unchanging stimulus leads to "neural adaptation." That is, under constant stimulation, visual neurons adjust their gain as to gradually stop responding. Neural adaptation saves energy but also limits sensory perception. Human neurons also adapt to sameness. However, the human visual system does much better than a frog's at detecting unmoving objects, because human eyes create their own motion, even during visual fixation. Fixational eye movements shift the visual scene across the retina, prodding visual neurons into action and counteracting neural adaptation. They thus prevent stationary objects from fading away.

The goal of oculomotor fixational mechanisms may not be retinal stabilization, but rather controlled image motion adjusted so as to overcome adaptation in an optimal fashion for visual processing [198].

In 1804, Troxler reported that precisely fixating the gaze on an object of interest causes stationary images in the surrounding region gradually to fade away. Thus, even a small reduction in the rate and size of fixational eye movements greatly impairs vision, even outside of the laboratory and for observers with healthy eyes and brains (Fig. 2.19c).

Eliminating *all* eye movements, however, can only be achieved in a laboratory. In the early 1950s, some research teams achieved this stilling effect with a tiny custom slide projector, mounted directly onto a contact lens that attached directly to the observer's eye with a suction device (Fig. 2.19d). In this setup, a person views the projected image through this lens, which moves with the eye. Using such a retinal stabilization technique, the image shifts every time the eye shifts. Thus it remains still with respect to the eye, causing the visual neurons to adapt and the image to fade away. Nowadays, researchers create this same result by measuring eye movements with a camera pointed at the eye. They transmit the eye-position data to a projection system that moves the image with the eye, thereby stabilizing the image on the retina.

Around the same time, three different types of fixational eye movements were characterized. *Microsaccades* are small, involuntary saccades that are produced when the subjects attempt to fixate their gaze on a visual target. They are the largest and fastest of the fixational eye movements, carrying an image across dozens to several hundreds of photoreceptors. *Drifts* are slow meandering motions that occur between the fast, linear microsaccades. *Tremor* is a tiny, very fast oscillation superimposed on drifts. Tremor is the smallest type of fixational eye movement, its motion no bigger than the size of one photoreceptor. See Martinez-Conde et al. [136, 139, 141] for some recent reviews of fixational eye movement parameters in humans, primates, and other vertebrates.

2.8.3   Microsaccades in Visual Physiology and Perception

Starting in the late 1990s, fixational eye movement research has focused on microsaccades. Physiological experiments found that microsaccades increase the firing of neurons in the visual cortex and lateral geniculate nucleus, by moving the images of stationary stimuli in and out of neuronal receptive fields. Firing rate increases following microsaccades were clustered in bursts of spikes, whereas individual spikes tended to occur in the periods between microsaccades. Moreover, bursts of spikes were better correlated with previous microsaccades than either single spikes or instantaneous firing rate. Bursts highly correlated with previous microsaccades had large spike numbers and short inter-spike intervals [137, 138]. Because microsaccades are related to maintaining visibility and counteracting fading (see further below), bursts that indicate previous microsaccades accurately must encompass the neural code for visibility. In area V1, optimal burst sizes following microsaccades tended to be three spikes or more. These bursts may be an important clue to the neural code or "language" that our brain uses to represent the visibility of the world [137]. The neural codes by which neurons, or neuronal populations, encode and transmit visual information are not only critical to our understanding of normal visual processing, but also to the development and refinement of neural prostheses.

Microsaccades could enhance spatial summation by synchronizing the activity of nearby neurons [137]. By generating bursts of spikes, microsaccades may also enhance temporal summation of responses from neurons with neighboring RFs [137]. Moreover, microsaccades may help disambiguate latency and brightness in visual perception, allowing us to use latency in our visual discriminations [137]. Changes in contrast can be encoded as changes in the latency of neuronal responses [2, 3, 77]. Since the brain knows when a microsaccade is generated, differential latencies in visual responses could be used by the brain to indicate differences in contrast and salience.

Despite several decades of debate (see [139] for a review), a direct link between microsaccade production and visual perception has only recently been demonstrated. Martinez-Conde et al. [140] found that increased microsaccade production during fixation resulted in enhanced visibility for visual targets. Conversely, decreased microsaccade production led to periods of visual fading. These results established a potential causal relationship between microsaccades and target visibility during fixation, and corroborated predictions from previous physiological studies in which microsaccades were found to increase the spiking rates in visual neurons [137, 138]. Microsaccade production has been subsequently linked to perceptual transitions in various other visual phenomena, such as binocular rivalry [215], filling-in of artificial scotomas [207], and illusory motion (perceived speed as well as subjective direction [108, 209]).

Fewer studies have addressed the neural and perceptual consequences of drifts and tremor. However, all fixational eye movements may contribute significantly to visual perception, depending on stimulation conditions. For example, receptive fields in the periphery may be so large that only microsaccades are large and fast enough – compared to drifts and tremor – to prevent visual fading, especially with low-contrast stimuli. Whereas foveal receptive fields may be so small that drifts and tremor can maintain vision in the absence of microsaccades. But even if drifts and/ or tremor can maintain foveal vision on their own, this does not rule out that microsaccades could also have a role. Thus, if one were to eliminate drifts and tremor, microsaccades alone might sustain foveal vision during fixation.

References

1. Ahnelt PK, Kolb H, Pflug R (1987), *Identification of a subtype of cone photoreceptor, likely to be blue sensitive, in the human retina.* J Comp Neurol, **255**(1): p. 18–34.
2. Albrecht DG (1995), *Visual cortex neurons in monkey and cat: effect of contrast on the spatial and temporal phase transfer functions.* Vis Neurosci, **12**(6): p. 1191–210.
3. Albrecht DG, Hamilton DB (1982), *Striate cortex of monkey and cat: contrast response function.* J Neurophysiol **48**: p. 217–37.
4. Albright TD (1984), *Direction and orientation selectivity of neurons in visual area MT of the macaque.* J Neurophysiol, **52**(6): p. 1106–30.
5. Alitto HJ, Usrey WM (2003), *Corticothalamic feedback and sensory processing.* Curr Opin Neurobiol, **13**(4): p. 440–5.
6. Alonso JM, Cudeiro J, Perez R, et al. (1993), *Influence of layer V of area 18 of the cat visual cortex on responses of cells in layer V of area 17 to stimuli of high velocity.* Exp Brain Res, **93**(2): p. 363–6.
7. Alonso JM, Cudeiro J, Perez R, et al. (1993), *Orientational influences of layer V of visual area 18 upon cells in layer V of area 17 in the cat cortex.* Exp Brain Res, **96**(2): p. 212–20.
8. Alonso JM, Martinez LM (1998), *Functional connectivity between simple cells and complex cells in cat striate cortex.* Nat Neurosci, **1**(5): p. 395–403.
9. Alonso JM, Usrey WM, Reid RC (2001), *Rules of connectivity between geniculate cells and simple cells in cat primary visual cortex.* J Neurosci, **21**(11): p. 4002–15.
10. Anderson JC, Martin KA, Whitteridge D (1993), *Form, function, and intracortical projections of neurons in the striate cortex of the monkey Macacus nemestrinus.* Cereb Cortex, **3**(5): p. 412–20.
11. Angelucci A, Levitt JB, Walton EJ, et al. (2002), *Circuits for local and global signal integration in primary visual cortex.* J Neurosci, **22**(19): p. 8633–46.
12. Attneave F (1954), *Some informational aspects of visual perception.* Psychol Rev, **61**(3): p. 183–93.
13. Baker JF, Petersen SE, Newsome WT, Allman JM (1981), *Visual response properties of neurons in four extrastriate visual areas of the owl monkey (Aotus trivirgatus): a quantitative comparison of medial, dorsomedial, dorsolateral, and middle temporal areas.* J Neurophysiol, **45**(3): p. 397–416.
14. Barlow HB (1961), *Possible principles underlying the transformation of sensory messages*, in *Sensory Communication*, Rosenblith WA, Editor. MIT Press: Cambridge, MA. p. 217–34.
15. Barlow HB (1989), *Unsupervised learning.* Neural Computation, **1**: p. 295–311.
16. Barlow HB, Fitzhugh R, Kuffler SW (1957), *Change of organization in the receptive fields of the cat's retina during dark adaptation.* J Physiol, **137**: p. 228–54.
17. Barone P, Batardiere A, Knoblauch K, Kennedy H (2000), *Laminar distribution of neurons in extrastriate areas projecting to visual areas V1 and V4 correlates with the hierarchical rank and indicates the operation of a distance rule.* J Neurosci, **20**(9): p. 3263–81.
18. Bauer R, Dow BM, Vautin RG (1980), *Laminar distribution of preferred orientations in foveal striate cortex of the monkey.* Exp Brain Res, **41**(1): p. 54–60.

19. Benevento LA, Standage GP (1982), *Demonstration of lack of dorsal lateral geniculate nucleus input to extrastriate areas MT and visual 2 in the macaque monkey.* Brain Res, **252**(1): p. 161–6.

20. Blasdel G, Obermayer K, Kiorpes L (1995), *Organization of ocular dominance and orientation columns in the striate cortex of neonatal macaque monkeys.* Vis Neurosci, **12**(3): p. 589–603.

21. Blasdel GG, Fitzpatrick D (1984), *Physiological organization of layer 4 in macaque striate cortex.* J Neurosci, **4**(3): p. 880–95.

22. Blasdel GG, Lund JS (1983), *Termination of afferent axons in macaque striate cortex.* J Neurosci, **3**(7): p. 1389–413.

23. Blasdel GG, Lund JS, Fitzpatrick D (1985), *Intrinsic connections of macaque striate cortex: axonal projections of cells outside lamina 4C.* J Neurosci, **5**(12): p. 3350–69.

24. Blasdel GG, Salama G (1986), *Voltage-sensitive dyes reveal a modular organization in monkey striate cortex.* Nature, **321**(6070): p. 579–85.

25. Bolz J, Gilbert CD (1986), *Generation of end-inhibition in the visual cortex via interlaminar connections.* Nature, **320**(6060): p. 362–5.

26. Bonhoeffer T, Grinvald A (1991), *Iso-orientation domains in cat visual cortex are arranged in pinwheel-like patterns.* Nature, **353**(6343): p. 429–31.

27. Bonin V, Mante V, Carandini M (2005), *The suppressive field of neurons in lateral geniculate nucleus.* J Neurosci, **25**(47): p. 10844–56.

28. Bowling DB (1980), *Light responses of ganglion cells in the retina of the turtle.* J Physiol, **299**: p. 173–96.

29. Boycott BB, Dowling JE (1969), *Organization of the primate retina: light microscopy.* Philos Trans R Soc Lond B Biol Sci, B, **255**: p. 109–84.

30. Boycott BB, Wassle H (1991), *Morphological Classification of Bipolar Cells of the Primate Retina.* Eur J Neurosci, **3**(11): p. 1069–88.

31. Brodmann K (1909), *Vergleichende Lokalisationlehre der Grosshirnrinde in ihren Prinzipien-Dargestellt auf Grund des Zellenbaues.* Leipzig: Barth.

32. Brown PK, Wald G (1963), *Visual pigments in human and monkey retinas.* Nature, **200**: p. 37–43.

33. Brown PK, Wald G (1964), *Visual Pigments In Single Rods And Cones Of The Human Retina.Direct Measurements Reveal Mechanisms Of Human Night And Color Vision.* Science, **144**: p. 45–52.

34. Bullier J, Henry GH (1980), *Ordinal position and afferent input of neurons in monkey striate cortex.* J Comp Neurol, **193**(4): p. 913–35.

35. Bullier J, Kennedy H (1983), *Projection of the lateral geniculate nucleus onto cortical area V2 in the macaque monkey.* Exp Brain Res, **53**(1): p. 168–72.

36. Callaway EM (1998), *Local circuits in primary visual cortex of the macaque monkey.* AnnuRev Neurosci, **21**: p. 47–74.

37. Callaway EM, Wiser AK (1996), *Contributions of individual layer 2–5 spiny neurons to local circuits in macaque primary visual cortex.* Vis Neurosci, **13**(5): p. 907–22.

38. Carandini M (2004), *Receptive fields and suppressive fields in the early visual system*, in *The cognitive neurosciences*, Gazzaniga MS, Editor. MIT Press: Cambridge, MA.

39. Carandini M, Heeger DJ, Movshon JA (1997), *Linearity and normalization in simple cells of the macaque primary visual cortex.* J Neurosci, **17**(21): p. 8621–44.

40. Chance FS, Nelson SB, Abbot LF (1999), *Complex cells as cortically amplified simple cells.* Nature Neurosciece, **2**: p. 277–82.

41. Cicerone CM, Nerger JL (1989), *The relative numbers of long-wavelength-sensitive to middle-wavelength-sensitive cones in the human fovea centralis.* Vision Res, **29**(1): p. 115–28.

42. Conley M, Fitzpatrick D (1989), *Morphology of retinogeniculate axons in the macaque.* Vis Neurosci, **2**(3): p. 287–96.

43. Crair MC, Ruthazer ES, Gillespie DC, Stryker MP (1997), *Ocular dominance peaks at pinwheel center singularities of the orientation map in cat visual cortex.* J Neurophysiol, **77**(6): p. 3381–5.

44. Curcio CA, Allen KA, Sloan KR, et al. (1991), *Distribution and morphology of human cone photoreceptors stained with anti-blue opsin.* J Comp Neurol, **312**(4): p. 610–24.

45. Curcio CA, Sloan KR, Jr., Packer O, et al. (1987), *Distribution of cones in human and monkey retina: individual variability and radial asymmetry.* Science, **236**(4801): p. 579–82.

46. Curcio CA, Sloan KR, Kalina RE, Hendrickson AE (1990), *Human photoreceptor topography.* J Comp Neurol, **292**(4): p. 497–523.

47. Dacey D, Packer OS, Diller L, et al. (2000), *Center surround receptive field structure of cone bipolar cells in primate retina.* Vision Res, **40**(14): p. 1801–11.

48. Dacey DM (1993), *The mosaic of midget ganglion cells in the human retina.* J Neurosci, **13**(12): p. 5334–55.

49. Dacey DM (1999), *Primate retina: cell types, circuits and color opponency.* Prog Retin Eye Res, **18**(6): p. 737–63.

50. Dacey DM (2000), *Parallel pathways for spectral coding in primate retina.* AnnuRev Neurosci, **23**: p. 743–75.

51. Dacey DM, Petersen MR (1992), *Dendritic field size and morphology of midget and parasol ganglion cells of the human retina.* Proc Natl Acad Sci USA, **89**(20): p. 9666–70.

52. Damasio AR, Benton AL (1979), *Impairment of hand movementsunder visual guidance.* Neurology, **29**(2): p. 170–4.

53. Damasio AR, Damasio H, Van Hoesen GW (1982), *Prosopagnosia: anatomic basis and behavioral mechanisms.* Neurology, **32**(4): p. 331–41.

54. Das A, Gilbert CD (1999), *Topography of contextual modulations mediated by short-range interactions in primary visual cortex.* Nature, **399**(6737): p. 655–61.

55. De Monasterio FM, Gouras P (1975), *Functional properties of ganglion cells of the rhesus monkey retina.* J Physiol, **251**(1): p. 167–95.

56. de Monasterio FM, Schein SJ (1982), *Spectral bandwidths of color-opponent cells of geniculocortical pathway of macaque monkeys.* J Neurophysiol, **47**(2): p. 214–24.

57. De Valois RL (1960), *Color vision mechanisms in the monkey.* J Gen Physiol, **43**(6): p. 115–28.

58. Dean P (1976), *Effects of inferotemporal lesions on the behavior of monkeys.* Psychol Bull, **83**(1): p. 41–71.

59. Desimone R, Fleming J, Gross CG (1980), *Prestriate afferents to inferior temporal cortex: an HRP study.* Brain Res, **184**(1): p. 41–55.

60. Desimone R, Gross CG (1979), *Visual areas in the temporal cortex of the macaque.* Brain Res, **178**(2–3): p. 363–80.

61. Desimone R, Schein SJ (1987), *Visual properties of neurons in area V4 of the macaque: sensitivity to stimulus form.* J Neurophysiol, **57**(3): p. 835–68.

62. Desimone R, Schein SJ, Moran J, Ungerleider LG (1985), *Contour, color and shape analysis beyond the striate cortex.* Vision Res, **25**(3): p. 441–52.

63. Desimone R, Ungerleider LG (1986), *Multiple visual areas in the caudal superior temporal sulcus of the macaque.* J Comp Neurol, **248**(2): p. 164–89.

64. Desimone R, Ungerleider LG (1989), *Neural mechanisims of visual processing in monkeys*, in *Handbook of neuropsychology*, Boller F, Graman J, Editors. Elsevier: Amsterdam. p. 267–99.

65. Donner KO, Reuter T (1965), *The dark-adaptation of singleunits in the frog's retina and its relation to the regeneration of rhodopsin.* Vision Res, **5**(11): p. 615–32.

66. Dowling JE, Boycott BB (1966), *Organization of the primate retina: electron microscopy.* Proc R Soc Lond B Biol Sci, **166**(2): p. 80–111.

67. Erisir A, Van Horn SC, Sherman SM (1997), *Relative numbers of cortical and brainstem inputs to the lateral geniculate nucleus.* Proc Natl Acad Sci USA, **94**(4): p. 1517–20.

68. Felleman DJ, Van Essen DC (1991),*Distributed hierarchical processing in the primate cerebral cortex.* Cereb Cortex, **1**(1): p. 1–47.

69. Ferster D, Chung S, Wheat H (1996), *Orientation selectivity of thalamic input to simple cells of cat visual cortex.* Nature, **380**(6571): p. 249–52.

70. Ferster D, Koch C (1987), *Neuronal connectionsunderlying orientation selectivity in cat visual cortex.* Trendes Neurosci, **10**: p. 487–92.

71. Ferster D, Miller KD (2000), *Neural mechanisms of orientation selectivity in the visual cortex.* Annu Rev Neurosci, **23**: p. 441–71.

72. Fitzpatrick D, Lund JS, Blasdel GG (1985), *Intrinsic connections of macaque striate cortex: afferent and efferent connections of lamina 4C*. J Neurosci, **5**(12): p. 3329–49.
73. Fitzpatrick D, Usrey WM, Schofield BR, Einstein G (1994), *The sublaminar organization of corticogeniculate neurons in layer 6 of macaque striate cortex*. Vis Neurosci, **11**(2): p. 307–15.
74. Gallant JL, Braun J, Van Essen DC (1993), *Selectivity for polar, hyperbolic, and Cartesian gratings in macaque visual cortex*. Science, **259**(5091): p. 100–3.
75. Gallant JL, Connor CE, Rakshit S, et al. (1996), *Neural responses to polar, hyperbolic, and Cartesian gratings in area V4 of the macaque monkey*. J Neurophysiol, **76**(4): p. 2718–39.
76. Gattass R, Gross CG (1981), *Visual topography of striate projection zone (MT) in posterior superior temporal sulcus of the macaque*. J Neurophysiol, **46**(3): p. 621–38.
77. Gawne TJ, Kjaer TW, Richmond BJ (1996), *Latency: another potential code for feature binding in striate cortex*. J Neurophysiol, **76**(2): p. 1356–60.
78. Gilbert CD (1977), *Laminar differences in receptive field properties of cells in cat primary visual cortex*. J Physiol, **268**(2): p. 391–421.
79. Gilbert CD, Das A, Ito M, et al. (1996), *Spatial integration and cortical dynamics*. Proc Natl Acad Sci USA, **93**(2): p. 615–22.
80. Gilbert CD, Wiesel TN (1979), *Morphology and intracortical projections of functionally characterised neurones in the cat visual cortex*. Nature, **280**(5718): p. 120–5.
81. Gilbert CD, Wiesel TN (1983), *Clustered intrinsic connections in cat visual cortex*. J Neurosci, **3**(5): p. 1116–33.
82. Gouras P (1968), *Identification of cone mechanisms in monkey ganglion cells*. J Physiol, **199**(3): p. 533–47.
83. Guillery RW, Sherman SM (2002), *Thalamic relay functions and their role in corticocortical communication: generalizations from the visual system*. Neuron, **33**(2): p. 163–75.
84. Gur M, Kagan I, Snodderly DM (2005), *Orientation and direction selectivity of neurons in V1 of alert monkeys: functional relationships and laminar distributions*. Cereb Cortex, **15**(8): p. 1207–21.
85. Hendrickson AE, Wilson JR, Ogren MP (1978), *The neuroanatomical organization of pathways between the dorsal lateral geniculate nucleus and visual cortex in Old World and New World primates*. J Comp Neurol, **182**(1): p. 123–36.
86. Hendry SH, Reid RC (2000), *The koniocellular pathway in primate vision*. AnnuRev Neurosci, **23**: p. 127–53.
87. Hendry SH, Yoshioka T (1994), *A neurochemically distinct third channel in the macaque dorsal lateral geniculate nucleus*. Science, **264**(5158): p. 575–7.
88. Hubel DH (1995), *Eye, brain and vision*.2ed. New York: Scientific American Library.242.
89. Hubel DH, Wiesel TN (1959), *Receptive fields of single neurones in the cat's striate cortex*. J Physiol, **148**: p. 574–91.
90. Hubel DH, Wiesel TN (1961), *Integrative action in the cat's lateral geniculate body*. J Physiol, **155**: p. 385–98.
91. Hubel DH, Wiesel TN (1962), *Receptive fields, binocular interaction and functional architecture in the cat's visual cortex*. J Physiol, **160**: p. 106–54.
92. Hubel DH, Wiesel TN (1965), *Receptive fields and functional architecture in two nonstriate visual areas (18 and 19) of the cat*. J Neurophysiol, **28**: p. 229–89.
93. Hubel DH, Wiesel TN (1968), *Receptive fields and functional architecture of monkey striate cortex*. J Physiol, **195**(1): p. 215–43.
94. Hubel DH, Wiesel TN (1972), *Laminar and columnar distribution of geniculo-cortical fibers in the macaque monkey*. J Comp Neurol, **146**(4): p. 421–50.
95. Hubel DH, Wiesel TN (1974), *Sequence regularity and geometry of orientation columns in the monkey striate cortex*. J Comp Neurol, **158**(3): p. 267–93.
96. Hupe JM, James AC, Payne BR, et al. (1998), *Cortical feedback improves discrimination between figure and background by V1, V2 and V3 neurons*. Nature, **394**(6695): p. 784–7.
97. Jones HE, Grieve KL, Wang W, Sillito AM (2001), *Surround suppression in primate V1*. J Neurophysiol, **86**(4): p. 2011–28.
98. Kandel ER, Schwartz JH, Jessell TM, eds (2000). *Principles of neural science*. 4th ed. McGraw Hill: New York.

99. Kaneko A (1970), *Physiological and morphological identification of horizontal, bipolar and amacrine cells in goldfish retina.* J Physiol, **207**(3): p. 623–33.

100. Kapadia MK, Westheimer G, Gilbert CD (1999),*Dynamics of spatial summation in primary visual cortex of alert monkeys.* Proc Natl Acad Sci USA, **96**(21): p. 12073–8.

101. Kaplan E, Shapley RM (1982), *X and Y cells in the lateral geniculate nucleus of macaque monkeys.* J Physiol, **330**: p. 125–43.

102. Kaplan E, Shapley RM (1986), *The primate retina contains two types of ganglion cells, with high and low contrast sensitivity.* Proc Natl Acad Sci USA, **83**(8): p. 2755–7.

103. Knierim JJ, van Essen DC (1992), *Neuronal responses to static texture patterns in area V1 of the alert macaque monkey.* J Neurophysiol, **67**(4): p. 961–80.

104. Kolb H, Dekorver L (1991), *Midget ganglion cells of the parafovea of the human retina: a study by electron microscopy and serial section reconstructions.* J Comp Neurol, **303**(4): p. 617–36.

105. Kolb H, Linberg KA, Fisher SK (1992), *Neurons of the human retina: a Golgi study.* J Comp Neurol, **318**(2): p. 147–87.

106. Kolb H, Marshak D (2003), *The midget pathways of the primate retina.*Doc Ophthalmol, **106**(1): p. 67–81.

107. Lachica EA, Beck PD, Casagrande VA (1992), *Parallel pathways in macaque monkey striate cortex: anatomically defined columns in layer III.* Proc Natl Acad Sci USA, **89**(8): p. 3566–70.

108. Laubrock J, Engbert R, Kliegl R (2008), *Fixational eye movements predict the perceived direction of ambiguous apparent motion.* J Vis, **8**(14): p. 1–17.

109. Leventhal AG, Rodieck RW, Dreher B (1981), *Retinal ganglion cell classes in the old world monkey: morphology and central projections.* Science, **213**(4512): p. 1139–42.

110. Levitt JB, Lund JS (2002), *The spatial extent over which neurons in macaque striate cortex pool visual signals.* Vis Neurosci, **19**(4): p. 439–52.

111. Livingstone M, Hubel D (1988), *Segregation of form, color, movement, and depth: anatomy, physiology, and perception.* Science, **240**(4853): p. 740–9.

112. Livingstone MS, Hubel DH (1982), *Thalamic inputs to cytochrome oxidase-rich regions in monkey visual cortex.* Proc Natl Acad Sci USA, **79**(19): p. 6098–101.

113. Livingstone MS, Hubel DH (1984), *Anatomy and physiology of a color system in the primate visual cortex.* J Neurosci, **4**(1): p. 309–56.

114. Lorente de Nó R (1949), *Cerebral cortex: architecture, intracortical connections, motor projections*, in *Physiology of the nervous system*, Fulton JF, Editor. Oxford University Press: Oxford. p. 288–330.

115. Lowel S, Schmidt KE, Kim DS, et al. (1998), *The layout of orientation and ocular dominance domains in area 17 of strabismic cats.* Eur J Neurosci, **10**(8): p. 2629–43.

116. Lund JS (1973), *Organization of neurons in the visual cortex, area 17, of the monkey (Macaca mulatta).* J Comp Neurol, **147**(4): p. 455–96.

117. Lund JS, Boothe RG, Lund RD (1977), *Development of neurons in the visual cortex (area 17) of the monkey (Macaca nemestrina): a Golgi study from fetal day 127 to postnatal maturity.* J Comp Neurol, **176**(2): p. 149–88.

118. Lund JS, Lund RD, Hendrickson AE, et al. (1975), *The origin of efferent pathways from the primary visual cortex, area 17, of the macaque monkey as shown by retrograde transport of horseradish peroxidase.* J Comp Neurol, **164**(3): p. 287–303.

119. Lund JS, Wu CQ (1997), *Local circuit neurons of macaque monkey striate cortex: IV. Neurons of laminae 1-3A.* J Comp Neurol, **384**(1): p. 109–26.

120. Macknik SL (2006), *Visual masking approaches to visual awareness.* Prog Brain Res, **155**: p. 177–215.

121. Macknik SL, Haglund MM (1999), *Optical images of visible and invisible percepts in the primary visual cortex of primates.* Proc Natl Acad Sci USA, **96**(26): p. 15208–10.

122. Macknik SL, Livingstone MS (1998), *Neuronal correlates of visibility and invisibility in the primate visual system.* Nat Neurosci, **1**(2): p. 144–9.

123. Macknik SL, Martinez-Conde S (2004), *The spatial and temporal effects of lateral inhibitory networks and their relevance to the visibility of spatiotemporal edges.* Neurocomputing, **58–60**: p. 775–82.

124. Macknik SL, Martinez-Conde S (2007), *The role of feedback in visual masking and visual processing*. Adv Cogn Psychol, **3**: p. 125–52.
125. Macknik SL, Martinez-Conde S (2009), *The role of feedback in visual attention and awareness*, in *The Cognitive Neurosciences, 4th edition*, Gazzaniga MS, Editor. MIT Press: Cambridege, MA, p. 1165–75.
126. Macknik SL, Martinez-Conde S, Haglund MM (2000), *The role of spatiotemporal edges in visibility and visual masking*. Proc Natl Acad Sci USA, **97**(13): p. 7556–60.
127. Macknik SL, Martinez-Conde S (2009), *Encyclopedia of Perception*, Ed. E. Bruce Goldstein, Sage Press, 522–24.
128. MacNeil MA, Masland RH (1998), *Extreme diversity among amacrine cells: implications for function*. Neuron, **20**(5): p. 971–82.
129. Maguire WM, Baizer JS (1984), *Visuotopic organization of the prelunate gyrus in rhesus monkey*. J Neurosci, **4**(7): p. 1690–704.
130. Malach R, Amir Y, Harel M, Grinvald A (1993), *Relationship between intrinsic connections and functional architecture revealed by optical imaging and in vivo targeted biocytin injections in primate striate cortex*. Proc Natl Acad Sci USA, **90**(22): p. 10469–73.
131. Marks WB, Dobelle WH, Macnichol EF, Jr. (1964), *Visual pigments of single primate cones*. Science, **143**: p. 1181–3.
132. Marr D, Hildreth E (1980), *Theory of edge detection*. Proc R Soc Lond Series B, **207**: p. 187–217.
133. Martinez-Conde S, Cudeiro J, Grieve KL, et al. (1999), *Effects of feedback projections from area 18 layers 2/3 to area 17 layers 2/3 in the cat visual cortex*. J Neurophysiol, **82**(5): p. 2667–75.
134. Martinez-Conde S, Macknik SL (2001). *Junctions are the most salient visual features in the early visual system*. in *Society for Neuroscience 31st Annual Meeting*. SanDiego, CA.
135. Martinez-Conde S, Macknik SL (2007), *Windows on the mind*. Sci Am, **297**(2): p. 56–63.
136. Martinez-Conde S, Macknik SL (2008), *Fixational eye movements across vertebrates: comparative dynamics, physiology, and perception*. J Vis, **8**(14): p. 1–16.
137. Martinez-Conde S, Macknik SL, Hubel DH (2000), *Microsaccadic eye movements and firing of single cells in the striate cortex of macaque monkeys*. Nature Neuroscience, **3**(3): p. 251–8.
138. Martinez-Conde S, Macknik SL, Hubel DH (2002), *The function of bursts of spikes during visual fixation in the awake primate lateral geniculate nucleus and primary visual cortex*. Proc Natl Acad Sci USA, **99**(21): p. 13920–5.
139. Martinez-Conde S, Macknik SL, Hubel DH (2004), *The role of fixational eye movements in visual perception*. Nat Rev Neurosci, **5**: p. 229–40.
140. Martinez-Conde S, Macknik SL, Troncoso XG, Dyar TA (2006), *Microsaccades counteract visual fading during fixation*. Neuron, **49**(2): p. 297–305.
141. Martinez-Conde S, Macknik SL, Troncoso XG, Hubel DH (2009), *Microsaccades: a neurophysiological analysis*. Trends Neurosci, **32**(9): p. 463–75.
142. Martinez LM, Alonso JM (2001), *Construction of complex receptive fields in cat primary visual cortex*. Neuron, **32**: p. 515–25.
143. Martinez LM, Wang Q, Reid RC, et al. (2005), *Receptive field structure varies with layer in the primary visual cortex*. Nat Neurosci, **8**(3): p. 372–9.
144. Masland RH, Ames A, 3rd (1976), *Responses to acetylcholine of ganglion cells in an isolated mammalian retina*. J Neurophysiol, **39**(6): p. 1220–35.
145. Maunsell JH, Newsome WT (1987), *Visual processing in monkey extrastriate cortex*. Annu Rev Neurosci, **10**: p. 363–401.
146. McGuire BA, Gilbert CD, Rivlin PK, Wiesel TN (1991), *Targets of horizontal connections in macaque primary visual cortex*. J Comp Neurol, **305**(3): p. 370–92.
147. Meadows JC (1974), *The anatomical basis of prosopagnosia*. J Neurol Neurosurg Psychiatry, **37**(5): p. 489–501.
148. Meadows JC (1974), *Disturbed perception of colours associated with localized cerebral lesions*. Brain, **97**(4): p. 615–32.
149. Merigan WH, Maunsell JH (1993), *How parallel are the primate visual pathways?* Annu Rev Neurosci, **16**: p. 369–402.
150. Miller RF, Slaughter MM (1986), *Excitatory amino acid receptors of the retina: diversity and subtype and conductive mechanisms*. TINS, **9**: p. 211–3.

151. Mishkin M, Ungerleider LG (1983), *Object vision and spatial vision: two cortical pathways.* Trendes Neurosci, **6**: p. 414–7.

152. Mollon JD, Bowmaker JK (1992), *The spatial arrangement of cones in the primate fovea.* Nature, **360**(6405): p. 677–9.

153. Mountcastle VB (1957), *Modality and topographic properties of single neurons of cat's somatic sensory cortex.* J Neurophysiol, **20**(4): p. 408–34.

154. Mountcastle VB, Berman AL, Davies PW (1955), *Topographic organization and modality representation in first somatic area of cat's cerebral cortex by method of singleunit analysis.* Am J Physiol, **183**: p. 646.

155. Movshon JA, Thompson ID, Tolhurst DJ (1978), *Spatial summation in the receptive fields of simple cells in the cat's striate cortex.* J Physiol, **283**: p. 53–77.

156. Muller JF, Dacheux RF (1997), *Alpha ganglion cells of the rabbit retina lose antagonistic surround responsesunder dark adaptation.* Vis Neurosci, **14**(2): p. 395–401.

157. Murphy PC, Duckett SG, Sillito AM (1999), *Feedback connections to the lateral geniculate nucleus and cortical response properties.* Science, **286**(5444): p. 1552–4.

158. Murphy PC, Sillito AM (1987), *Corticofugal feedback influences the generation of length tuning in the visual pathway.* Nature, **329**(6141): p. 727–9.

159. Nawy S, Copenhagen DR (1987), *Multiple classes of glutamate receptor on depolarizing bipolar cells in retina.* Nature, **325**(6099): p. 56–8.

160. Nelson R, Famiglietti EV, Jr., Kolb H (1978), *Intracellular staining reveals different levels of stratification for on- and off-center ganglion cells in cat retina.* J Neurophysiol, **41**(2): p. 472–83.

161. Nelson R, Kolb H (1983), *Synaptic patterns and response properties of bipolar and ganglion cells in the cat retina.* Vision Res, **23**(10): p. 1183–95.

162. Obermayer K, Blasdel GG (1993), *Geometry of orientation and ocular dominance columns in monkey striate cortex.* J Neurosci, **13**(10): p. 4114–29.

163. Olavarria JF, Van Essen DC (1997), *The global pattern of cytochrome oxidase stripes in visual area V2 of the macaque monkey.* Cereb Cortex, **7**(5): p. 395–404.

164. Østerberg G (1935), *Topography of the layer of rods and cones in the human retina.* Acta Ophthalmologica, **6**: p. 1–103.

165. Pack CC, Livingstone MS, Duffy KR, Born RT (2003), *End-stopping and the aperture problem: two-dimensional motion signals in macaque V1.* Neuron, **39**(4): p. 671–80.

166. Pasupathy A, Connor CE (1999), *Responses to contour features in macaque area V4.* J Neurophysiol, **82**(5): p. 2490–502.

167. Pearlman AL, Birch J, Meadows JC (1979), *Cerebral color blindness: an acquired defect in hue discrimination.* Ann Neurol, **5**(3): p. 253–61.

168. Peichl L, Wassle H (1983), *The structural correlate of the receptive field centre of alpha ganglion cells in the cat retina.* J Physiol, **341**: p. 309–24.

169. Perkel DJ, Bullier J, Kennedy H (1986), *Topography of the afferent connectivity of area 17 in the macaque monkey: a double-labelling study.* J Comp Neurol, **253**(3): p. 374–402.

170. Perry VH, Cowey A (1981), *The morphological correlates of X- and Y-like retinal ganglion cells in the retina of monkeys.* Exp Brain Res, **43**(2): p. 226–8.

171. Perry VH, Oehler R, Cowey A (1984), *Retinal ganglion cells that project to the dorsal lateral geniculate nucleus in the macaque monkey.* Neuroscience, **12**(4): p. 1101–23.

172. Poggio GF, Doty RW, Jr., Talbot WH (1977), *Foveal striate cortex of behaving monkey: single-neuron responses to square-wave gratings during fixation of gaze.* J Neurophysiol, **40**(6): p. 1369–91.

173. Polyak S (1941), *The retina.* Chicago: University of Chicago Press.

174. Powell TP, Mountcastle VB (1959), *Some aspects of the functional organization of the cortex of the postcentral gyrus of the monkey: a correlation of findings obtained in a singleunit analysis with cytoarchitecture.* Bull Johns Hopkins Hosp, **105**: p. 133–62.

175. Ramón y Cajal S (1893), *La rétine des vertébrés.* Cellule, **9**: p. 117–257.

176. Ramón y Cajal S (1900), *Structure of the Mammalian Retina.* Madrid.

177. Rao RPN, Olshausen BA, Lewicki MS (2002), *Probabilistic models of the brain: perception and neural function.* Cambridge, MA: MIT Press.
178. Ratcliff G, Davies-Jones GA (1972), *Defective visual localization in focal brain wounds.* Brain, **95**(1): p. 49–60.
179. Ratliff F (1965), *Mach bands: Quantitative studies on neural networks in the retina.* San Francisco: Holden-Day, Inc.
180. Reid RC, Alonso JM (1995), *Specificity of monosynaptic connections from thalamus to visual cortex.* Nature, **378**(6554): p. 281–4.
181. Ringach DL (2002), *Orientation selectivity in macaque V1: diversity and laminar dependence.* J Neurosci, **22**(13): p. 5639–51.
182. Ringach DL (2002), *Spatial structure and symmetry of simple cell receptive fields in macaque primary visual cortex.* J Neurophysiol, **88**: p. 455–463.
183. Rockland KS, Lund JS (1983), *Intrinsic laminar lattice connections in primate visual cortex.* J Comp Neurol, **216**(3): p. 303–18.
184. Rockland KS, Saleem KS, Tanaka K (1994),*Divergent feedback connections from areas V4 and TEO in the macaque.* Vis Neurosci, **11**(3): p. 579–600.
185. Rockland KS, Virga A (1989), *Terminal arbors of individual "feedback" axons projecting from area V2 to V1 in the macaque monkey: a study using immunohistochemistry of anterogradely transported Phaseolus vulgaris-leucoagglutinin.* J Comp Neurol, **285**(1): p. 54–72.
186. Rodieck RW (1998), *The first steps in seeing.* Sunderland, Massachusetts: Sinauer Associates. 562.
187. Roorda A, Williams DR (1999), *The arrangement of the three cone classes in the living human eye.* Nature, **397**(6719): p. 520–2.
188. Sceniak MP, Hawken MJ, Shapley R (2001), *Visual spatial characterization of macaque V1 neurons.* J Neurophysiol, **85**(5): p. 1873–87.
189. Schiller PH, Malpeli JG (1978), *Functional specificity of lateral geniculate nucleus laminae of the rhesus monkey.* J Neurophysiol, **41**(3): p. 788–97.
190. Schultze M (1866), *Zur Anatomie und Physiologie der Retina.* Arch Mikrosk Anat Entwicklungsmech, **2**: p. 165–286.
191. Shapley R, Hawken M, Ringach DL (2003), *Dynamics of orientation selectivity in the primary visual cortex and the importance of cortical inhibition.* Neuron, **38**(5): p. 689–99.
192. Shapley R, Perry JS (1986), *Cat and monkey retinal ganglion cells and their visual functional roles.* Trendes Neurosci, **9**: p. 229–35.
193. Sherman SM, Guillery RW (1998), *On the actions that one nerve cell can have on another: distinguishing "drivers" from "modulators".* Proc Natl Acad Sci USA, **95**(12): p. 7121–6.
194. Sherman SM, Guillery RW (2001), *Exploring the thalamus.* SanDiego: Academic Press.
195. Shipp S, Zeki S (1985), *Segregation of pathways leading from area V2 to areas V4 and V5 of macaque monkey visual cortex.* Nature, **315**(6017): p. 322–5.
196. Shipp S, Zeki S (1989), *The organization of connections between areas V5 and V1 in macaque monkey visual cortex.* Eur J Neurosci, **1**(4): p. 309–32.
197. Sincich LC, Horton JC (2005), *The circuitry of V1 and V2: integration of color, form, and motion.* Annu Rev Neurosci, **28**: p. 303–26.
198. Skavenski AA, Hansen RM, Steinman RM, Winterson BJ (1979), *Quality of retinal image stabilization during small natural and artificial body rotations in man.* Vision Res, **19**(6): p. 675–83.
199. Slaughter MM, Miller RF (1981), *2-amino-4-phosphonobutyric acid: a new pharmacological tool for retina research.* Science, **211**(4478): p. 182–5.
200. Slaughter MM, Miller RF (1983), *An excitatory amino acid antagonist blocks cone input to sign-conserving second-order retinal neurons.* Science, **219**(4589): p. 1230–2.
201. Steriade M, Jones EG, McCormick DA, eds (1997). *Thalamus.* Elsevier: New York.
202. Stone J, Dreher B, Leventhal A (1979), *Hierarchical and parallel mechanisms in the organization of visual cortex.* Brain Res, **180**(3): p. 345–94.
203. Suzuki W, Saleem KS, Tanaka K (2000),*Divergent backward projections from the anterior part of the inferotemporal cortex (area TE) in the macaque.* J Comp Neurol, **422**(2): p. 206–28.

204. Tomita T (1965), *Electrophysiological study of the mechanisms subserving color coding in the fish retina.* Cold Spring Harb Symp Quant Biol, **30**: p. 559–66.
205. Trifonov YA (1968), *Study of synaptic transmission between the photoreceptor and the horizontal cellusing electrical stimulation of the retina.* Biofizika, **10**: p. 673–80.
206. Troncoso XG, Macknik SL, Martinez-Conde S (2005), *Novel visual illusions related to Vasarely's 'nested squares' show that corner salience varies with corner angle.* Perception, **34**(4): p. 409–20.
207. Troncoso XG, Macknik SL, Martinez-Conde S (2008), *Microsaccades counteract perceptual filling-in.* J Vis, **8**(14): p. 1–9.
208. Troncoso XG, Macknik SL, Martinez-Conde S (2009), *Corner salience varies linearly with corner angle during flicker-augmented contrast: a general principle of corner perception based on Vasarely's artworks.* Spat Vis, **22**(3): p. 211–24.
209. Troncoso XG, Macknik SL, Otero-Millan J, Martinez-Conde S (2008), *Microsaccades drive illusory motion in the Enigma illusion.* Proc Natl Acad Sci USA, **105**(41): p. 16033–8.
210. Troncoso XG, Tse PU, Macknik SL, et al. (2007), *BOLD activation varies parametrically with corner angle throughout human retinotopic cortex.* Perception, **36**(6): p. 808–20.
211. Ts'o DY, Frostig RD, Lieke EE, Grinvald A (1990), *Functional organization of primate visual cortex revealed by high resolution optical imaging.* Science, **249**(4967): p. 417–20.
212. Ts'o DY, Gilbert CD, Wiesel TN (1986), *Relationships between horizontal interactions and functional architecture in cat striate cortex as revealed by cross-correlation analysis.* J Neurosci, **6**(4): p. 1160–70.
213. Ungerleider LG, Desimone R (1986), *Cortical connections of visual area MT in the macaque.* J Comp Neurol, **248**(2): p. 190–222.
214. Ungerleider LG, Desimone R (1986), *Projections to the superior temporal sulcus from the central and peripheral field representations of V1 and V2.* J Comp Neurol, **248**(2): p. 147–63.
215. Ungerleider LG, Mishkin M (1982), *Two cortical visual systems*, in *Analysis of visual behavior*, Ingle DG, Goodale MA, Mansfield JQ, Editors. MIT Press: Cambridge, MA. p. 549–86.
216. Usrey WM, Alonso JM, Reid RC (2000), *Synaptic interactions between thalamic inputs to simple cells in cat visual cortex.* J Neurosci, **20**(14): p. 5461–7.
217. vanDam LC, van Ee R (2006), *Retinal image shifts, but not eye movements per se, cause alternations in awareness during binocular rivalry.* J Vis, **6**(11): p. 1172–9.
218. Van Essen DC, Anderson CH, Felleman DJ (1992), *Information processing in the primate visual system: an integrated systems perspective.* Science, **255**(5043): p. 419–23.
219. Van Essen DC, Gallant JL (1994), *Neural mechanisms of form and motion processing in the primate visual system.* Neuron, **13**(1): p. 1–10.
220. Van Essen DC, Zeki SM (1978), *The topographic organization of rhesus monkey prestriate cortex.* J Physiol, **277**: p. 193–226.
221. Vasarely V (1970), *Vasarely II.* Plastic arts of the 20th century, ed. Joray M. Switzerland: Éditions duGriffon Neuchâtel.
222. Verweij J, Dacey DM, Peterson BB, Buck SL (1999), *Sensitivity and dynamics of rod signals in H1 horizontal cells of the macaque monkey retina.* Vision Res, **39**(22): p. 3662–72.
223. Wässle H, Boycott BB (1991), *Functional architecture of the mammalian retina.* Physiol Rev, **71**(2): p. 447–80.
224. Watanabe M, Rodieck RW (1989), *Parasol and midget ganglion cells of the primate retina.* J Comp Neurol, **289**(3): p. 434–54.
225. Werblin FS, Dowling JE (1969), *Organization of the retina of the mudpuppy, Necturus maculosus. II. Intracellular recording.* J Neurophysiol, **32**(3): p. 339–55.
226. Wiesel TN, Hubel DH, Lam DM (1974), *Autoradiographic demonstration of ocular-dominance columns in the monkey striate cortex by means of transneuronal transport.* Brain Res, **79**(2): p. 273–9.
227. Wiser AK, Callaway EM (1996), *Contributions of individual layer 6 pyramidal neurons to local circuitry in macaque primary visual cortex.* J Neurosci, **16**(8): p. 2724–39.
228. Yarbus AL (1967), *Eye movements and vision.* New York: Plenum Press.

229. Zeki SM (1974), *Cells responding to changing image size and disparity in the cortex of the rhesus monkey.* J Physiol, **242**(3): p. 827–41.
230. Zeki SM (1974), *Functional organization of a visual area in the posterior bank of the superior temporal sulcus of the rhesus monkey.* J Physiol, **236**(3): p. 549–73.
231. Zeki SM (1978), *Functional specialisation in the visual cortex of the rhesus monkey.* Nature, **274**(5670): p. 423–8.
232. Zeki SM (1978), *Uniformity and diversity of structure and function in rhesus monkey prestriate visual cortex.* J Physiol, **277**: p. 273–90.
233. Zhaoping L (2005), *The primary visual cortex creates a bottom-up saliency map*, in *Neurobiology of Attention*, Itti L, Rees G, Tsotsos JK, Editors. Elsevier: Oxford. p. 570–75.
234. Zihl J, von Cramon D, Mai N (1983), *Selective disturbance of movement vision after bilateral brain damage.* Brain, **106** (Pt2): p. 313–40.

Chapter 3
Retinal Remodeling and Visual Prosthetics

Bryan W. Jones, Robert E. Marc, and Carl B. Watt

Abstract Retinal degenerative disease induces a cascade of events that ultimately result in phased revision of neuronal populations and circuitry of the retina. These changes reveal plasticity in the retina that mimics that seen during development and in instances of neural deafferentation in other central nervous system (CNS) systems, involving neuronal as well as glial cell populations. These retinal remodeling changes occur across the spectrum of retinal degenerative disease and are observed in defects of the retinal pigment epithelium (RPE), rhodopsin packaging and transport defects as well as other non-retinitis pigmentosa (RP) related diseases with the final result being fundamental revision of neuronal populations and circuitry. These revisions impact potential biological and bionic rescues of visual function and must be overcome before vision restoration strategies can be viable.

Abbreviations

AC	Amacrine cell
AMD	Age-related macular degeneration
BC	Bipolar cell
CMP	Computational molecular phenotyping
CNS	Central nervous system
IPL	Inner plexiform layer
RP	Retinitis pigmentosa
RPE	Retinal pigment epithelium
TEM	Transmission electron microscopy

B.W. Jones (✉)
Moran Eye Center, University of Utah, 65 Mario Capecchi Drive,
Salt Lake City, UT 84132, USA
e-mail: bryan.jones@m.cc.utah.edu

G. Dagnelie (ed.), *Visual Prosthetics: Physiology, Bioengineering, Rehabilitation*,
DOI 10.1007/978-1-4419-0754-7_3, © Springer Science+Business Media, LLC 2011

3.1 Introduction

Our understanding of retinal structure and function has been a 150 year journey through biological science with the goal of understanding precisely how the retina is anatomically composed and how that structure interacts physiologically. Unfortunately this goal while close to completion in some areas, remains woefully lacking in complete detail of development, participants, connectivity, physiology and pathology, particularly in disease processes. This chapter examines the retina in disease and will attempt to clarify some of the long misunderstood aspects of retinal pathology, discussing how those pathologies impact bionic and/or biological strategies in the rescue of vision.

Other chapters in this book will discuss bionic approaches to "curing" vision loss using various prosthetics, leaving this chapter to function as a biological primer of sorts to introduce some of the biological realities that any therapeutic intervention will have to deal with as all of the inherited retinal degenerations studied to date reveal a biological moving target that must be considered prior to therapy.

The question of whether or not the neural retina is receptive to bionic or biological intervention is one that historically has been investigated without consideration of the actual disease process neural systems proceed through when they experience loss of photoreceptors. Though the neural retina grossly appears to survive photoreceptor loss in diseases such as retinitis pigmentosa (RP) and age-related macular degeneration (AMD), the reality is that the retina is no different from other CNS pathways when their afferent inputs are lost. When photoreceptor inputs are lost, the retina engages in a wide variety of remodeling events driven by loss of signaling inputs. These transformations include glial hypertrophy and possible hyperplasia, neuronal translocations, neuronal loss and the emergence of retinal circuit alteration with the formation of novel synaptically active neuronal processes. These new processes are perhaps the most significant impediment to prosthetic retinal rescue through bionic or biological interventions as the disease process corrupts and modifies the normal visual information processing so as to make it indecipherable by the visual cortex. If we are to proceed, vision rescue strategies need to contend with the biological realities of retinal remodeling.

3.2 Background

The effort to build, design and implement neuroprosthetic devices has been challenging due to not just the complexity of the retina, but also the difficulty of dealing with a complex, reactive biological tissue that changes its fundamental connectivity in disease processes. Efforts from a variety of labs now make it clear that the neural retina does not adopt a passive role with respect to photoreceptor degenerations and that extensive alterations and remodeling occur from the molecular scale up through the synaptic, cellular and tissue levels [2, 18, 19, 21, 31, 33, 49, 51, 52, 54, 58, 60, 68–71, 80, 85, 86, 88–90, 92]. Regardless of whether the intervention is survival factor delivery [29, 30], genetic [8, 61], cellular [36, 97, 105] or bionic [27, 46, 56, 102, 110], approaches will have to address the ongoing process of neuronal death, alterations to gene expression, neuronal circuit rewiring and migration and the elaboration of novel glial barriers. While these prospects may appear daunting, prosthetic devices may in fact be the ideal intervention with which to rescue and reconfigure neural retinas altered by disease.

3.3 Retinal Disease and Its Diversity

Retinal disease including the well characterized retinitis pigmentosa (Fig. 3.1) with an incidence of 1–4,000 [11] and the less well understood, yet far more prevalent age-related macular degeneration affect millions of people world wide. While RP affects a significant portion of the working age population, AMD is far more common overall, with an incidence in the United States alone estimated to reach three million by 2020 [35]. Indeed, it has been estimated that AMD is the leading cause of new cases of blindness in Americans over 60 with an estimated 18% of Americans between 65 and 74 and 30% of Americans older than 74 showing signs of possible future AMD [106].

Regardless of the form retinal degenerative disease takes, the final common pathway of photoreceptor loss followed by downstream reactive biological processes results in a system that has proven difficult to rescue. While rescues of

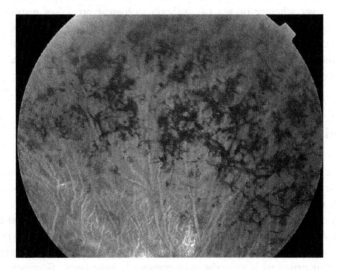

Fig. 3.1 Funduscopic image from 46 year old male with a diagnosis of X-linked retinitis pigmentosa, showing "pigmented bone spicules," accumulations of pigment epithelium that are formed by migration of the pigment epithelium into the neural retina along glial columns. These clinically pathologic findings are often seen in the peripheral retina in patients with RP

vision targeted towards the anterior eye have been possible for a great many years due to the accessibility and amenability of the tissues involved to pharmacological and surgical interventions, retinal disease presents a significantly larger challenge that has proven more difficult to combat due to its complex and progressive nature and the number of potential gene loci involved. These diverse pathological insults currently number close to 200 gene defects associated with various retinal diseases http://www.sph.uth.tmc.edu/RetNet/ including AMD, RP, diabetic retinopathy and glaucoma. These disease loci are located on 23 different genes in addition to mito-chondrial gene loci and result in vision loss through diverse mechanisms including defects in retinal pigment epithelium cells seen in recessive Leber congenital amau-rosis [1, 39, 77], defects in the ATP binding cassette transporter seen in recessive Stargardt disease [3–6, 17], the c-mer proto-oncogene receptor tyrosine kinase [20, 26, 39], alterations in cilia function and intraflagellar transport [57, 104], arrestin [83, 84], and transducin defects [24, 107], rod cGMP phosphodiesterase defects [45, 74, 75], metabotropic glutamate receptor defects [25, 108], peripherin defects [16], fatty acid biosynthetic enzymes [96, 109], and a diverse assortment of other gene loci encoding proteins responsible for signaling [14, 43, 44, 101] and rhodopsin gene mutations [34, 47, 60].

Defects for AMD are likely as numerous and complex as the RP causes [5, 6, 9, 12, 15, 17, 23, 28, 37, 41, 48, 53, 62, 76, 87, 98, 103], yet are dependent upon a number of potential gene defect interactions that over time and with the accumulation of other risk factors result in retinal degeneration of the central portion of the retina responsible for high acuity vision [103]. Additionally, because no one specific cause of AMD has been identified, there is some difficulty defining a precise definition complicated by significant overlap of clinical manifestations. Indeed there is even some degree of controversy over whether or not the pathophysiological processes responsible for many of the sequellae of AMD including drusen accumulation, geographic atrophy, pigmentary changes and alterations in the vascular network are even directly related. Whatever the mechanism(s) involved, the end result of AMD is likely the same; pho-toreceptor cell death followed by retinal remodeling.

3.4   Retinal Remodeling

While work prior to the last decade assumed that retinal degenerative disease only affected the sensory retina, it is now commonly understood that these diseases also involve the neural retina to dramatic fashion [2, 8, 49, 52, 69, 71, 85, 89, 91, 92]. The reality of retinal degenerative disease and the subsequent changes that occur to the anatomy and physiology of the retina present profound difficulties to prospects of rescue, whether that rescue is biologically based or bionic in nature. Retinas that have lost their principal inputs, the photoreceptors, have been effectively deaffer-ented and undergo changes to their circuitry early and likely initially clinically occult. Regardless of the initial molecular or environmental insult, the proverb "omnes viae Romam ducunt" or "all roads lead to Rome" summarizes where these mechanisms take us with respect to retinal remodeling. All defects resulting in loss of photoreceptor input to the neural retina initiate a series of events that change the fundamental ground truth of the retinal neural circuitry. This alteration in how the retina processes signals presents a significant challenge to retinal rescue through bionic prosthetic devices or biological interventions and it can be argued that most approaches to intervention have waited far too long in the degenerative process to hope for any substantive visual rescue. By the time photoreceptors are gone (Fig. 3.1), the changes to wiring are well underway.

Most implant strategies presume substantial survival of retinal outflow architectures and while it has long been claimed that the neural retina remains unchanged after the death of the sensory retina, this perspective is incorrect. In retinal degenerations, the neural retina undergoes a series of phases initiated by a period of photoreceptor or retinal pigment epithelial cell stress (Fig. 3.2). The standard metabolic phenotypes of some cells (Müller cells) become altered possibly indicating fundamental changes in the abilities of these cells to maintain their function and viability, but neuronal metabolic profiles appear to be maintained until cell death. Initially clinically occult changes also occur to the circuitry of the neural retina as well in even early stages of retinal degeneration. Subsequent to phase one, the neural retina enters into a phase of outer nuclear layer modification that includes photoreceptor cell death, apparent death of bystander neurons, phagocytic consumption of dying neurons and the walling off or entombment of the remnant neural retina beneath Müller cell processes. The final tertiary phase of retinal degeneration occurs as the retina enters a protracted period of remodeling characterized by disruption of topology by glial hypertrophy and continued neuronal migration, continued neuronal cell death and extensive rewiring with elaboration of de novo neurite and synaptic formation [49, 51, 52, 70]. Late in the course of retinal degeneration, neuronal death becomes extensive. Though many neurons persist after death of the sensory retina, all are susceptible to cell death in varying fractions and patterns. Focal depletion of the inner nuclear layer is common and some genetic types of photoreceptor degenerations express massive ganglion cell loss in large patches of retina. In the most extreme cases, the Müller cell seal breaks down and neurons do in fact emigrate from the retina into the remnant choroid [50].

These three phases of retinal remodeling culminate in the rewiring of all cell classes and essentially reprogram the retina rendering the circuitry incapable of processing visual data and delivering those data to visual cortex [69]. It should be noted that even though the first report of aberrant circuitry in the human RP retina goes back to 1974 [54], the concept of neural remodeling events are abundant in the epilepsy literature [55, 79, 93] and the vision research community is coming late to the game. These alterations in retinal morphology and physiology are seen across the spectrum of retinal degenerations from inherited [33] to engineered [51, 70] and induced photoreceptor degenerations [21] with changes occurring relatively early after photoreceptor cell stress and death [69].

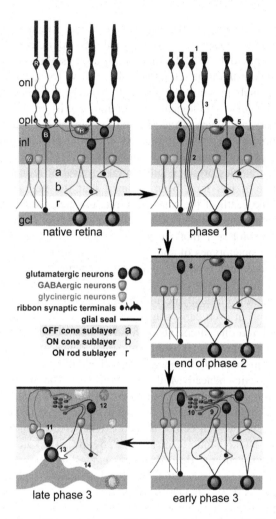

onl

ople

inl

a
b
r

gcl

native retina   phase 1

glutamatergic neurons
GABAergic neurons
glycinergic neurons
ribbon synaptic terminals
glial seal
OFF cone sublayer   a
ON cone sublayer   b
ON rod sublayer   r

end of phase 2

late phase 3   early phase 3

Fig. 3.2 (1) Truncation of photoreceptors. (2) Rod axon extension. (3) Cone axon extension. (4) Rod bipolar cell dendrite retraction. (5) Cone bipolar cell dendrite retraction. (6) Horizontal Cell axon remodeling. (7) Glial seal. (8) Phenotype revisions. (9) Neurite fascicle formation. (10) Microneuroma formation. (11) Neuronal migration. (12) Neuronal death. (13) IPL rewiring. (14) Laminar deformation. A schematic representation of the three stages of retinal degeneration showing both rod and cone photoreceptors, rod and cone bipolar cells, ganglion cells, a horizontal cell, GABAergic amacrine and glycinergic amacrine cells. The two nuclear layers are illustrated as *horizontal bands*. The first frame, native retina shows normal lamination and connectivity of cell classes in the retina. Phase 1 reveals early photoreceptor stress and outer segment shortening (1) along with rod and cone neurite extensions projecting down into inner nuclear layer and ganglion cell layer (2, 3). Horizontal cells are also seen contributing to the neurite projections (6) along with rod and cone bipolar cells undergoing dendrite retraction (4,5). Müller cells may also begin to hypertrophy in this stage. By the end of phase 2 there is a complete loss of photoreceptors and elaboration of a Müller cell seal over the neural retina (7), sealing it-off away from the remnant choroid. Neuronal phenotypic revisions are underway or complete at this time (8). Early phase 3 events ensue with the elaboration of neurite extensions from glycinergic and

3.5 Retinal Circuitry

Though analysis of the neural retina and its circuitry goes back over 100 years ago to Ramón y Cajal's work, most work examining the anatomy of circuitry in retinal disease is more recent, encompassing efforts in the last three decades to understand the components and their function. This work has revealed the retina to be a bi-laminar device with sensory and computational layers.

The sensory retina is composed of photoreceptors and is the photon transduction layer, while the neural retina is composed of the remaining neuron classes that comprise the image-processing layer. Even in a simple retina like the mammalian, the retinal circuitry is complex, comprising approximately 14 patterned outflow channels, realized as ganglion cells. The number of cell classes in the mammalian retina includes one rod class, one rod horizontal cell, one rod bipolar cell, two to three cone classes, one to three cone horizontal cells, 9+ cone bipolar cells, 27 amacrine cells, and about 15–20 ganglion cells. Thus, about 60–70 cellular devices form the outflow channels [66, 73, 99]. These outflow channels involve the flow of information through a set of stereotypical circuits from photoreceptors to bipolar cells to ganglion and amacrine cells with amacrine cells providing both feedback and feedforward control [65, 72, 73, 99]. It should be noted however that even two bipolar cells providing input to two separate ganglion cells, interconnected by a single amacrine cell provides a combinatorial 90 distinct and separate motifs assuming lumped-parameter circuitry. Assuming distributed parameter circuitry [100] expands the number of combinatorials to over 2,000 potential motifs. This approximation of a circuit diagram does not include any weightings for differential synaptic strength, cell class diversity, and coupling by gap junctions. Nor does this approximation include the most common form of synaptic connection in the retina between cells, the amacrine–amacrine cell serial synaptic chain. However, we know that the outflow of signals from the mammalian retina is represented by only 15–20 ganglion cell classes [67, 82], greatly simplifying the number of possible outputs, though we do not know what the total network topology is.

Even rich models [42] that mimic physiologic data acquired over limited spatiotemporal domains predict little about network topology or emergent features. Despite a broad view of the bounds of biophysical performance provided by physiology, models derived from physiology are essentially degenerate: not unique to any one network topology. In addition, remodeling and reprogramming of neural networks in retinal disease strongly argues that network scrambling is a key pathology [69]. Network motif diversity is analogous to genetic diversity: many connective motifs (gene sequences) are

---

Fig. 3.2 (continued) GABAergic amacrine cells along with contributions from bipolar cells and ganglion cells forming complex tangles of processes called microneuromas (9, 10) that form outside the normal lamination of the inner plexiform layer, sometimes merging with the inner plexiform layer. These microneuromas possess active synaptic elements corruptive of normal signaling. By late phase 3, retinal degeneration is advanced with neuronal migration or translocation events occurring in a bi-directional fashion (11) along with neuronal death of many cell classes (12). IPL rewiring and laminar deformation of the plexiform layers can also be observed

possible, but only a subset form good filters (proteins), and mutating motifs generates neural malfunction (genetic disease).

In theory, subretinal implants drive remnant circuits with cone-like inputs and epiretinal implants drive ganglion cell channels by mimicking bipolar-amacrine cell networks. Both schemes require survival of retinal neurons to drive perceptual and oculomotor systems, and presume no alterations in cell patterning or connectivity, nor any corruptive signal invasion into retinal networks. Subretinal strategies uniquely require positioning within the subretinal space. *These presumptions (preservation of topology, cell numbers and wiring) are false for most retinal degenerations.*

3.6 Retinal Circuitry Revision

Neuronal translocations in the remodeling retina are complex and do not just involve migrations of cell somata that leave their dendrites and axons in the original locations, preserving connectivity. The reality is far more insidious as the elaboration of new neurite, axonal, and synaptic structures occurs before gross cellular migration ensues (Figs. 3.3 and 3.4). These structures occur individually and may assemble into fascicles and microneuromas that may run for many microns underneath the Müller cell seal, forever changing and corrupting completely the neuronal circuitry of the retina [49–52, 68–71]. Modeling of new circuits demonstrates that all observed circuits are corruptive and many form resonant circuits, rendering the remnant neural retina no longer effective as an image processor [49, 51, 69].

While most analyses of retinal degeneration have focused on events surrounding phase 2 and photoreceptor death, rewiring of the neural retina occurs in all phases of retinal degeneration, and likely begins prior to photoreceptor death during the stress phase [69, 70]. Early in phase 2, ganglion cell light responses are altered, resulting in the loss of ON responses with the simultaneous preservation of OFF responses [80]. Once the photoreceptors are completely lost, ganglion cells spike throughout the retina of the Pde6b^{rd1} mouse retina [86], possibly providing a mechanism behind the scintillating scotomas reported by many patients with RP [22].

Therefore, passive anatomy alone does not reveal the scope of neural change in response to retinal degenerative disease. The growing evidence supports retinal rewiring as a common feature in retinal degenerations that involve photoreceptor loss and recent work [69] indicates profound changes in physiology through the use of excitation mapping [63, 64] and mapping cellular identity across disease states with single-cell resolution [67] along with in vivo and in vitro ligand activation in wild-type mice and rdcl and hrhoG mutant mice exhibiting rapid photoreceptor degeneration. In addition, the Marc 2007 study included in vitro excitation mapping in a sample of human RP retina revealing reprogramming events in bipolar cells that likely impact all forms of proposed retinal rescue strategies as the remodeling goes beyond rewiring and morphological change to include molecular reprogramming.

These findings are perhaps not surprising in that changes in circuitry have been documented in the literature for years. Other than the previously noted 1974 study

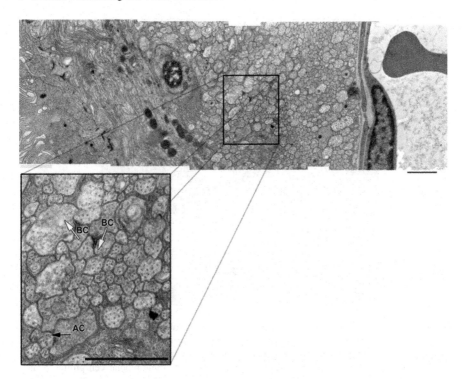

Fig. 3.3 EM image of a microneuroma underneath a distal retinal Müller cell seal on the left with a blood vessel and portion of an erythrocyte on the right. One hundred to four hundred nanometer diameter processes are running parallel together, perpendicularly through this plane of section comprised of five transmission electron microscopy (TEM) images mosaiced together. Synaptic profiles are present in this microneuroma with apparent bipolar cell (BC) like synapses with dyads, yet bereft of ribbons as well as conventional synapses from amacrine cells (AC) shown in the inset indicating that microneuromas are potentially not passive structures with respect to circuitry. Efforts to reconstruct microneuromas are underway to define the pathology of circuitry in these retinas. Scale bars = 1 μm

by Kolb, some of the earliest indications of retinal rewiring or connectivity defects can be seen in a paper by Li et al. in 1995 [58] where the authors documented aberrantly sprouting rod photoreceptors. Fei in 2002 [32] documented sprouting cones, Machida et al. [60] found abnormal sprouting of photoreceptors and horizontal cells in the degenerating retinas of the P23H transgenic rat and Fariss et al. [31] documented anomalous extension of rod, horizontal and amacrine cell neurites throughout the neural retina while Gregory-Evans et al. identified abnormal cone synapses in human cone-rod dystrophy [38]. Peng et al. identified ectopic synapses in the RCS rat [78] while other investigators working concurrently in mouse models of RP identified some of the earliest changes in the second order neurons, with dendritic retraction of rod bipolar and horizontal cells after photoreceptor cell loss [88, 90, 95]. Documentation of neuronal migration [49, 51, 52, 70], the identification of corruptive synaptic machinery in rod and cone bipolar cells as well as

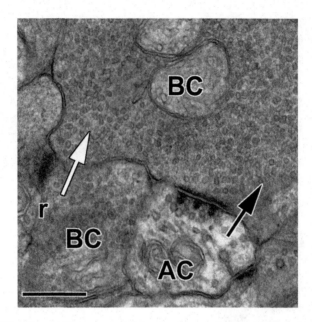

Fig. 3.4 Additional synaptic structures often present in microneuromas, though with immature forms. This example shows a aberrant presynaptic multi-projection amacrine cell (AC) making synaptic contact onto a bipolar cell in parallel with another bipolar cell (BC) profile making a simultaneous synapse complete with synaptic ribbon, onto the same bipolar cell profile. Scale bar = 200 nm

horizontal cells [18, 19] and most significantly, formation of new neuronal connectivities and reprogramming [69] have made for a compelling literature that will absolutely impact the implementation and success of rescues designed to preserve or restore vision.

3.7   Implications for Bionic Rescue

Because of the vast diversity of potential insults in both RP and AMD, any one, targeted intervention is likely to be useful for only a small percentage of potential individuals. Therefore, approaches designed to replace entire systems with bionic and biological solutions may appear attractive. However, as noted all retinal degenerations lead to problems of access and alterations to the fundamental image processing circuitry. The problem of how to rescue vision is further compounded by the issue of when to intervene. Current therapies or interventions are limited to those patients who have lost a considerable portion of their vision and are legally blind. These patients often present at late stage with advanced retinal degeneration (Fig. 3.1) and already likely exhibit profound alterations to the retinal circuitry that corrupt any surrogate inputs.

While modern engineering has allowed significant advancements in miniaturization of circuitry combined with the ability to power potentially prosthetic devices [7, 59], we still are lacking in our development and implementation of visual system interfaces to those devices. Additionally, the design and implementation will depend upon where in the visual system we intend to attempt an intervention and at what stage of retinal degeneration the subject might be in. Specifically, intervening in a degenerative retina will present an entirely different set of engineering difficulties than intervening at the optic nerve or the visual cortex and it could be argued that until we understand how each component of the visual system processes information, we will not be successful in the implementation of vision rescue prosthetics that attempt a simulation of the retinotopically organized flow of information to the visual cortex where properly patterned inputs result in spatiotemporally correct percepts.

Furthermore, how to actually stimulate the retina is one consideration, but unless one knows the circuitry, or can model the circuitry, there is no predicting the possible output of the neural retina. Additionally, given that neurons in the retina appear to be relatively promiscuous with respect to contacts on cell classes they make during the degenerative process and that those contacts appear to be impoverished, predicting the output of the retina irrespective of the type or methodology of stimulus will be difficult at best. Some modeling [69] predicts that circuits may "ring" for many seconds, essentially leaving the visual cortex no option but to filter these inputs out. Additionally, since retinal degeneration is a progressive disease, one might suspect that the neural retina will continue to remodel, possibly even recruiting and corrupting interventions or rescues into a continued degenerative process.

3.8 Implications for Biological Rescue

These critiques and findings can also be applied to biological rescues in that certain transplantation schemes may slow some forms of retinal degeneration when implemented before degeneration of the sensory retina is complete and remodeling becomes dominant. This however, is not a viable strategy for most human disease. Further, it is likely that the outcomes of most transplants will be impacted by at least three factors including cell fusion, improper rewiring, and co-opting of transplanted cells into defective or non-functional forms by resident neurons and glia. Many reports of transplanted stem cells assuming phenotypes of host cells are now known to be instances of cell fusion [94]. It is also a distinct possibility that when delivery of exogenous cells induces trauma from the surgery, aberrant protein and DNA uptake can also alter host and guest phenotypes, confounding analysis.

In addition to the corruptive local and global rewiring that occurs in retinal remodeling, retinas appear to have lost patterning restrictions as well. Naïve cells do not carry the normally present developmental structuring that occurs during retinal maturation and do not induce re-patterning. Moreover, transplanted photoreceptors or any other fragments of retina will certainly engage in wide-area neurite extensions if they survive, and degenerating retinas already engage in profuse generation of aberrant neurites. There is no evidence that any of these processes make proper connections.

This of course also begs the question: What phenotype should an uncommitted stem cell assume and how will it be transcriptionally guided in forming that phenotype? Additionally, properly phenotyping transplanted cells [13] is critical and most transplant studies fail in this regard. Any emergent phenotypes, if informed by local signaling from negatively remodeling cells, will most likely be co-opted into an aberrant phenotype. Additionally, most transplanted cells are rejected [10] or slowly lose their own mature phenotypes after transplantation. In short, the key error in transplant designs is a belief that the neural retina is normal. It is not normal and in the degenerate retina, there is hardly a cell type that demonstrates normality. The basic assumptions of transplant technologies (intactness, receptivity and instructional capacity of the host neural retina) are false for most retinal degenerations. Moreover, expectations that cells transplanted into negatively remodeling environments will restore normalcy to host cells, maintain mature phenotypes or assume proper phenotypes seem baseless and are, as yet, untested.

It should be noted that biological approaches should not necessarily be thought of as impossible as there are a number of "lower" organisms that possess retinas far more complex than mammalian retinas. Yet, these organisms with more complex retinas are able to restore or repair to some extent damage incurred to their retinas through stem cell dedifferentiation and recapitulation of an approximate structure and function of the retina [81]. However, these appraisals are gross as there have been no efforts that these authors are aware of that describes the nature of the circuitry in repair zones.

3.9   Final Remarks

The diversity of potential defects is impressive because of the complex specialization and highly optimized function of the mammalian retina. Because of this complexity, it may seem tempting to attempt a rescue or target a solution that would bypass all of the potential defects through a straightforward bionic approach, solving all potential blinding diseases with a single solution. Fifty years from now, this may in fact be how history records a cure for vision loss, but any intervention, bionic or otherwise is going to have to deal with a progressive disease that exhibits a plastic, reactive neural retina with likely downstream visual alterations in the circuitry of visual elements in cortex that display their own ability to adapt [40] and potentially remodel in response to retinal deafferentation or alteration in efferent retinal signals resulting from retinal rewiring. These diseases are not focal and will spread, possibly even involving interventions designed to rescue the retina.

Acknowledgments Supported in part by an Unrestricted Grant from Research to Prevent Blindness, Inc., New York, NY, to the Department of Ophthalmology & Visual Sciences, University of Utah. Dr. Bryan William Jones is a recipient of a Research to Prevent Blindness Career Development Award (BWJ). Edward N. & Della L. Thome Memorial Foundation, Bank of

America, N.A. Trustee, (BWJ) NEI R01 EY02576, R01 EY015128, P01 EY014800 (REM); support from the Cal and JeNeal Hatch Presidential Endowed Chair (REM).

Competing Interests

Robert E. Marc is a principal of Signature Immunologics. All other authors declare no other competing interests.

References

1. Aguirre GD, Baldwin V, Pearce-Kelling S, et al. (1998), *Congenital stationary night blindness in the dog: common mutation in the RPE65 gene indicates founder effect.* Mol Vis, **4**: p. 23.
2. Aleman TS, Cideciyan AV, Sumaroka A, et al. (2007), *Inner retinal abnormalities in X-linked retinitis pigmentosa with RPGR mutations.* Invest Ophthalmol Vis Sci, **48**(10): p. 4759–65.
3. Allikmets R (1997), *A photoreceptor cell-specific ATP-binding transporter gene (ABCR) is mutated in recessive Stargardt macular dystrophy.* Nat Genet, **17**(1): p. 122.
4. Allikmets R (2000), *Simple and complex ABCR: genetic predisposition to retinal disease.* Am J Hum Genet, **67**(4): p. 793–9.
5. Allikmets R, Shroyer NF, Singh N, et al. (1997), *Mutation of the Stargardt disease gene (ABCR) in age-related macular degeneration.* Science, **277**(5333): p. 1805–7.
6. Allikmets R, Singh N, Sun H, et al. (1997), *A photoreceptor cell-specific ATP-binding transporter gene (ABCR) is mutated in recessive Stargardt macular dystrophy.* Nat Genet, **15**(3): p. 236–46.
7. Asher A, Segal WA, Baccus SA, et al. (2007), *Image processing for a high-resolution optoelectronic retinal prosthesis.* IEEE Trans Biomed Eng, **54**(6 Pt 1): p. 993–1004.
8. Bainbridge JW, Smith AJ, Barker SS, et al. (2008), *Effect of gene therapy on visual function in Leber's congenital amaurosis.* N Engl J Med, **358**(21): p. 2231–9.
9. Boon CJ, Klevering BJ, Hoyng CB, et al. (2008), *Basal laminar drusen caused by compound heterozygous variants in the CFH gene.* Am J Hum Genet, **82**(2): p. 516–23.
10. Bull ND, Limb GA, Martin KR (2008), *Human Müller stem cell (MIO-M1) transplantation in a rat model of glaucoma: survival, differentiation, and integration.* Invest Ophthalmol Vis Sci, **49**(8): p. 3449–56.
11. Bunker CH, Berson EL, Bromley WC, et al. (1984), *Prevalence of retinitis pigmentosa in Maine.* Am J Ophthalmol, **97**(3): p. 357–65.
12. Cameron DJ, Yang Z, Gibbs D, et al. (2007), *HTRA1 variant confers similar risks to geographic atrophy and neovascular age-related macular degeneration.* Cell Cycle, **6**(9): p. 1122–5.
13. Canola K, Angenieux B, Tekaya M, et al. (2007), *Retinal stem cells transplanted into models of late stages of retinitis pigmentosa preferentially adopt a glial or a retinal ganglion cell fate.* Invest Ophthalmol Vis Sci, **48**(1): p. 446–54.
14. Chen CK, Burns ME, He W, et al. (2000), *Slowed recovery of rod photoresponse in mice lacking the GTPase accelerating protein RGS9-1.* Nature, **403**(6769): p. 557–60.
15. Chen H, Yang Z, Gibbs D, et al. (2008), *Association of HTRA1 polymorphism and bilaterality in advanced age-related macular degeneration.* Vision Res, **48**(5): p. 690–4.
16. Clarke G, Goldberg AF, Vidgen D, et al. (2000), *Rom-1 is required for rod photoreceptor viability and the regulation of disk morphogenesis.* Nat Genet, **25**(1): p. 67–73.
17. Cremers FP, van de Pol DJ, van Driel M, et al. (1998), *Autosomal recessive retinitis pigmentosa and cone-rod dystrophy caused by splice site mutations in the Stargardt's disease gene ABCR.* Hum Mol Genet, **7**(3): p. 355–62.

18. Cuenca N, Pinilla I, Sauve Y, et al. (2004), *Regressive and reactive changes in the connectivity patterns of rod and cone pathways of P23H transgenic rat retina.* Neuroscience, **127**(2): p. 301–17.

19. Cuenca N, Pinilla I, Sauve Y, Lund R (2005), *Early changes in synaptic connectivity following progressive photoreceptor degeneration in RCS rats.* Eur J Neurosci, **22**(5): p. 1057–72.

20. D'Cruz PM, Yasumura D, Weir J, et al. (2000), *Mutation of the receptor tyrosine kinase gene Mertk in the retinal dystrophic RCS rat.* Hum Mol Genet, **9**(4): p. 645–51.

21. de Raad S, Szczesny PJ, Munz K, Reme CE (1996), *Light damage in the rat retina: glial fibrillary acidic protein accumulates in Müller cells in correlation with photoreceptor damage.* Ophthalmic Res, **28**(2): p. 99–107.

22. Delbeke J, Pins D, Michaux G, et al. (2001), *Electrical stimulation of anterior visual pathways in retinitis pigmentosa.* Invest Ophthalmol Vis Sci, **42**(1): p. 291–7.

23. Dewan A, Liu M, Hartman S, et al. (2006), *HTRA1 promoter polymorphism in wet age-related macular degeneration.* Science, **314**(5801): p. 989–92.

24. Dryja TP, Berson EL, Rao VR, Oprian DD (1993), *Heterozygous missense mutation in the rhodopsin gene as a cause of congenital stationary night blindness.* Nat Genet, **4**(3): p. 280–3.

25. Dryja TP, McGee TL, Berson EL, et al. (2005), *Night blindness and abnormal cone electroretinogram ON responses in patients with mutations in the GRM6 gene encoding mGluR6.* Proc Natl Acad Sci USA, **102**(13): p. 4884–9.

26. Duncan JL, Yang H, Vollrath D, et al. (2003), *Inherited retinal dystrophy in Mer knockout mice.* Adv Exp Med Biol, **533**: p. 165–72.

27. Eckhorn R, Wilms M, Schanze T, et al. (2006), *Visual resolution with retinal implants estimated from recordings in cat visual cortex.* Vision Res, **46**(17): p. 2675–90.

28. Edwards AO, Ritter R, III, Abel KJ, et al. (2005), *Complement factor H polymorphism and age-related macular degeneration.* Science, **308**(5720): p. 421–4.

29. Faktorovich EG, Steinberg RH, Yasumura D, et al. (1990), *Photoreceptor degeneration in inherited retinal dystrophy delayed by basic fibroblast growth factor.* Nature, **347**(6288): p. 83–6.

30. Faktorovich EG, Steinberg RH, Yasumura D, et al. (1992), *Basic fibroblast growth factor and local injury protect photoreceptors from light damage in the rat.* J Neurosci, **12**(9): p. 3554–67.

31. Fariss RN, Li ZY, Milam AH (2000), *Abnormalities in rod photoreceptors, amacrine cells, and horizontal cells in human retinas with retinitis pigmentosa.* Am J Ophthalmol, **129**(2): p. 215–23.

32. Fei Y (2002), *Cone neurite sprouting: an early onset abnormality of the cone photoreceptors in the retinal degeneration mouse.* Mol Vis, **8**: p. 306–14.

33. Fletcher EL, Kalloniatis M (1996), *Neurochemical architecture of the normal and degenerating rat retina.* J Comp Neurol, **376**(3): p. 343–60.

34. Frederick JM, Krasnoperova NV, Hoffmann K, et al. (2001), *Mutant rhodopsin transgene expression on a null background.* Invest Ophthalmol Vis Sci, **42**(3): p. 826–33.

35. Friedman DS, O'Colmain BJ, Munoz B, et al. (2004), *Prevalence of age-related macular degeneration in the United States.* Arch Ophthalmol, **122**(4): p. 564–72.

36. Gias C, Jones M, Keegan D, et al. (2007), *Preservation of visual cortical function following retinal pigment epithelium transplantation in the RCS rat using optical imaging techniques.* Eur J Neurosci, **25**(7): p. 1940–8.

37. Gold B, Merriam JE, Zernant J, et al. (2006), *Variation in factor B (BF) and complement component 2 (C2) genes is associated with age-related macular degeneration.* Nat Genet, **38**(4): p. 458–62.

38. Gregory-Evans K, Fariss RN, Possin DE, et al. (1998), *Abnormal cone synapses in human cone-rod dystrophy.* Ophthalmology, **105**(12): p. 2306–12.

39. Gu SM, Thompson DA, Srikumari CR, et al. (1997), *Mutations in RPE65 cause autosomal recessive childhood-onset severe retinal dystrophy.* Nat Genet, **17**(2): p. 194–7.

40. Gutnisky DA, Dragoi V (2008), *Adaptive coding of visual information in neural populations.* Nature, **452**(7184): p. 220–4.
41. Hageman GS, Anderson DH, Johnson LV, et al. (2005), *A common haplotype in the complement regulatory gene factor H (HF1/CFH) predisposes individuals to age-related macular degeneration.* Proc Natl Acad Sci USA, **102**(20): p. 7227–32.
42. Hennig MH, Funke K, Worgotter F (2002), *The influence of different retinal subcircuits on the nonlinearity of ganglion cell behavior.* J Neurosci, **22**(19): p. 8726–38.
43. Hu G, Wensel TG (2002), *R9AP, a membrane anchor for the photoreceptor GTPase accelerating protein, RGS9-1.* Proc Natl Acad Sci USA, **99**(15): p. 9755–60.
44. Hu G, Zhang Z, Wensel TG (2003), *Activation of RGS9-1GTPase acceleration by its membrane anchor, R9AP.* J Biol Chem, **278**(16): p. 14550–4.
45. Huang SH, Pittler SJ, Huang X, et al. (1995), *Autosomal recessive retinitis pigmentosa caused by mutations in the alpha subunit of rod cGMP phosphodiesterase.* Nat Genet, **11**(4): p. 468–71.
46. Humayun MS, de Juan E, Jr., Dagnelie G, et al. (1996), *Visual perception elicited by electrical stimulation of retina in blind humans.* Arch Ophthalmol, **114**(1): p. 40–6.
47. Humphries MM, Rancourt D, Farrar GJ, et al. (1997), *Retinopathy induced in mice by targeted disruption of the rhodopsin gene.* Nat Genet, **15**(2): p. 216–9.
48. Jakobsdottir J, Conley YP, Weeks DE, et al. (2005), *Susceptibility genes for age-related maculopathy on chromosome 10q26.* Am J Hum Genet, **77**(3): p. 389–407.
49. Jones BW, Marc RE (2005), *Retinal remodeling during retinal degeneration.* Exp Eye Res, **81**(2): p. 123–37.
50. Jones BW, Marc RE, Watt CB, et al. (2006), *Neural plasticity revealed by light-induced photoreceptor lesions.* Adv Exp Med Biol, **572**: p. 405–10.
51. Jones BW, Watt CB, Frederick JM, et al. (2003), *Retinal remodeling triggered by photoreceptor degenerations.* J Comp Neurol, **464**(1): p. 1–16.
52. Jones BW, Watt CB, Marc RE (2005), *Retinal remodelling.* Clin Exp Optom, **88**(5): p. 282–91.
53. Kaplan J, Gerber S, Larget-Piet D, et al. (1993), *A gene for Stargardt's disease (fundus flavimaculatus) maps to the short arm of chromosome 1.* Nat Genet, **5**(3): p. 308–11.
54. Kolb H, Gouras P (1974), *Electron microscopic observations of human retinitis pigmentosa, dominantly inherited.* Invest Ophthalmol, **13**(7): p. 487–98.
55. Koyama R, Yamada MK, Fujisawa S, et al. (2004), *Brain-derived neurotrophic factor induces hyperexcitable reentrant circuits in the dentate gyrus.* J Neurosci, **24**(33): p. 7215–24.
56. Lakhanpal RR, Yanai D, Weiland JD, et al. (2003), *Advances in the development of visual prostheses.* Curr Opin Ophthalmol, **14**(3): p. 122–7.
57. Li JB, Gerdes JM, Haycraft CJ, et al. (2004), *Comparative genomics identifies a flagellar and basal body proteome that includes the BBS5 human disease gene.* Cell, **117**(4): p. 541–52.
58. Li ZY, Kljavin IJ, Milam AH (1995), *Rod photoreceptor neurite sprouting in retinitis pigmentosa.* J Neurosci, **15**(8): p. 5429–38.
59. Loudin JD, Simanovskii DM, Vijayraghavan K, et al. (2007), *Optoelectronic retinal prosthesis: system design and performance.* J Neural Eng, **4**(1): p. S72–84.
60. Machida S, Kondo M, Jamison JA, et al. (2000), *P23H rhodopsin transgenic rat: correlation of retinal function with histopathology.* Invest Ophthalmol Vis Sci, **41**(10): p. 3200–9.
61. Maguire AM, Simonelli F, Pierce EA, et al. (2008), *Safety and efficacy of gene transfer for Leber's congenital amaurosis.* N Engl J Med, **358**(21): p. 2240–8.
62. Maller JB, Fagerness JA, Reynolds RC, et al. (2007), *Variation in complement factor 3 is associated with risk of age-related macular degeneration.* Nat Genet, **39**(10): p. 1200–1.
63. Marc RE (1999), *Kainate activation of horizontal, bipolar, amacrine, and ganglion cells in the rabbit retina.* J Comp Neurol, **407**(1): p. 65–76.
64. Marc RE (1999), *Mapping glutamatergic drive in the vertebrate retina with a channel-permeant organic cation.* J Comp Neurol, **407**(1): p. 47–64.

65. Marc RE (2004), *Retinal Neurotransmitters*, in *The Visual Neurosciences*, Chalupa LM, Werner J, Editors. MIT Press: Cambridge, MA. p. 315–30.
66. Marc RE (2008), Functional Neuroanatomy of the Retina, in *Albert and Jakobiec's Principles and Practice of Ophthalmology*, 3rd edition, Editors. Albert and Miller, Elsevier, p. 1565–92.
67. Marc RE, Jones BW (2002), *Molecular phenotyping of retinal ganglion cells*. J Neurosci, **22**(2): p. 413–27.
68. Marc RE, Jones BW (2003), *Retinal remodeling in inherited photoreceptor degenerations*. Mol Neurobiol, **28**(2): p. 139–47.
69. Marc RE, Jones BW, Anderson JR, et al. (2007), *Neural reprogramming in retinal degeneration*. Invest Ophthalmol Vis Sci, **48**(7): p. 3364–71.
70. Marc RE, Jones BW, Watt CB, Strettoi E (2003), *Neural remodeling in retinal degeneration*. Prog Retin Eye Res, **22**(5): p. 607–55.
71. Marc RE, Jones BW, Watt CB, et al. (2008), *Extreme retinal remodeling triggered by light damage: implications for age related macular degeneration*. Mol Vis, **14**: p. 782–806.
72. Masland RH (2001), *Neuronal diversity in the retina*. Curr Opin Neurobiol, **11**(4): p. 431–6.
73. Masland RH (2001), *The fundamental plan of the retina*. Nat Neurosci, **4**(9): p. 877–86.
74. McLaughlin ME, Ehrhart TL, Berson EL, Dryja TP (1995), *Mutation spectrum of the gene encoding the beta subunit of rod phosphodiesterase among patients with autosomal recessive retinitis pigmentosa*. Proc Natl Acad Sci USA, **92**(8): p. 3249–53.
75. McLaughlin ME, Sandberg MA, Berson EL, Dryja TP (1993), *Recessive mutations in the gene encoding the beta-subunit of rod phosphodiesterase in patients with retinitis pigmentosa*. Nat Genet, **4**(2): p. 130–4.
76. Molday LL, Rabin AR, Molday RS (2000), *ABCR expression in foveal cone photoreceptors and its role in Stargardt macular dystrophy*. Nat Genet, **25**(3): p. 257–8.
77. Morimura H, Fishman GA, Grover SA, et al. (1998), *Mutations in the RPE65 gene in patients with autosomal recessive retinitis pigmentosa or leber congenital amaurosis*. Proc Natl Acad Sci USA, **95**(6): p. 3088–93.
78. Peng YW, Senda T, Hao Y, et al. (2003), *Ectopic synaptogenesis during retinal degeneration in the royal college of surgeons rat*. Neuroscience, **119**(3): p. 813–20.
79. Pollard H, Khrestchatisky M, Moreau J, et al. (1994), *Correlation between reactive sprouting and microtubule protein expression in epileptic hippocampus*. Neuroscience, **61**(4): p. 773–87.
80. Pu M, Xu L, Zhang H (2006), *Visual response properties of retinal ganglion cells in the royal college of surgeons dystrophic rat*. Invest Ophthalmol Vis Sci, **47**(8): p. 3579–85.
81. Raymond PA, Barthel LK, Bernardos RL, Perkowski JJ (2006), *Molecular characterization of retinal stem cells and their niches in adult zebrafish*. BMC Dev Biol, **6**: p. 36.
82. Rockhill RL, Daly FJ, MacNeil MA, et al. (2002), *The diversity of ganglion cells in a mammalian retina*. J Neurosci, **22**(9): p. 3831–43.
83. Sommer ME, Farrens DL (2006), *Arrestin can act as a regulator of rhodopsin photochemistry*. Vision Res, **46**(27): p. 4532–46.
84. Sommer ME, Smith WC, Farrens DL (2006), *Dynamics of arrestin–rhodopsin interactions: acidic phospholipids enable binding of arrestin to purified rhodopsin in detergent*. J Biol Chem, **281**(14): p. 9407–17.
85. Specht D, Tom Dieck S, Ammermuller J, et al. (2007), *Structural and functional remodeling in the retina of a mouse with a photoreceptor synaptopathy: plasticity in the rod and degeneration in the cone system*. Eur J Neurosci, **26**(9): p. 2506–15.
86. Stasheff SF (2008), *Emergence of sustained spontaneous hyperactivity and temporary preservation of OFF responses in ganglion cells of the retinal degeneration (rd1) mouse*. J Neurophysiol, **99**(3): p. 1408–21.
87. Stone EM, Braun TA, Russell SR, et al. (2004), *Missense variations in the fibulin 5 gene and age-related macular degeneration*. N Engl J Med, **351**(4): p. 346–53.
88. Strettoi E, Pignatelli V (2000), *Modifications of retinal neurons in a mouse model of retinitis pigmentosa*. Proc Natl Acad Sci USA, **97**(20): p. 11020–5.

89. Strettoi E, Pignatelli V, Rossi C, et al. (2003), *Remodeling of second-order neurons in the retina of rd/rd mutant mice.* Vision Res, **43**(8): p. 867–77.
90. Strettoi E, Porciatti V, Falsini B, et al. (2002), *Morphological and functional abnormalities in the inner retina of the rd/rd mouse.* J Neurosci, **22**(13): p. 5492–504.
91. Sullivan R, Penfold P, Pow DV (2003), *Neuronal migration and glial remodeling in degenerating retinas of aged rats and in nonneovascular AMD.* Invest Ophthalmol Vis Sci, **44**(2): p. 856–65.
92. Sullivan RK, Woldemussie E, Pow DV (2007), *Dendritic and synaptic plasticity of neurons in the human age-related macular degeneration retina.* Invest Ophthalmol Vis Sci, **48**(6): p. 2782–91.
93. Sutula T (2002), *Seizure-induced axonal sprouting: assessing connections between injury, local circuits, and epileptogenesis.* Epilepsy Curr, **2**(3): p. 86–91.
94. Terada N, Hamazaki T, Oka M, et al. (2002), *Bone marrow cells adopt the phenotype of other cells by spontaneous cell fusion.* Nature, **416**(6880): p. 542–5.
95. Varela C, Igartua I, De la Rosa EJ, De la Villa P (2003), *Functional modifications in rod bipolar cells in a mouse model of retinitis pigmentosa.* Vision Res, **43**(8): p. 879–85.
96. Vasireddy V, Uchida Y, Salem N, Jr., et al. (2007), *Loss of functional ELOVL4 depletes very long-chain fatty acids (> or =C28) and the unique omega-O-acylceramides in skin leading to neonatal death.* Hum Mol Genet, **16**(5): p. 471–82.
97. Vugler A, Lawrence J, Walsh J, et al. (2007), *Embryonic stem cells and retinal repair.* Mech Dev, **124**(11–12): p. 807–29.
98. Wang QL, Chen S, Esumi N, et al. (2004), *QRX, a novel homeobox gene, modulates photoreceptor gene expression.* Hum Mol Genet, **13**(10): p. 1025–40.
99. Wassle H (2004), *Parallel processing in the mammalian retina.* Nat Rev Neurosci, **5**(10): p. 747–57.
100. Weiss TF (1996), *Cellular Biophysics: Electrical Properties.* Vol. 2. MIT Press: Cambridge, MA. p. 557.
101. Wensel TG (2008), *Signal transducing membrane complexes of photoreceptor outer segments.* Vision Res, **48**(20): p. 2052–61.
102. Yanai D, Weiland JD, Mahadevappa M, et al. (2007), *Visual performance using a retinal prosthesis in three subjects with retinitis pigmentosa.* Am J Ophthalmol, **143**(5): p. 820–27.
103. Yates JR, Sepp T, Matharu BK, et al. (2007), *Complement C3 variant and the risk of age-related macular degeneration.* N Engl J Med, **357**(6): p. 553–61.
104. Yen HJ, Tayeh MK, Mullins RF, et al. (2006), *Bardet–Biedl syndrome genes are important in retrograde intracellular trafficking and Kupffer's vesicle cilia function.* Hum Mol Genet, **15**(5): p. 667–77.
105. Young MJ, Ray J, Whiteley SJ, et al. (2000), *Neuronal differentiation and morphological integration of hippocampal progenitor cells transplanted to the retina of immature and mature dystrophic rats.* Mol Cell Neurosci, **16**(3): p. 197–205.
106. Zarbin MA (2004), *Current concepts in the pathogenesis of age-related macular degeneration.* Arch Ophthalmol, **122**(4): p. 598–614.
107. Zeitz C, Gross AK, Leifert D, et al. (2008), *Identification and functional characterization of a novel rhodopsin mutation associated with autosomal dominant CSNB.* Invest Ophthalmol Vis Sci, **49**(9): p. 4105–14.
108. Zeitz C, van Genderen M, Neidhardt J, et al. (2005), *Mutations in GRM6 cause autosomal recessive congenital stationary night blindness with a distinctive scotopic 15-Hz flicker electroretinogram.* Invest Ophthalmol Vis Sci, **46**(11): p. 4328–35.
109. Zhang K, Kniazeva M, Han M, et al. (2001), *A 5-bp deletion in ELOVL4 is associated with two related forms of autosomal dominant macular dystrophy.* Nat Genet, **27**(1): p. 89–93.
110. Zrenner E (2002), *Will retinal implants restore vision?* Science, **295**(5557): p. 1022–5.

Chapter 4
Cortical Plasticity and Reorganization in Severe Vision Loss

Eduardo Fernández and Lotfi B. Merabet

Abstract Blind individuals make striking adjustments to their loss of sight. Current experimental evidence suggests that these behavioral adaptations are based on dramatic neurophysiological changes at the level of the brain. In particular, is the fact that the occipital cortex (the area of the brain normally ascribed with visual processing) is functionally recruited to process non-visual sensory modalities. The impact of these neuroplastic changes on the success of implementing a rehabilitative strategy such as a visual based neuroprosthesis remains unknown. Here we discuss several factors such as potential limits of plasticity, potential mechanisms and methods to modulate neuroplasticity so as to promote rehabilitative potential. We should thus remain aware that some of the impediments to future progress in visual neuroprosthesis development are not only technical, engineering and surgical issues, but are also related to the development and implementation of strategies designed to interface with the visually deprived brain. New evidence regarding experience-dependent plasticity in the adult brain together with the achievements in other neuroprosthesis efforts allows cautious optimism that some degree of functional vision can be restored in profoundly blind individuals. However, it is essential that future research explore the mechanisms underlying brain plasticity following the loss of vision. These new findings should be integrated in order to enhance the development of suitable rehabilitative strategies for each particular type of visual neuroprosthesis and achieve the best possible behavioral outcome for a given person using these devices.

Abbreviations

AMD	Age-related macular degeneration
D-AMPH	D-Amphetamine
rTMS	Repetitive transcranial magnetic stimulation

E. Fernández (✉)
Instituto de Bioingeniería, Universidad Miguel Hernández, Avda de la Universidad s/n, 03202 Elche (Alicante), Spain
e-mail: e.fernandez@umh.es

G. Dagnelie (ed.), *Visual Prosthetics: Physiology, Bioengineering, Rehabilitation,*
DOI 10.1007/978-1-4419-0754-7_4, © Springer Science+Business Media, LLC 2011

SPECT   Single photon emission computerized tomography
tDCS   Transcranial direct current stimulation
TMS   Transcranial magnetic stimulation

4.1   Introduction

The question of what happens to the brain following the loss of sight is of seminal importance for any rehabilitative strategy for the blind. In order to interact effectively with their environment, blind individuals have to make striking adjustments to their loss of sight. Growing experimental evidence now suggests that these behavioral adaptations are reflected by dramatic neurophysiological changes at the level of the brain and specifically, with regions of the brain responsible for processing vision itself. These changes may represent the exploitation of spatial and temporal processing inherent within occipital visual cortex that allow a blind individual to adapt to the loss of sight and remain integrated in highly visually-dependent society.

Over the past 25 years, great strides have been made in understanding the neurophysiological mechanisms underlying visual perception. What is less known are the changes associated with how the brain adapts to the loss of sight. For example, what is the physiological and functional fate of cortical areas normally associated with the processing of visual information once vision is lost (e.g. from ocular disease or trauma)? Would this have an impact on the success of implementing a rehabilitative strategy such as a visual based neuroprosthesis in the hope of restoring functional vision? As research and development continues, we should be aware that some of the impediments to future progress in implementing a visual neuroprosthesis approach are not only technical, engineering and surgical issues, but are also related to the development and implementation of strategies designed to interface with the visually deprived brain.

In this chapter, we will review recent advances in the knowledge about brain plasticity and emphasize its importance in order to achieve optimal and desired behavioral outcomes with respect to neuroprosthesis development. Other important questions that will be reviewed relate to the time course of the plastic changes and whether cortical areas deprived of their normal sensory input can still process the lost sensory modality.

4.2   Current Concepts on Brain Plasticity and Implications for Visual Rehabilitation

Classical thought has held the view that the bulk of brain development occurred during childhood and that thereafter there was little opportunity for dramatic adaptive change. It was understood however, that adult brains must display some form of adaptation or "plasticity" since we are capable of sensory and motor learning throughout life. Furthermore, it has been postulated that the site of these ongoing changes was limited to "higher order" perceptual processing areas as opposed to primary sensory and motor cortices. Thus, primary visual, auditory and somatosensory cortical areas were strictly implicated with seeing, hearing and touch respectively. Currently, these concepts can now be viewed as an oversimplification as research in the field of neuroplasticity has expanded rapidly to suggest that sensory modalities are not as inherently distinct and independent as was previously believed and that the adult brain has a remarkable capacity to change and adapt throughout a developmental lifetime [3, 7, 28, 34, 37, 42, 61, 63, 72, 76].

With respect to the discussion here, there is also considerable evidence that adaptive and compensatory changes occur within the brain following the loss of sight [12, 24, 33, 60, 61, 73, 76, 82, 83]. Current evidence suggests that in response to the loss of sight, regions of the occipital cortex (areas normally ascribed to the processing of visual information) are functionally recruited to process tactile and auditory stimuli and even higher order cognitive functions such as verbal memory (Fig. 4.1).

One important question to address would be to uncover the underlying nature of this functional recruitment of occipital cortex to process other sensory modalities. Is it possible that this recruitment is related to the ability of blind subjects to extract greater information from the remaining sensory modalities for which they are so highly dependent? In this context, plasticity can be viewed as an active component of sensory processing capable of altering processing patterns and cortical topography. However, it is important to note that not all neuroplastic changes following sensory

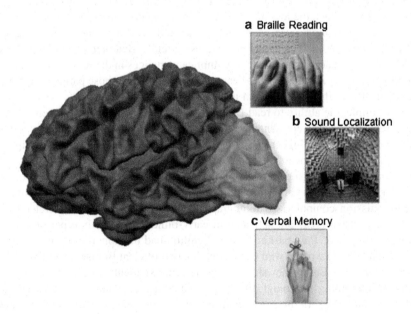

a Braille Reading

b Sound Localization

c Verbal Memory

Fig. 4.1 Recruitment of occipital cortical areas in the blind in response to different tasks. (**a**) Braille reading. (**b**) Sound localization. (**c**) A verbal memory task. See text for references and more details

loss should be assumed to be beneficial or necessarily lead to functional recovery. In reality, neuroplasticity can be viewed as both a positive and negative response. On one hand, it can contribute to changes that are functionally adaptive when a sensory modality is lost. On the other, neuroplasticity can also constrain the degree of adaptation. Therefore, the consequences of neuroplastic change need to be considered not only as a consequence of sensory loss, but also with respect to an individual's own experiences.

Plasticity is not only essential to allow the brain to adjust to its ever-changing sensory environment and experiences and improving perceptual skills, but also plays a crucial role in the recovery from damage and insult. This is also true with regard to the visual system, the adaptation to blindness and ultimately, the restoration of sight. In the case of restoring sight through neuroprosthetic means, it would be a great over-simplification to believe that re-introduction of the lost sensory input by itself will immediately restore the lost sense. Specific strategies have to be developed to modulate information processing by the brain and to extract relevant and functionally meaningful information from neuroprosthetic inputs.

4.3 Clinical Evidence for Reorganization of Cortical Networks in the Blind and Visually Impaired

It would seem reasonable to presume that in the setting of visual deprivation, the brain would reorganize itself to exploit the sensory inputs at its disposal [5, 17, 38, 40, 54, 61] and in fact, the loss of sight has been associated with superior non-visual perception in the blind, such as auditory and tactile abilities, and even in higher cognitive functions such as linguistic processing and verbal memory [4, 5, 12, 60, 79, 81, 92]. These adjustments not only implicate changes in the remaining sensory modalities (for example, touch and hearing) but also involve those parts of the brain once dedicated to the task of vision itself.

For example, the ability to read Braille is associated with an enlargement of the somatosensory cortical representation of the reading index finger (but not the corresponding non-reading finger) in association with the recruitment of the occipital visual cortex for the processing of tactile information. The functional significance of this cross-modal plasticity is supported by a variety of additional converging data. For example, Uhl et al. using event-related electroencephalography and single photon emission computerized tomography (SPECT) [90, 91] demonstrated that the primary visual areas are activated in early-blind subjects while performing a Braille reading task. Pascual-Leone and Hamilton had the opportunity of studying a congenitally blind woman who was an extremely proficient Braille reader (working as an editor for a Braille newsletter) who became suddenly Braille alexic while otherwise remaining neurologically intact, following a bilateral occipital stroke [40]. The interesting finding was contrary to expectations as the lesion did not affect the somatosensory cortex, but rather damaged the occipital pole bilaterally. Although she was well aware of the presence of the dot elements contained in the

Braille text, she was unable to extract enough information to determine the meaning contained in the dot patterns. Further support demonstrating the functional role of occipital cortex comes from the fact that reversible disruption of occipital cortex function, for example by transcranial magnetic stimulation (TMS), impairs Braille reading ability in the blind [24].

While initial work focused on the task of Braille reading [82, 83], increasing evidence has also demonstrated activation of occipital cortical areas in congenitally blind subjects during tasks of auditory localization [51, 93]. This issue was further addressed by assessing language-related brain activity implicated in speech processing and auditory verb-generation. It has been shown that speech comprehension activates not only parts of the brain associated with language (as with sighted adult controls) [80], but also striate and extra-striate regions of the visual cortex [5]. As with tactile processing, reversible functional disruption of the occipital cortex by TMS impairs verb-general performance only in blind subjects [4] providing further evidence that the recruitment of the occipital cortex in high-level cognitive processing is functionally relevant.

The work demonstrating the functional recruitment of occipital cortex for the processing of non-visual information may reveal only the "tip of the iceberg" in terms of the brain reorganization that follows visual deprivation. Nevertheless these neuroplastic changes define a specific time window for the success of any visual neuroprosthesis (before full cross-modal adaptation) that probably is influenced by factors such as the onset and duration of visual deprivation and the mechanisms and profile of the visual loss (Fig. 4.2).

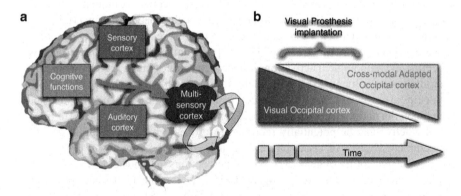

Fig. 4.2 Adaptive and compensatory changes at the occipital cortex after visual deprivation. (**a**) Following visual deafferentation, inputs from other sensory processing areas reach the occipital cortex via connections through multisensory cortical areas (and possible through direct connections). These adaptive changes include the functional recruitment of visual cortical areas for the processing of non-visual information such as tactile information, auditory information and higher-order cognitive functions (e.g. verbal memory). (**b**) Over time, these neuroplastic changes may eventually lead to the establishment of new connections and functional roles with clear implications on the right time of implantation and the likelihood of success in recreating functional vision with a visual neuroprosthetic device

4.4 What Are the Limits to This Cortical Plasticity?

It is important to clarify that evidence of cortical reorganization exists in both congenital and late blind individuals. However, the neuroplastic changes associated with partial blindness remain less understood [19, 22, 27, 58]. For example, Baker et al. [8] found evidence of cortical reorganization in individuals with partial visual field loss due to age-related macular degeneration (AMD). Dilks et al. [27] reported converging behavioral and neuroimaging evidence from a stroke patient consistent with reorganization after deafferentation of human visual cortex, but other studies have been unable to demonstrate the same results [58, 86]. It is also worth noting, if not at least anecdotally, that some blind or severely visually deprived subjects do not adapt well to the loss of sight. In these subjects, preliminary studies suggest that the occipital cortex is not or at most, only partially reorganized [20, 33, 34, 61, 62, 77]. For these individuals, a neuroprosthetic approach might be a functionally desirable solution but it is important to understand their capacity for processing visual information.

The variability in both the behavioral and neurophysiological adaptations could be related not only to the profoundness of the visual loss but also the amount of time spent visually deprived. Thus, the time course associated with neuroplastic changes and whether or not cortical areas can still process visual information after longstanding visual deprivation are important questions to be considered. In this context, we have recently reported the case of a late blind patient (at the time of study, he had no light perception in either eye for at least 12 years), who suddenly experienced elementary and complex visual hallucinations located in his right visual field [2]. Neuro-radiological examination revealed that the patient had an arteriovenous malformation located in his left striate cortex. Although the under- lying pathophysiology of visual hallucinations in this patient is uncertain, the presence of visual hallucinations could represent an increased level of excitability, possibly related to cortical "release" [23, 57] or from an "irritative" phenomenon reflecting the pathological activation of neural ensembles in the regions where the occipital lesion is located [6]. Nonetheless, these findings provide evidence that the occipital cortex of some blind subjects can still generate visual perceptions and strengthen the notion that long-term deafferented occipital cortex can remain func- tionally active and be potentially recruited to process visual information despite the complete absence of visual input. Of course, it will be very useful to know in advance the feasibility of generating visual perceptions in blind subjects before the implantation of any visual neuroprosthetic device. In this context it has been proposed that image-guided transcranial magnetic stimulation (TMS), can be used as a non- invasive method to systematically map the visual sensations induced by focal stimulation of the human occipital cortex and there are already clinical protocols which allow to relate the localization of the real site of stimulation to characteristic positions in the visual field [32]. This procedure has the potential to improve our understanding of physiologic organization and plastic changes in the human visual system and to establish the degree and extent of remaining functional visual cortex in blind subjects (Fig. 4.3).

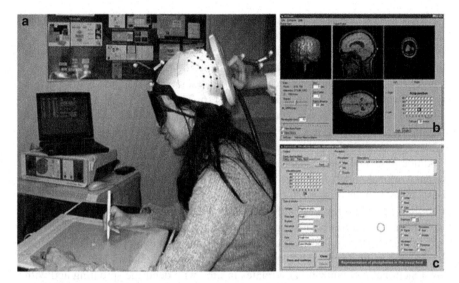

Fig. 4.3 Frameless interface to aid in the positioning of the transcranial magnetic stimulator coil over a subject's brain. This technique enables use of anatomical information as an interactive navigational guide for stimulator coil position and allows recording of the position and orientation of the coil at the instant of stimulation for later correlation with the response data. (**a**) An example of the live display. (**b**) Interface to identify the position of the TMS coil over the subject's MRI data at several customized locations (modified from [32]). (**c**) Customized interface to facilitate the recording of phosphene data

Evidence of cortical reorganization within a time frame as short as months has been observed in animal studies [47, 48]. However human research on this matter remains equivocal [8, 27] and more conclusive data and measures of these functional changes need to be obtained. Furthermore, although rapid reorganization seems possible, it may be dependent on the interaction of several factors (i.e. age, time since onset of blindness, disease severity, etc.). At the same time, the cerebral organization between two given blind persons may be substantially different depending on the pathology, the age of blindness, personal experience, etc. Clearly, much more evidence is needed about the extent of cortical reorganization in the adult human visual system and the conditions under which reorganization occurs [26].

4.5   Possible Mechanisms Behind Brain Plasticity

As stated previously, brain plasticity refers to the neurophysiological changes that occur in relation to the organization of the brain and in response to alterations of neural activity and sensory experience. These changes are especially evident during development and after neurologic injury and can be best conceptualized with encompassing multiple levels of operation from molecular to cellular levels on to neural systems and ultimately, on to behavior.

As it was first suggested by Santiago Ramón y Cajal in his *Textura del Sistema Nervioso del Hombre y los Vertebrados* [18], plasticity can be regarded as a type of "structural modification that implies the formation of new neural pathways through ramification and progressive growth of the dendritic terminals". In general, these changes can involve rapid reinforcement of pre-established synaptic pathways and even the formation of new neuronal pathways [16, 24, 37, 55, 72, 73]. Consequently, and as a result of experience neurotransmitters are modulated, synapses change their morphology, dendrites and spines grow and contract, axons change their trajectory and cortical representational maps are altered.

Although several mechanisms are likely to be involved in all these processes, the current working model emphasizes the selective strengthening of synapses following classical Hebbian learning rules that have guided much of the work in both the field of cortical synaptic plasticity and cortical representational reorganization [41]. At the cellular level, it has been suggested that functional reorganization may rely on mechanisms involving modifications in the excitatory/inhibitory neurotransmission [30, 45, 95] and/or a increased synthesis of neurotrophic factors [46, 49]. In this way, functional reorganization correlates in time with dendritic sprouting and with changes in the excitatory/inhibitory neurotransmission.

There also is evidence that the mechanisms involved in synaptic plasticity can vary between cortical regions [84, 85] and involve glial cells [13]. Thus, the last few years have provided extraordinary evidence regarding the molecular mechanisms underlying neuroplastic changes. The exciting future in this context involves the possibility of developing new approaches such as specific rehabilitation strategies and pharmacological interventions to modulate these processes and ultimately optimize rehabilitative outcomes.

4.6 Modulation of Brain Plasticity: Recent Developments

Enhancement of function-enabling plasticity and prevention of function-disabling plasticity can be accomplished through several approaches such as specific rehabilitation procedures, pharmacological interventions and/or exogenous electrical/magnetic stimulation. For example, a number of studies have demonstrated that cortical maps can be contracted or expanded by loss of peripheral inputs or by enhanced use [3, 9, 21, 29, 37, 55, 56, 72, 78, 96]. Furthermore, as demonstrated by mapping studies after micro-infarcts, it is clear that behavior is one of the most powerful modulators of post-injury recovery and therefore, behavioral intervention to enhance recovery is becoming increasingly popular [3, 70, 72, 75]. These rehabilitation therapies have significantly improved the quality of life of many of patients after brain damage and suggest that this ability of the brain to reorganize itself by experience-dependent neural plasticity could be also used to develop new training strategies to accelerate learning and maximize the adaptation to prosthetic vision devices. Learning electrically stimulated visual patterns can be a new and difficult experience. Although the continued use of the visual implant may, by itself, restore some visual perception, carefully guided visual rehabilitation may be necessary to maximize the adaptation and get the most from these devices.

There are other possible methods to modulate plasticity that are currently being evaluated and could potentially act as adjuvant therapy. Pharmacological approaches [43], for example, the coupling of D-amphetamine (D-AMPH) and rehabilitative training seems to be useful in promoting behavioral function as well as neurotrophic and neuroplastic responses in animal studies [10, 88] and these effects have also been established during language learning in the human [15]. However, clinical studies on the use of D-AMPH as a pharmacological adjunct to post-stroke rehabilitation have yielded mixed results probably because of type, dosage and timing of drug delivery [87, 94]. Nonetheless, it seems that practice-dependent changes of cortical plasticity can be facilitated by pharmacological strategies that rely on adrenergic and cholinergic mechanisms and show a trend towards decrease of antagonists to these neurotransmitter systems [43]. These findings provide preliminary evidence that the pharmacological approach combined with appropriate rehabilitation strategies may be beneficial to promote post-injury recovery facilitating the "re-learning" processes and could potentially extend to the case of visual rehabilitation.

Another possibility to enhance training effects is the use of intracortical micro-stimulation techniques [1, 68, 69, 71, 75]. This procedure appears to induce dendritic growth in a frequency specific manner, however direct cortical microstimulation requires extensive neurosurgical procedures (e.g. a craniotomy) thus questioning its overall feasibility within the clinical setting. Alternative approaches that could induce similar effects and that are being currently tested include noninvasive brain stimulation techniques such as transcranial magnetic stimulation (TMS) and transcranial direct current stimulation (tDCS). TMS is a non-invasive and painfree technique for cortical stimulation in humans [11] that consists of a magnetic field emanating from a wire coil held outside the head. tDCS consists of applying a weak electrical current (1–2 mA) to modulate the activity of neurons. Both techniques are able to induce electrical currents in nearby regions of the brain that can influence brain plasticity and reorganization [25, 44, 89]. These techniques have recently been tested in studies with stroke patients with very promising results [3, 66] and show potential for future application in the rehabilitation of persons with a visual neuroprosthesis. For example, high frequency repetitive transcranial magnetic stimulation (rTMS) over the occipital cortex could be used prior to the training sessions with any visual prosthetic device to enhance excitability and facilitate the interpretation of the visual percepts. Such as strategy may be especially appropriate during early stages of training and learning.

4.7 Neuroplasticity and Other Neuroprostheses Efforts

In terms of restoring lost sensory function, cochlear implant research has arguably been the most successful and has translated into a viable therapeutic option for over 110,000 deaf adults and children worldwide [14, 31, 53, 59, 74]. After some time,

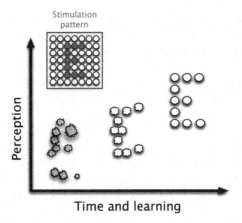

Fig. 4.4 The neural plasticity of the visual cortex can contribute to ever-improving correlation between the physical world and evoked phosphenes. Immediately after implantation the evoked phosphenes are likely to induce a poor perception of an object (the letter "E" in this example). However, appropriate learning and rehabilitation strategies will contribute to provide concordant perceptions (modified from [67])

and with adequate and intensive training, deaf individuals can learn to comprehend and in some cases, even acquire speech. By analogy, after the surgical implantation of any visual prosthesis, the right learning and rehabilitation strategies could potentially help to modulate the plasticity of the brain and contribute to ever-improving performance and more concordant perceptions [67] (Fig. 4.4).

It is remarkable that studies investigating neuroplasticity in the deaf and following cochlear implantation bear striking parallels with those seen in visual neuropros theses development. For example, similar to activation of the occipital cortex by auditory stimuli in the blind, the auditory cortex (Brodmann's areas 41, 42 and 22) is activated in deaf subjects in response to visual stimuli [36]. Thus, like in the case of the blind, the removal of one sensory modality leads to neural reorganization of the remaining ones. Visual–auditory plasticity in the deaf has also been found in patients with cochlear implants [31, 50, 64, 65]. Neuroimaging studies indicate that after implantation of a cochlear device, primary auditory cortex is activated by the sound of spoken words in deaf patients who had lost hearing before the development of language. Interestingly, it appears that if metabolism in the auditory cortex is restored by cross-modal plasticity changes before implantation, the auditory cortex can no longer respond to signals from a cochlear implant installed afterwards and patients do not show improvement in language capabilities [52]. It is important to note that many changes are invariably linked to chronic electrical stimulation that in turn, complicate the interpretation of the neuroplastic changes that ensue. Again, drawing from cochlear research experience, chronic stimulation can lead to a significant reduction in spiral ganglion neurons; de-myelination of residual ganglion neurons; shrinkage of the perikaryon of neurons throughout the auditory pathway and reduced spontaneous activity throughout the auditory pathway [31]. These neuroanatomical and physiological changes also need to be taken into consideration in terms of the effects of chronic electrical stimulation and the potential long term benefits.

4.8 A Look at What Is Ahead

The lessons to be learned are that simple re-introduction of the lost sensory input by itself might not be sufficient to restore the lost sense. For restoring functional vision in the blind, we must first understand how the brain adapts to blindness and uncover adaptive resources such as cross-modal representations. There is no doubt that plasticity will contribute to the success of any visual neuroprostheses, but specific strategies will then be necessary to modulate information processing by the brain and to extract relevant and functionally meaningful information from the electrical stimulation patterns [33, 61, 62] (Fig. 4.5).

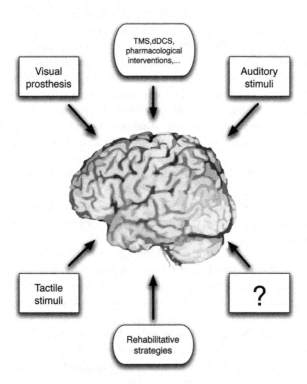

Fig. 4.5 Some possible experimental strategies proposed to enhance functional vision and the adaptation to a visual neuroprosthetic device. It should be taken into account that the rehabilitation of the blind is a very complex problem, requiring intimate collaborations among clinicians, basic scientists, engineers, educators and rehabilitative experts

Several studies have highlighted that following the loss of vision the brain undergoes profound neuroplastic changes. This plasticity takes place at a variety of levels, from the synaptic interactions among single neurons and the circuits in which neurons interact, to large-scale systems comprising those circuits. Furthermore it has been also suggested that glial cells could have central roles in the adaptation to blindness [13]. The precise understanding of these changes will be crucial in developing and projecting the success of novel visual neuroprosthetic strategies will certainly have implications for rehabilitative training and device development. This endeavor will require strong interactions between basic scientists, clinicians, engineers and rehabilitation experts to help make decisions about (a) whether potentially residual capacity for vision exists; (b) how this plasticity can be driven and (c) what the inputs should be to maximize this restitution. These issues are central to the development of any visual neuroprosthesis approach and will provide a mechanistic rationale for understanding therapeutic interventions and teaching strategies for the blind.

New evidence about experience-dependent plasticity of the adult brain together with the achievements of other neuroprosthesis efforts allows cautious optimism about the possibility to restore some functional vision to profoundly blind individuals, but there are still several important issues that should be taken into account. Case studies of surgical sight restoration following long-term visual deprivation [35, 39] provide a relevant insight. For example, patients blinded for many years experience profound difficulty in various visual tasks, particularly those requiring the identification and recognition of objects following ocular surgical procedures aimed at regaining some degree of functional vision. Interestingly, if these patients were allowed to explore the same object through touch, they can recognize it immediately as to register their newly acquired visual percepts with their existing senses. These results suggest that the simple restoration of the lost sensory input may not itself suffice for achieving a functional sense. One possibility to overcome this problem might be to develop a patient controlled system that coordinates and registers the visual perceptions generated by a visual prosthesis with the identification of objects perceived through other senses (such as touch and audition). Patients could then learn to integrate these concordant sources of sensory stimuli into meaningful percepts [61].

Finally, although the effects of neural plasticity are prominent in the context of any visual neuroprosthesis, they are usually unrecognized or greatly underestimated. Therefore, it is essential that future research explore the mechanisms that underlie brain plasticity following the loss of vision and that research studies in the field of visual prosthesis learn to integrate these new findings to enhance the translation of this knowledge to clinical research and practice. We have now an unprecedented number of tools for the restoration of sight through artificial means but we have to use these tools to select appropriate candidates for implantation, to develop suitable rehabilitative strategies for each particular type of visual neuroprosthesis and to achieve the best possible behavioral outcome for a given person using these devices.

References

1. Adachi K, Lee JC, Hu JW, et al. (2007), *Motor cortex neuroplasticity associated with lingual nerve injury in rats*. Somatosens Mot Res, **24**(3): p. 97–109.
2. Alfaro A, Concepcion L, Merabet L, Fernandez E (2006), *An atypical presentation of visual hallucinatory experiences following prolonged blindness*. Neurocase, **12**(4): p. 212–5.
3. Alonso-Alonso M, Fregni F, Pascual-Leone A (2007), *Brain stimulation in poststroke rehabilitation*. Cerebrovasc Dis, **24**(Suppl 1): p. 157–66.
4. Amedi A, Floel A, Knecht S, et al. (2004), *Transcranial magnetic stimulation of the occipital pole interferes with verbal processing in blind subjects*. Nat Neurosci, **7**(11): p. 1266–70.
5. Amedi A, Raz N, Pianka P, et al. (2003), *Early 'visual' cortex activation correlates with superior verbal memory performance in the blind*. Nat Neurosci, **6**(7): p. 758–66.
6. Anderson SW, Rizzo M (1994), *Hallucinations following occipital lobe damage: the pathological activation of visual representations*. J Clin Exp Neuropsychol, **16**(5): p. 651–63.
7. Bach-y-Rita P (2004), *Tactile sensory substitution studies*. Ann N Y Acad Sci, **1013**: p. 83–91.
8. Baker CI, Peli E, Knouf N, Kanwisher NG (2005), *Reorganization of visual processing in macular degeneration*. J Neurosci, **25**(3): p. 614–8.
9. Bao S, Chan VT, Merzenich MM (2001), *Cortical remodelling induced by activity of ventral tegmental dopamine neurons*. Nature, **412**(6842): p. 79–3.
10. Barbay S, Zoubina EV, Dancause N, et al. (2006), *A single injection of d-amphetamine facilitates improvements in motor training following a focal cortical infarct in squirrel monkeys*. Neurorehabil Neural Repair, **20**(4): p. 455–8.
11. Barker AT, Jalinous R, Freeston IL (1985), *Non-invasive magnetic stimulation of human motor cortex*. Lancet, **1**: p. 1106–7.
12. Bavelier D, Neville HJ (2002), *Cross-modal plasticity: where and how?* Nat Rev Neurosci, **3**(6): p. 443–52.
13. Bernabeu A, Alfaro A, Garcia M, Fernandez E (2009), *Proton magnetic resonance spectroscopy (1H-MRS) reveals the presence of elevated myo-inositol in the occipital cortex of blind subjects*. Neuroimage, **47**(4): p. 1172–6.
14. Bouccara D, Avan P, Mosnier I, et al. (2005), *Auditory rehabilitation*. Med Sci (Paris), **21**(2): p. 190–7.
15. Breitenstein C, Wailke S, Bushuven S, et al. (2004), *d-Amphetamine boosts language learning independent of its cardiovascular and motor arousing effects*. Neuropsychopharmacology, **29**(9): p. 1704–14.
16. Buonomano DV, Merzenich MM (1998), *Cortical plasticity: from synapses to maps*. Annu Rev Neurosci, **21**: p. 149–86.
17. Burton H, Snyder AZ, Conturo TE, et al. (2002), *Adaptive changes in early and late blind: a fMRI study of Braille reading*. J Neurophysiol, **87**(1): p. 589–607.
18. Cajal SR (1904), *Textura del sistema nervioso del hombre y de los vertebrados*. Imprenta y librería de Nicolás Moya: Madrid.
19. Calford MB, Chino YM, Das A, et al. (2005), *Neuroscience: rewiring the adult brain*. Nature, **438**(7065): p. E3; discussion E3–4.
20. Celesia GG (2005), *Visual plasticity and its clinical applications*. J Physiol Anthropol Appl Human Sci, **24**(1): p. 23–7.
21. Celnik P, Hummel F, Harris-Love M, et al. (2007), *Somatosensory stimulation enhances the effects of training functional hand tasks in patients with chronic stroke*. Arch Phys Med Rehabil, **88**(11): p. 1369–76.
22. Cheung SH, Legge GE (2005), *Functional and cortical adaptations to central vision loss*. Vis Neurosci, **22**(2): p. 187–201.
23. Cogan DG (1973), *Visual hallucinations as release phenomena*. Albrecht Von Graefes Arch Klin Exp Ophthalmol, **188**(2): p. 139–50.

24. Cohen LG, Celnik P, Pascual-Leone A, et al. (1997), *Functional relevance of cross-modal plasticity in blind humans*. Nature, **389**(6647): p. 180–3.
25. Cohen LG, Ziemann U, Chen R, et al. (1998), *Studies of neuroplasticity with transcranial magnetic stimulation*. J Clin Neurophysiol, **15**(4): p. 305–24.
26. Dagnelie G (2006), *Visual prosthetics 2006: assessment and expectations*. Expert Rev Med Devices, **3**(3): p. 315–25.
27. Dilks DD, Serences JT, Rosenau BJ, et al. (2007), *Human adult cortical reorganization and consequent visual distortion*. J Neurosci, **27**(36): p. 9585–94.
28. Dreher B, Burke W, Calford MB (2001), *Cortical plasticity revealed by circumscribed retinal lesions or artificial scotomas*. Prog Brain Res, **134**: p. 217–46.
29. Duque J, Mazzocchio R, Stefan K, et al. (2008), *Memory formation in the motor cortex ipsilateral to a training hand*. Cereb Cortex, **18**: p. 1395–406.
30. Dyck RH, Chaudhuri A, Cynader MS (2003), *Experience-dependent regulation of the zincergic innervation of visual cortex in adult monkeys*. Cereb Cortex, **13**(10): p. 1094–109.
31. Fallon JB, Irvine DR, Shepherd RK (2008), *Cochlear implants and brain plasticity*. Hear Res, **238**(1–2): p. 110–7.
32. Fernandez E, Alfaro A, Tormos JM, et al. (2002), *Mapping of the human visual cortex using image-guided transcranial magnetic stimulation*. Brain Res Protoc, **10**: p. 115–24.
33. Fernandez E, Pelayo F, Romero S, et al. (2005), *Development of a cortical visual neuroprosthesis for the blind: the relevance of neuroplasticity*. J Neural Eng, **2**(4): p. R1–12.
34. Ferrandez JM, Alfaro A, Bonomini P, et al. (2003), *Brain plasticity: feasibility of a cortical visual prosthesis for the blind*. In *Engineering in Medicine and Biology Society. Proceedings of the 25th Annual International Conference of the IEEE*.
35. Fine I, Wade AR, Brewer AA, et al. (2003), *Long-term deprivation affects visual perception and cortex*. Nat Neurosci, **6**(9): p. 915–6.
36. Finney EM, Fine I, Dobkins KR (2001), *Visual stimuli activate auditory cortex in the deaf*. Nat Neurosci, **4**(12): p. 1171–3.
37. Froemke RC, Merzenich MM, Schreiner CE (2007), *A synaptic memory trace for cortical receptive field plasticity*. Nature, **450**(7168): p. 425–9.
38. Gizewski ER, Gasser T, de Greiff A, et al. (2003), *Cross-modal plasticity for sensory and motor activation patterns in blind subjects*. Neuroimage, **19**(3): p. 968–75.
39. Gregory RL (2003), *Seeing after blindness*. Nat Neurosci, **6**(9): p. 909–10.
40. Hamilton R, Keenan JP, Catala M, Pascual-Leone A (2000), *Alexia for Braille following bilateral occipital stroke in an early blind woman*. Neuroreport, **11**(2): p. 237–40.
41. Hebb DO (1947), *The effects of early experience on problem solving at maturity*. Am Psychol, **2**: p. 737–45.
42. Hernandez Muela S, Mulas F, Mattos L (2004), *Functional neuronal plasticity*. Rev Neurol, **38**(Suppl 1): p. 58–68.
43. Hummel FC, Cohen LG (2005), *Drivers of brain plasticity*. Curr Opin Neurol, **18**(6): p. 667–74.
44. Hummel FC, Cohen LG (2006), *Non-invasive brain stimulation: a new strategy to improve neurorehabilitation after stroke?* Lancet Neurol, **5**(8): p. 708–12.
45. Iwai Y, Fagiolini M, Obata K, Hensch TK (2003), *Rapid critical period induction by tonic inhibition in visual cortex*. J Neurosci, **23**(17): p. 6695–702.
46. Johansson BB (2000), *Brain plasticity and stroke rehabilitation. The Willis lecture*. Stroke, **31**(1): p. 223–30.
47. Kaas JH (2002), *Sensory loss and cortical reorganization in mature primates*. Prog Brain Res, **138**: p. 167–76.
48. Kaas JH, Krubitzer LA, Chino YM, et al. (1990), *Reorganization of retinotopic cortical maps in adult mammals after lesions of the retina*. Science, **248**(4952): p. 229–31.
49. Komitova M, Johansson BB, Eriksson PS (2006), *On neural plasticity, new neurons and the postischemic milieu: an integrated view on experimental rehabilitation*. Exp Neurol, **199**(1): p. 42–55.

50. Kral A, Tillein J, Heid S, et al. (2006), *Cochlear implants: cortical plasticity in congenital deprivation.* Prog Brain Res, **157**: p. 283–313.
51. Kujala T, Alho K, Paavilainen P, et al. (1992), *Neural plasticity in processing of sound location by the early blind: an event-related potential study.* Electroencephalogr Clin Neurophysiol, **84**(5): p. 469–72.
52. Lee DS, Lee JS, Oh SH, et al. (2001), *Cross-modal plasticity and cochlear implants.* Nature, **409**(6817): p. 149–50.
53. Loeb GE (1990), *Cochlear prosthetics.* Annu Rev Neurosci, **13**: p. 357–71.
54. Maeda K, Yasuda H, Haneda M, Kashiwagi A (2003), *Braille alexia during visual hallucination in a blind man with selective calcarine atrophy.* Psychiatry Clin Neurosci, **57**(2): p. 227–9.
55. Mahncke HW, Bronstone A, Merzenich MM (2006), *Brain plasticity and functional losses in the aged: scientific bases for a novel intervention.* Prog Brain Res, **157**: p. 81–109.
56. Mahncke HW, Connor BB, Appelman J, et al. (2006), *Memory enhancement in healthy older adults using a brain plasticity-based training program: a randomized, controlled study.* Proc Natl Acad Sci USA, **103**(33): p. 12523–8.
57. Manford M, Andermann F (1998), *Complex visual hallucinations. Clinical and neurobiological insights.* Brain, **121**(Pt 10): p. 1819–40.
58. Masuda Y, Dumoulin SO, Nakadomari S, Wandell BA (2008), *V1 projection zone signals in human macular degeneration depend on task, not stimulus.* Cereb Cortex, **18**: p. 2483–93.
59. McDermott HJ (2004), *Music perception with cochlear implants: a review.* Trends Amplif, **8**(2): p. 49–82.
60. Merabet LB, Pascual-Leone A (2010), *Neural reorganization following sensory loss: the opportunity of change.* Nat Rev Neurosci, **11**(1): p. 44–52.
61. Merabet LB, Rizzo JF, Amedi A, et al. (2005), *What blindness can tell us about seeing again: merging neuroplasticity and neuroprostheses.* Nat Rev Neurosci, **6**(1): p. 71–7.
62. Merabet LB, Rizzo JF, III, Pascual-Leone A, Fernandez E (2007), *'Who is the ideal candidate?': decisions and issues relating to visual neuroprosthesis development, patient testing and neuroplasticity.* J Neural Eng, **4**(1): p. S130–5.
63. Merabet L, Thut G, Murray B, et al. (2004), *Feeling by sight or seeing by touch?* Neuron, **42**(1): p. 173–9.
64. Middlebrooks JC, Bierer JA, Snyder RL (2005), *Cochlear implants: the view from the brain.* Curr Opin Neurobiol, **15**(4): p. 488–93.
65. Mitchell TV, Maslin MT (2007), *How vision matters for individuals with hearing loss.* Int J Audiol, **46**(9): p. 500–11.
66. Nair DG, Hutchinson S, Fregni F, et al. (2007), *Imaging correlates of motor recovery from cerebral infarction and their physiological significance in well-recovered patients.* Neuroimage, **34**(1): p. 253–63.
67. Normann RA, Maynard E, Guillory KS, Warren DJ (1996), *Cortical implants for the blind.* IEEE Spectrum, **33**(5): p. 54–9.
68. Nudo RJ (2003), *Adaptive plasticity in motor cortex: implications for rehabilitation after brain injury.* J Rehabil Med, **41**(Suppl): p. 7–10.
69. Nudo RJ (2003), *Functional and structural plasticity in motor cortex: implications for stroke recovery.* Phys Med Rehabil Clin N Am, **14**(1 Suppl): p. S57–76.
70. Nudo RJ (2006), *Plasticity.* NeuroRx, **3**(4): p. 420–7.
71. Nudo RJ, Jenkins WM, Merzenich MM (1990), *Repetitive microstimulation alters the cortical representation of movements in adult rats.* Somatosens Mot Res, **7**(4): p. 463–83.
72. Pascual-Leone A, Amedi A, Fregni F, Merabet LB (2005), *The plastic human brain cortex.* Annu Rev Neurosci, **28**: p. 377–401.
73. Pascual-Leone A, Hamilton R, Tormos JM, et al. (1999), *Neuroplasticity in the adjustment to blindness.* In *Neuronal Plasticity: Building a Bridge from the Laboratory to the Clinic*, J. Grafman, Christen Y, Editors. Springer: Berlin.
74. Pena C, Bowsher K, Samuels-Reid J (2004), *FDA-approved neurologic devices intended for use in infants, children, and adolescents.* Neurology, **63**(7): p. 1163–7.

75. Plautz EJ, Barbay S, Frost SB, et al. (2003), *Post-infarct cortical plasticity and behavioral recovery using concurrent cortical stimulation and rehabilitative training: a feasibility study in primates*. Neurol Res, **25**(8): p. 801–10.

76. Ptito M, Kupers R (2005), *Cross-modal plasticity in early blindness*. J Integr Neurosci, **4**(4): p. 479–88.

77. Ptito M, Moesgaard SM, Gjedde A, Kupers R (2005), *Cross-modal plasticity revealed by electrotactile stimulation of the tongue in the congenitally blind*. Brain, **128**(Pt 3): p. 606–14.

78. Ramos-Estebanez C, Merabet LB, Machii K, et al. (2007), *Visual phosphene perception modulated by subthreshold crossmodal sensory stimulation*. J Neurosci, **27**(15): p. 4178–81.

79. Rauschecker JP (1995), *Compensatory plasticity and sensory substitution in the cerebral cortex*. Trends Neurosci, **18**(1): p. 36–43.

80. Roder B, Stock O, Bien S, et al. (2002), *Speech processing activates visual cortex in congenitally blind humans*. Eur J Neurosci, **16**(5): p. 930–6.

81. Roder B, Teder-Salejarvi W, Sterr A, et al. (1999), *Improved auditory spatial tuning in blind humans*. Nature, **400**(6740): p. 162–6.

82. Sadato N, Pascual-Leone A, Grafman J, et al. (1996), *Activation of the primary visual cortex by Braille reading in blind subjects*. Nature, **380**(6574): p. 526–8.

83. Sadato N, Pascual-Leone A, Grafman J, et al. (1998), *Neural networks for Braille reading by the blind*. Brain, **121**(Pt 7): p. 1213–29.

84. Shaw KN, Commins S, O'Mara SM (2003), *Deficits in spatial learning and synaptic plasticity induced by the rapid and competitive broad-spectrum cyclooxygenase inhibitor ibuprofen are reversed by increasing endogenous brain-derived neurotrophic factor*. Eur J Neurosci, **17**(11): p. 2438–46.

85. Shaw CA, Lanius RA, van den Doel K (1994), *The origin of synaptic neuroplasticity: crucial molecules or a dynamical cascade?* Brain Res Brain Res Rev, **19**(3): p. 241–63.

86. Smirnakis SM, Brewer AA, Schmid MC, et al. (2005), *Lack of long-term cortical reorganization after macaque retinal lesions*. Nature, **435**(7040): p. 300–7.

87. Sonde L, Lokk J (2007), *Effects of amphetamine and/or l-dopa and physiotherapy after stroke – a blinded randomized study*. Acta Neurol Scand, **115**(1): p. 55–9.

88. Stroemer RP, Kent TA, Hulsebosch CE (1998), *Enhanced neocortical neural sprouting, synaptogenesis, and behavioral recovery with d-amphetamine therapy after neocortical infarction in rats*. Stroke, **29**(11): p. 2381–93; discussion 2393–5.

89. Thickbroom GW (2007), *Transcranial magnetic stimulation and synaptic plasticity: experimental framework and human models*. Exp Brain Res, **180**(4): p. 583–93.

90. Uhl F, Franzen P, Lindinger G, et al. (1991), *On the functionality of the visually deprived occipital cortex in early blind persons*. Neurosci Lett, **124**(2): p. 256–9.

91. Uhl F, Franzen P, Podreka I, et al. (1993), *Increased regional cerebral blood flow in inferior occipital cortex and cerebellum of early blind humans*. Neurosci Lett, **150**(2): p. 162–4.

92. Van Boven RW, Hamilton RH, Kauffman T, et al. (2000), *Tactile spatial resolution in blind Braille readers (1)*. Am J Ophthalmol, **130**(4): p. 542.

93. Weeks R, Horwitz B, Aziz-Sultan A, et al. (2000), *A positron emission tomographic study of auditory localization in the congenitally blind*. J Neurosci, **20**(7): p. 2664–72.

94. Whiting E, Chenery HJ, Chalk J, Copland DA (2007), *Dexamphetamine boosts naming treatment effects in chronic aphasia*. J Int Neuropsychol Soc, **13**(6): p. 972–9.

95. Zepeda A, Sengpiel F, Guagnelli MA, et al. (2004), *Functional reorganization of visual cortex maps after ischemic lesions is accompanied by changes in expression of cytoskeletal proteins and NMDA and GABA(A) receptor subunits*. J Neurosci, **24**(8): p. 1812–21.

96. Zhou X, Merzenich MM (2007), *Intensive training in adults refines A1 representations degraded in an early postnatal critical period*. Proc Natl Acad Sci USA, **104**(40): p. 15935–40.

Chapter 5
Visual Perceptual Effects of Long-Standing Vision Loss

Ava K. Bittner and Janet S. Sunness

Abstract This chapter focuses on the changes in vision experienced by patients with RP and AMD. The specific aspects of vision that are reviewed include progressive changes in central acuity, contrast sensitivity, visual field, color vision, night vision, glare, and light and dark adaptation. Emphasis is on patients' perspectives, including the impact on functioning and performance of activities of daily living, as well as rates, patterns of vision loss, and day-to-day visual fluctuations experienced by those with retinal degenerative diseases. Several types of visual phenomena are presented, including Charles Bonnet Syndrome hallucinations in AMD, perceptual completion or filling-in of scotomas in AMD, remapping visual cortex in AMD, the preferred retinal locus in AMD, and photopsias or light show type flashes in RP. The proposed implications of these visual changes and phenomena as they apply to retinal prosthetic vision are discussed.

Abbreviations

AIBSE	Acute idiopathic blind spot enlargement
AMD	Age-related macular degeneration
AZOOR	Acute zonal occult outer retinopathy
CBS	Charles Bonnet syndrome
fMRI	Functional magnetic resonance imaging
GA	Geographic atrophy
MEWDS	Multiple evanescent white dot syndrome
PIC	Punctate inner choroidopathy
PRL	Preferred retinal locus
RP	Retinitis pigmentosa
VEGF	Vascular endothelial growth factor

A.K. Bittner (✉)
Lions Vision Research & Rehabilitation Center, Wilmer Eye Institute, Johns Hopkins University School of Medicine, 550 N. Broadway, 6th floor, Baltimore, MD 21205, USA
e-mail: abittne1@jhmi.edu

G. Dagnelie (ed.), *Visual Prosthetics: Physiology, Bioengineering, Rehabilitation*, 93
DOI 10.1007/978-1-4419-0754-7_5, © Springer Science+Business Media, LLC 2011

5.1 Introduction

Retinal degenerations are characterized by a loss of vision. The loss of photoreceptors leads to the development of blind spots (scotomas) or reduction in the visual field area. The features of the scotomas (whether they are peripheral or central, the time course of development, etc) are characteristic of the particular retinal disease, but all have in common the loss of vision.

However, as a consequence of the retinal disease, whether at the level of "sick" retinal cells, changes in the optic nerve, or changes in the brain, there is the generation of new visual phenomena. Patients may report flashing lights (photopsias), positive scotomas (perceived as blurry or missing areas of vision), filling-in phenomena, and visual hallucinations. While the basis of these phenomena is not clearly understood, they are reported by a large number of patients, and must be taken into account both in the design of visual prostheses and when interpreting visual responses from patients implanted with such devices.

5.2 Vision Changes Experienced by RP Patients

5.2.1 Overview

The most prominent and earliest symptoms of RP are progressive night blindness and field loss, though central vision may also be reduced. The vision loss is bilateral and symmetrical. There are two patterns of night vision loss [35]. In type 1 rod-cone degeneration, there is reduced night vision from birth. In these patients, early rod dysfunction may be demonstrated by dark-adapted two-color static perimetry. In type 2 (sometimes called regional), night vision is normal until field loss begins. In type 2 patients, dark-adapted visual field perimetry shows rod photoreceptor function in non-scotomatous retinal areas. Usually the patients with type 2 degeneration once had the ability to see stars at night, while patients with type 1 were never able to see stars. In both forms, however, the initial symptoms typically include either mobility problems in dim or dark illumination or an inability to change quickly from one light level to another.

Most individuals are first symptomatic between the ages of 5–30 years, although some cases have been reported to emerge later in life [39]. The age of onset of RP varies for different genetic mutations, but across all patients, the average age at which RP was diagnosed by an ophthalmologist was reported as 35 years [56]. As a generalization, patients with X-linked RP begin having visual field loss the earliest (typically during the teenage years), patients with autosomal recessive RP are in the middle, and patients with autosomal dominant RP may not develop significant field loss until the 40s or later. The proportion of RP patients with autosomal recessive inheritance is approximately 30–40%, autosomal dominant is presumed to occur in about 50–60%, and x-linked inheritance is estimated in

5–15% [22]. For patients without a prior family history of RP, it is often diagnosed incidentally during a routine eye examination, and sometimes on the basis of subjectively reported reduced night vision.

5.2.2   Visual Field Loss in RP

Peripheral visual field loss is universal in RP. It typically starts in the midperipheral region of the retina and spreads both out toward the periphery and in toward the macula [39]. Patients may develop a full or partial mid-peripheral ring scotoma, which then expands outward and inward. Since the nasal retina extends more peripherally than the temporal retina, the far periphery of the temporal visual field may be spared when the scotoma reaches the edge of the nasal, superior and inferior fields. Some RP patients may retain far peripheral, temporal islands of vision later in the course of the degeneration, even when the central field is <20° or in some cases when there is no remaining central vision. If the peripheral spared areas are large enough, they can enable patients to detect moving objects from the side or give valuable information during mobility to avoid bumping into objects or people. Early in the course of the disease, individuals with RP may be labeled as being clumsy or careless in terms of mobility, bumping into people and obstacles hidden by their (as yet unknown) scotomas. As peripheral visual field loss progresses, they are increasingly prone to bumps, bruises and falls.

The rate of visual field progression in RP is typically slow, with estimates of about 5–14% lost per year [5, 20, 27, 34]. Figures 5.1 and 5.2 show examples of visual field progression in retinitis pigmentosa measured by Goldmann perimetry over 13 and 16 years, respectively [21]. For most individuals, the progression is steady, but some report that their rate of visual field loss is variable over time, with occasional lengthy periods of perceived stabilization. The slow rate of progression enables RP patients to adapt well to their vision loss, and they often do not seek mobility training or assistance until late in the disease when only a few degrees of central vision remain. A previous survey indicated that about 23% of RP patients were not aware that they had visual field loss, although they showed constriction of their field [24]. Often patients who have good central acuity but substantial field loss, who would be characterized as legally blind on the basis of a visual field diameter <20°, are surprised to learn the extent of the loss through visual field testing. This is because they have adjusted gradually to the progressive field loss and are still fully functional in their daily activities. Due to constrictions in the visual field, RP patients tend to use scanning to survey their environment for orientation and mobility. When walking in unfamiliar areas, instead of gazing straight ahead at a distant target, RP patients tend to direct their gaze at nearby objects on the walls, downward, or at the layout (i.e., edge-lines or boundaries between walls) [57]. The smaller the horizontal visual field extent, the more they tend to use downward-directed fixations, which are important to detect changes in the walking surface and avoid low-lying obstacles.

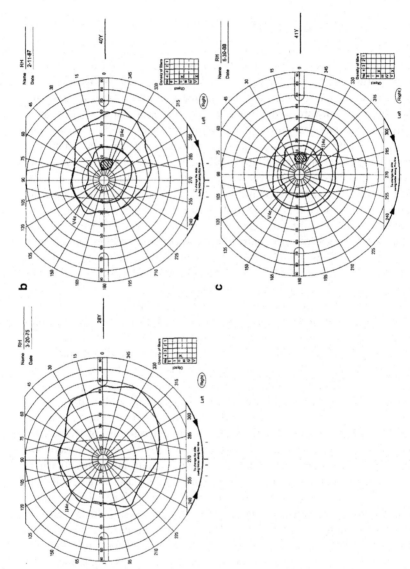

Fig. 5.1 (a–c) Goldmann visual fields obtained on the right eye of a patient with retinitis pigmentosa, with targets as marked, showing a pattern IA loss of visual fields over a period of 13 years. Reprinted from [21], with permission

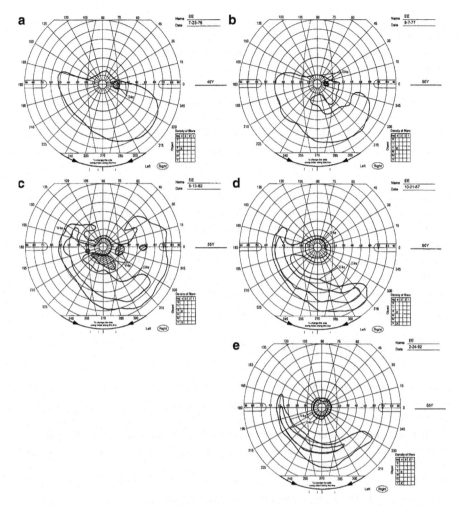

Fig. 5.2 (**a–e**) Goldmann visual fields obtained on the right eye of a patient with retinitis pigmentosa, with targets as marked, showing a pattern IIB loss of visual fields over a period of 16 years. Reprinted from [21], with permission

5.2.3 Changes in Color Vision and Glare Sensitivity in RP

Early in RP, color vision is typically normal since the central visual field where the vast majority of cone photoreceptors are located is not affected by the initial rod photoreceptor degeneration. However, as the disease progresses, abnormalities in color vision are highly correlated with the extent of visual field loss. Among those with a visual acuity of 20/30 or better, autosomal dominant cases are less likely to show extensive color defects when compared to other genetic types of RP [15]. As central visual acuity is initially lost, the development of dyschromatopsia to pale, desaturated and similar colors occurs first. Then, defects in blue color discrimination are common in RP. The color matches of RP patients, are on average more protanomalous (i.e., requiring a greater than normal red/green mixture ratio during color matching) than those in normally sighted individuals [60]. As the retinal degeneration progresses, bright red and orange colors are the last hues that are typically lost.

Most RP patients become increasingly sensitive to light, which can include bright sunlight or diffuse glare, as in white cloudy weather. Many RP patients complain of visual impairment or of discomfort in bright light, independent of cataract. The amount of intraocular light scatter in RP has been correlated to visual field area [2]. It is possible that when vision is reduced due to the retinal degeneration, even a minimal further reduction due to bright light may move the patient into a range of functional disability [18]. Also, RP patients require a longer time to recover visual acuity following transitions between areas with different levels of light. Therefore, they experience difficulty when transitioning from a bright sunny day outdoors to dimmer indoor lighting, or vice versa [26]. To help with light sensitivity and glare, the majority of RP patients always wear sunglasses or tinted lenses when outdoors on sunny days. Most wear them only sometimes on cloudy days, and rarely if ever indoors [58].

5.2.4 Vision Fluctuations in RP

There are visual phenomena present in RP patients that are unexplained at present. The most striking is the presence of "good" and "bad" days. Many patients report having good and bad days, without any clear correlation with ambient lighting or weather. Visual acuity and contrast sensitivity measures are two to three times more variable in legally blind RP patients when compared to normally-sighted individuals [31]. Variability in visual acuity or visual field appears to increase as visual acuity or visual field is reduced in RP; however, contrast sensitivity does not appear to vary according to the level of remaining contrast sensitivity.

Periodic shifts or changes in the way the retinal degeneration affects patients' ability to function and accomplish important tasks leads to experiences of increased disability at potentially critical times. Some RP patients indicate that stress or fatigue decreases vision temporarily, and that their vision improves when these factors are alleviated. Day-to-day decreases in visual field test results appear to be related to corresponding periodic increases in perceived stress or decreases in general health. Research in this area, based on patient feedback and focused on investigating the concerns pertinent to patients, is currently being conducted to understand and mitigate RP patients' visual fluctuations that can result in significant distress, morbidity, and reduced quality of life [32]. Attempts to understand and manage these aspects of retinal disease processes may also help identify therapies and improve the reliability of outcome measures in clinical trials. Day-to-day fluctuations in retinal sensitivity in response to electrical stimulation with prostheses and resulting short-term variations in visual function are also likely to occur with retinal prostheses.

5.3   Visual Changes in Patients with Advanced Macular Degeneration

Unlike RP, patients with macular disease experience visual loss within the central field of vision, and generally have preserved peripheral vision throughout the course of the disease. The most common type of advanced macular degeneration is that associated with age-related macular degeneration (AMD). Two types of advanced AMD lead to loss of central vision, and are commonly referred to as the wet and dry forms of AMD. In patients with advanced AMD, about two-thirds have the wet form.

5.3.1   Changes Due to Wet AMD or Choroidal Neovascularization

Wet AMD (or choroidal neovascularization) is characterized by the growth of abnormal new blood vessels underneath the retina, with a predilection for development in the foveal region (i.e. the very center of the vision, specialized and required for fine vision such as reading). These new blood vessels leak, bleed, and scar, leading to an acute drop of vision and severe visual acuity loss with progression [7]. Despite a variety of treatments (including various types of laser and photodynamic therapy), until 2005 most patients with wet AMD eventually lost vision to the 20/200 or worse level, because of recurrence of the new blood vessels. Only in this decade have there been treatments developed that address the root cause of the development of new blood vessels. Vascular endothelial growth factor (VEGF) has been identified as one of the factors stimulating new blood vessel growth. Several anti-VEGF medications have been developed. These require injection into the eye, on a monthly basis. Clinical trials using Ranibizumab (Lucentis) have shown that about 90% of patients achieve stabilization of their visual acuity, and 35% improve their visual acuity [42]. Thus, while wet AMD has been the leading cause of severe visual loss in the population over age 60, the frequency of severe visual loss from this condition should decrease markedly in the future. Patients may still be left with scotomas, with features similar to those described below.

5.3.2   Changes Due to Dry AMD or Geographic Atrophy

The second type of advanced AMD is known as the dry AMD form or geographic atrophy (GA). In this condition, there is gradual death of the retinal pigment epithelium, the layer of cells beneath the retina, with consequent death of the overlying photoreceptors and scotoma development. These areas of atrophy and scotoma may

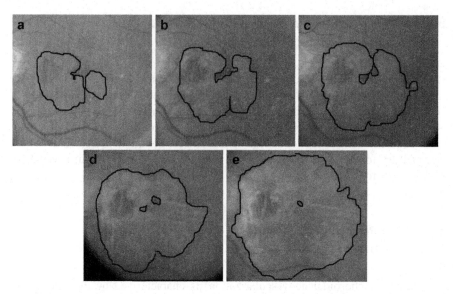

Fig. 5.3 Progression of advanced dry age-related macular degeneration (geographic atrophy) over a 5-year period. The *black* outline shows the border of the atrophy and the presence of spared areas within it. (**a**) Baseline: there were two areas of geographic atrophy, partly surrounding the fovea. Visual acuity was 20/40. (**b**) One-year follow-up: the atrophy enlarged and coalesced into a horseshoe-shaped area of atrophy and corresponding dense scotoma. The fovea was spared, as was the area immediately above the fovea. Visual acuity was still 20/40, but the reading rate dropped by more than 50%. (**c**) Two-year follow-up: the atrophy has enlarged and the spared vertical region has gotten narrower. The fovea is still spared. Visual acuity was 20/60. (**d**) Three-year follow-up: the atrophy has enlarged and coalesced superiorly, so that there is a ring of atrophy, with two small spared areas within it. Visual acuity was still 20/60. (**e**) Five-year follow-up: the atrophy has enlarged further, and there is only a tiny slit of fovea spared within the atrophy. Visual acuity was 20/200. The patient now used eccentric retina for reading text that was 20/600 or greater in size

first develop in areas near, but not involving, the fovea (Fig. 5.3a). As these atrophic areas enlarge and coalesce, the patient may develop a horseshoe of blind area around the fovea, and then a ring of scotoma surrounding the fovea (Fig. 5.3b) [45]. These patients may have good visual acuity, but they have difficulty reading and recognizing faces because the whole word or face does not "fit" in the spared seeing area that is surrounded by scotoma (Fig. 5.3c and 5.3d) [52]. Eventually, the fovea itself becomes atrophic and severe visual acuity loss occurs (Fig. 5.3e). Unlike wet AMD, there are no treatments available to slow or prevent vision loss in GA at the present time. Once an area becomes atrophic, the photoreceptor cells die, so that a potential treatment that prevents cell death will not restore vision to an atrophic area.

Patients with GA and other forms of advanced AMD have profound reductions in vision in dim illumination due to reduced contrast sensitivity, and require increased lighting in order to read [53]. They are also more sensitive to the effects of glare and bright sunlight, requiring solar shields with tinted lenses outdoors, and in some cases indoors as well. Unlike advanced RP patients, the vast majority of individuals with advanced AMD do not require use of a white cane or mobility training since the peripheral vision remains intact even in late stages of the disease.

Patients with other forms of macular disease, such as Stargardt disease (the most common form of macular degeneration in young people) and diabetic retinopathy, have similar symptoms to those described above. Optimizing the visual performance of patients with macular disease generally involves improving the lighting, increasing contrast, and teaching the patient how to adapt to the presence of scotomas in the central visual field.

5.4 Charles Bonnet Syndrome

5.4.1 Overview

Hallucinations are organized perceptions in the absence of an external stimulus. Charles Bonnet Syndrome (CBS) is a condition first described by Charles Bonnet in 1760, in which visual hallucinations occur in visually impaired individuals. The patient is aware that the hallucination is not reality and there is an absence of cognitive impairment. CBS is usually associated with impaired vision due to retinal degeneration, and can be present in either AMD or RP patients. It is possible for CBS to occur with any type of ocular pathology, but it is less common in patients with glaucoma, optic neuritis or cataracts than in those with retinal disease. It is more common in the elderly. A diagnosis of CBS can be made when alternative conditions known to give rise to hallucinations such as migraine, occipital lobe epilepsy, and psychiatric disease have been excluded. If other senses are involved (e.g. hearing, smell), then it is unlikely to be CBS.

The prevalence of CBS has been estimated up to 15% [44], but likely varies due to patients' concern to hide their symptoms for fear of being labeled psychiatrically unstable. CBS is very rare in those without vision loss, with only monocular loss or in those with no light perception in both eyes. The onset of some cases has been reported prior to vision loss. Some patients have found a reduction in the occurrence of CBS when their vision loss progressed significantly or led to complete blindness. Cases of CBS have also been reported following enucleation or removal of the eye [43].

The visual hallucinations can involve photopsias, colors, moving parts, patterns or shapes (simple) or well-defined forms, images and scenes (complex) [1]. They tend to appear centrally, within the areas of vision loss, and about half of AMD patients report that the visual hallucinations are clearer than their current vision. Examples of the most common types of visual hallucinations include images of people or faces, followed by geometric patterns, and plants, flowers or trees [1, 30]. They are normally not frightening. For example, one of our patients reported that she sees people in her apartment. She checks that her door is still double-locked, goes over and says hello, and then gets back to what she was doing. Some actually give great pleasure. One woman with advanced macular degeneration reported she saw little girls with beautifully embroidered aprons. She enjoyed watching them, and knew that it must be a visual hallucination because if she were truly using her visually-impaired eyes, she would not be able to see the fine embroidery. One woman lost 25 pounds because she kept seeing bugs in her food. One man thought he was seeing saints, because he saw many different faces. However, the persistence or frequent recurrence of episodes of CBS can become annoying to patients.

5.4.2   Complexity of Visual Hallucinations in CBS

Using functional magnetic resonance imaging (fMRI) of the visual cortex, simple visual hallucinations have been found to originate early in the visual pathway (V1 and/or V2), whereas more complex visual hallucinations were generated in the higher visual areas [9]. Only a very small proportion, 8%, of the reported visual hallucinations was restricted to the area of binocular field loss [1].

In patients with bilateral scotomas secondary to AMD, the likelihood or complexity of CBS is not predicted by the extent of visual acuity loss [1]. The rate of vision loss may be a predictor of CBS, as cases of AMD patients with rapid loss of vision due to significant exudation or laser photocoagulation have been reported. Cognitive factors, such as state of arousal, may play a central role in the development of CBS once the vision loss has reached a critical threshold level, either in terms of the extent of visual field loss or amount of reduction in visual cortical processing.

The neural basis underlying CBS remains under debate, with many hypotheses proposed [37, 40]. Initially CBS was thought to be related to epileptic discharges, but imaging did not confirm this hypothesis. Another etiological theory is that it may be a deafferentation phenomenon, such as a visual analogue to a phantom limb in amputees, in which brain activity occurs without sensory input. The significant biochemical changes in the areas of the deafferented synapses may result in a release phenomenon or hyperexcitability response.

5.4.3   Predictors and Alleviating Factors for CBS

The mean age of onset of CBS ranges from 75 to 84 years [37], however, cases have also been reported in much younger individuals, including children. Fewer cases among children may reflect an increased plasticity of the immature afferent pathway and/or an inability of the patient to understand or describe the visual experiences. Younger age among AMD patients has been identified as a potential predictor for CBS [1]. CBS is more common among women; however this likely reflects the female bias of an elderly population, as well as a greater willingness among female patients to report visual phenomena that may be considered "abnormal" by caregivers.

CBS is more common in those who live alone and have limited social contacts. Other risk factors for CBS include loss of energy, low extroversion, shyness, use of beta-blocker medications, loneliness or bereavement [55]. Typically an episode of CBS is brief, lasting only a few seconds to minutes, and they tend to occur more often in the evening. In one study, about half of the patients reported that their visual hallucinations typically lasted between 1 and 60 min [44]. About a quarter to a third of patients with CBS experience visual hallucinations daily, whereas about half experience them weekly or monthly. The frequency and durations of the visual hallucinations can vary within and between individuals. They tend to last longer when the individual is drowsy or tired, indicating a relationship with the patient's state of arousal [54]. As with photopsias, some potential triggers for CBS include stress and fatigue. CBS is more likely to occur in low levels of illumination, whereas photopsias are associated with either absence of light or bright light, but more often with bright light [37].

Patients report that they have little control over the appearance or duration of the images. Some patients are able to stop their hallucinations through rapid closing and opening the eyes, blinking, sustained eye closure, turning on a light, looking at something else for distraction, walking away, or hitting or shouting at the hallucination when alone [12]. Potential treatments may include optical devices, such as prisms or telescopes, tinted lenses, use of night lights in the bedroom or increasing social contacts [37]. There are some case reports indicating that some of the atypical antipsychotic or antiepileptic medications can alleviate symptoms; however their effectiveness in clinical trials has not been established [13, 44].

5.5   Filling-In Phenomena (Perceptual Completion)

In order to explain the visual deficit in macular degeneration to people with normal vision, an image is often presented showing a black splotch in the middle of the picture. While this conveys the fact that it is the central vision that is involved, most patients do not see a black splotch. Instead, they report that things are blurry. Patients seem to be able to distinguish this type of blurriness from the difficulty of seeing small print, for example. The blurriness is a result of the patient's brain and visual system trying to "fill-in" what is not seen [41]. The area corresponding to the scotoma cannot look clear, because in truth the image is not being seen. But the image is completed, and the patients are often not aware that the reason they cannot recognize a face is that parts are really missing.

The filling-in phenomenon or perceptual completion has been associated with the difficulty appreciating a scotoma on Amsler grid testing. The Amsler grid is a square of graph paper, subtending 20° horizontally and vertically at the defined viewing distance. There is a large dot in the middle of the grid. The way in which this test is generally used is to try to have the patient center the eye on the grid (for example, by seeing the four corners of the grid), and then report if the central dot can be seen. (Alternatively, the patient is told to look directly at the dot). The task is to determine if any of the lines of the grid are distorted, wavy, or missing. The Amsler grid often is used to try to help patients with AMD detect the early development of wet AMD by detecting a distortion in the grid. However, patients who have definite blind spots on visual field testing often cannot detect any part of the grid to be missing or distorted. One study found that 40% of patients with absolute scotomas in their central field, from the heavy laser treatment used for macular degeneration in the past, could not detect a defect on Amsler grid testing [14]. This was further examined by placing the grid directly over the scotoma in a scanning laser ophthalmoscope, in which the examiner could see that it was directly over the blind area. The patients still could not detect the defect on the Amsler grid testing [46].

There are ways to make the patient aware of where filling-in is taking place. In a technique using face fields [49], the patient is instructed to cover one eye. The patient is told to look at the nose of the examiner, so that the nose is seen as clearly as possible. While the patient is looking at the nose, he/she is asked if there is any part of the face that is blurry, distorted, or missing. The face is a very salient stimulus, and patients are often able to say that an eye is missing, or a piece of the cheek is blurred, etc. This is very helpful for defining the location of the preferred retinal locus of fixation (see section 5.7), to allow for more effective low vision training.

5.6   Remapping of Primary Visual Cortex in Patients with Central Scotomas from Macular Disease

Much of the primary visual cortex is devoted to representing the macula [28, 48]. When there are central scotomas, so that much of the macula is blind, what happens to the corresponding areas of primary visual cortex? [25]. Do they remain silent, are they somehow recruited by surrounding retinal areas, or are they stimulated by other cerebral areas? Certainly, cortical areas corresponding to other sensory modalities do not remain silent. For example, there is the phantom limb phenomenon following amputation [17]. For vision, there is the additional question as to whether the potential remapping of the visual cortex could contribute to the use of an eccentric retinal locus of fixation as a "pseudofovea" [59].

Functional MRI (fMRI) can be used to perform retinotopic mapping of the visual cortex [12]. One can have the patient observe a given pattern on a monitor and measure which cortical areas are stimulated by the differential amount of oxygen consumption in each area. Interest has focused on whether there remain silent areas in the primary visual cortex, corresponding to the macular scotomas, or whether there is remapping and therefore activity in these areas that previously subserved the now scotomatous macular region. Work in this area has been ongoing for only the past 5 years. Current information suggests that cortical remapping occurs only when the fovea itself is scotomatous. When the fovea is seeing, the visual cortex corresponding to a scotoma in the macular region near it does not appear to have remapping [3, 4, 36, 51]. One paper reported more stimulation of the so-called lesion projection zone when the eccentric preferred retinal locus of fixation was stimulated [47]. The stimulation of the lesion projection zone may be task-related [36], and may be from higher cortical levels. When patients with central scotomas were shown scrambled patterns, there were silent areas in the primary visual cortex. When the same patients were shown faces, the formerly silent areas were stimulated, and the face-specific extrastriate cortex was stimulated as well (Rosenau BJ, Greenberg AS, Sunness JS, Yantis S. Cortical Lesion Projection Zone activity in Retinal Disease Patients is Caused by Object-Specific Feedback, not Plasticity, presented at the 2008 VSS Meeting). The hypothesis is that there may be top-down stimulation of primary visual cortex. This may relate to the filling-in phenomenon described above.

5.7   The Preferred Retinal Locus for Fixation

When the fovea is no longer functional, the patient must use an eccentric retinal area to fixate and view the object of interest. Patients vary widely in their ability to create a "pseudofovea", that is an eccentric preferred retinal locus (PRL) to which the oculomotor system can direct the object of interest [59]. It is not understood why some patients are able to adopt an eccentric PRL effectively and read a chart by looking up or to a side, while other patients with similar scotomas must use repeated scanning movements, report letters coming in and out of view, and have much greater difficulty in reading.

There is evidence that adoption of an eccentric PRL can improve visual acuity. In a prospective study of patients with bilateral advanced dry AMD (geographic atrophy) followed for 3 years, 17% improved their visual acuity by at least 0.2 logMAR (i.e., seeing letters 2/3 the size or smaller) over the course of the follow-up [50]. This improvement occurred only in the worse-seeing eye at baseline, and occurred despite continual enlargement of the central scotoma over the 3-year period. Scanning laser ophthalmoscope analysis showed that at baseline, these worse-seeing eyes could not use peripheral retina effectively; the fixation cross was placed within the scotomatous area where it was not seen, and it could not be stably placed on seeing eccentric retina. At 3 years, these worse-seeing eyes had acquired the ability to use an eccentric retinal locus for fixation, with consequent improvement in visual function. The same improvement did not happen in the better-seeing eyes, presumably because these eyes already were using eccentric retinal fixation loci at baseline. With both eyes open, the better-seeing eye's fixation pattern likely dominated the two eyes, perhaps interfering with the development of a monocular PRL in the worse-seeing eye. By 3 years, with worsening of the better eye as well, more attention was now directed to the worse-seeing eye, with improved use of peripheral seeing retina and consequent improvement in visual function.

This spontaneous improvement in visual acuity in the worse-seeing eye has important implications for future clinical trials for retinal prostheses. In all likelihood, the eye to be treated initially for each patient will be the one with worse visual acuity. The attention itself that is directed to this eye by virtue of the intervention may improve the patient's ability to use peripheral retina and thereby improve the visual acuity. This phenomenon may be addressed in part by providing some low vision training prior to the clinical trial.

5.8 Photopsias

5.8.1 Photopsias in RP

The basis of photopsias, or light flashes, in RP and other conditions is not well-understood. They may be manifestations of spontaneous activity in compromised retinal cells, or in retinal microneuromas, triggered through inner plexiform layer connections, possibly due to remodeling and/or ganglion cell and axon loss in the degenerating retina. Photopsias may be linked in important ways to the processes occurring during retinal implant stimulation, and their characterization may be helpful for the future development of prosthetic vision. Photopsias may interfere with visual function testing during clinical trials, as well as RP patients' vision while performing daily activities, underscoring the importance of their characterization among this patient population.

A survey of RP patients conducted in the clinic indicated that 35% reported flashes of light [24]. A more recent internet-based anonymous survey of photopsias in RP patients found that 93% of those who completed the survey had experienced photopsias. The photopsias in this survey were described as phosphenes (slow, localized dots or shapes) by 71%, flashes (all or most of the field at once) by 58%, static noise (like on a television without reception) by 31%, and fluorescence (a background glow) by 20% of those who noted photopsias [6]. Photopsias were most commonly reported to have a shape of a crescent, arc or semi-circle by over half of the respondents. The factors that were most commonly reported to be associated with an increase in photopsias were bright light, fatigue, stress, exercise and absence of light.

Photopsias are commonly noted by RP patients in both the earlier stages of the disease, as well as in those with end-stage retinal degeneration. Nearly half of those who have photopsias experienced them before they were diagnosed with RP, and 60% stated that they first noticed photopsias when they were less than 30 years of age [6].

RP patients who were able to read normal or small sized font without magnification, were driving currently, or who could easily navigate or had only some difficulty with mobility in unfamiliar areas, were two to three times more likely to note photopsias mostly or only peripherally versus in their central vision. Therefore, the extent and location of photopsias appear to be related to residual photoreceptor function assessed by self-reported vision and performance of daily living activities. Photopsias tend to start in the periphery early in RP and then later occur more centrally and in areas with vision as deficits in visual function occur. Therefore RP patients may become more aware of photopsias as vision loss becomes more advanced.

The majority of RP patients indicated that photopsias interfere with their vision, and interference was more likely when photopsias occurred daily, increased in frequency over time, or were located across a larger area over time. About half of RP patients who report photopsias experience them daily. The increased frequency of photopsias in RP appears to be related to increased perceived stress and decreased positive mood. About a quarter experienced photopsias constantly, and over half experienced photopsias for only a few seconds at a time. The location or frequency of photopsias in later RP stages may obstruct vision at times, and is a potential issue for patients' function or when obtaining vision measures.

5.8.2 Photopsias in AMD and Other Ocular Diseases

One report indicated that photopsias are common in patients with macular choroidal neovascularization [8], occurring in 59%, and of those, 59% experienced white colored photopsias and described them as typically lasting several seconds. Subretinal fluid, cicatrix formation and larger disciform scars were more common among individuals who noted photopsias than in those who did not. The occurrence of photopsias among those with macular choroidal neovascularization may potentially be due to sensory deprivation; when normal input to the visual system is repressed and the activity of other neural tissue may become more apparent.

Rare retinal diseases with scotomata and the possible presence of photopsias are acute zonal occult outer retinopathy (AZOOR) [61], multiple evanescent white dot syndrome (MEWDS) [29], acute idiopathic blind spot enlargement (AIBSE) affecting the retina around the optic nerve without optic nerve head swelling or choroiditis [16], autoimmune retinopathy (including cancer- and melanoma-associated retinopathies) [23], photoreceptor dysfunction due to digitalis toxicity [38], and punctate inner choroidopathy (PIC) [19]. Some patients with optic neuritis secondary to multiple sclerosis [11, 33] and restrictive thyroid ophthalmopathy with tight inferior recti eye muscles [10] have also reported photopsias associated with eye movements, likely related to compression or traction.

5.9 Concluding Remarks

In addition to the anatomical and functional changes that occur in the retinal and visual cortex, it is important to consider and address the various types of changes in vision experienced by patients with RP and AMD that will impact both the objective and subjective outcomes with prostheses. In particular, emphasis should be placed on the patients' perspectives of functioning with a retinal degenerative disease, considering the disability and uncertainty associated with the performance of activities of daily living and day-to-day visual fluctuations. In the assessment and rehabilitation of prosthetic vision, several types of visual phenomena that may ordinarily occur in those with advanced AMD or end-stage RP will potentially interfere with the visual percepts produced by prosthetic devices. These aspects need to be better understood and managed by researchers and clinicians working in the area of prosthetic vision.

References

1. Abbott EJ, Connor GB, Artes PH, Abadi RV (2007), *Visual loss and visual hallucinations in patients with age-related macular degeneration (Charles Bonnet syndrome).* Invest Ophthalmol Vis Sci, **48**(3): p. 1416–23.
2. Alexander KR, Fishman GA, Derlacki DJ (1996), *Intraocular light scatter in patients with retinitis pigmentosa.* Vision Res, **36**(22): p. 3703–9.
3. Baker CI, Dilks DD, et al. (2008), *Reorganization of visual processing in macular degeneration: replication and clues about the role of foveal loss.* Vision Res, **48**(18): p. 1910–9.
4. Baker CI, Peli E, Knouf N, Kanwisher NG (2005), *Reorganization of visual processing in macular degeneration.* J Neurosci, **25**(3): p. 614–8.
5. Berson EL, Sandberg MA, Rosner B, et al. (1985), *Natural course of retinitis pigmentosa over a three-year period.* Am J Ophthalmol, **99**: p. 24–51.
6. Bittner AK, Diener-West M, Dagnelie G (2009), *A survey of photopsias in self-reported retinitis pigmentosa: location of photopsias is related to disease severity.* Retina, **29**(10): p. 1513–21.
7. Bressler NM, Bressler SB, Fine SL (1988), *Age-related macular degeneration.* Surv Ophthalmol, **32**: p. 375–413.
8. Brown GC, Murphy RP (1992), *Visual symptoms associated with choroidal neovascularization. Photopsias and the Charles Bonnet syndrome.* Arch Ophthalmol, **110**(9): p. 1251–6.
9. Burke W (2002), *The neural basis of Charles Bonnet hallucinations: a hypothesis.* J Neurol Neurosurg Psychiatry, **73**: p. 535–41.
10. Danks JJ, Harrad RA (1998), *Flashing lights in thyroid eye disease: a new symptom described and (possibly) explained.* Br J Ophthalmol, **82**(11): p. 1309–11.
11. Davis FA, Bergen D, Schauf C, et al. (1976), *Movement phosphenes in optic neuritis: a new clinical sign.* Neurology, **26**: p. 1100–4.
12. Engel S, Glover G, Wandell BA, et al. (1997), *Retinotopic organization in human visual cortex and the spatial precision of functional MRI.* Cereb Cortex, **7**: p. 181–92.
13. Eperjesi F, Akbarali N (2004), *Rehabilitation in Charles Bonnet syndrome: a review of treatment options.* Clin Exp Optom, **87**(3): p. 149–52.
14. Fine AM, Elman MJ, Ebert JE, et al. (1986), *Earliest symptoms caused by neovascular membranes in the macula.* Arch Ophthalmol, **104**: p. 513–4.
15. Fishman GA, Young RS, Vasquez V, Lourenço P (1981), *Color vision defects in retinitis pigmentosa.* Ann Ophthalmol, **13**(5): p. 609–18.
16. Fletcher WA, Imes RK, Goodman D, Hoyt WF (1988), *Acute idiopathic blind spot enlargement: a big blind spot syndrome without optic disc edema.* Arch Ophthalmol, **106**: p. 44–9.
17. Flor H, Elbert T, Knecht S, et al. (1995), *Phantom-limb pain as a perceptual correlate of cortical reorganization following arm amputation.* Nature, **375**: p. 482–4.
18. Gawande AA, Donovan WJ, Ginsburg AP, Marmor MF (1989), *Photoaversion in retinitis pigmentosa.* Br J Ophthalmol, **73**(2): p. 115–20.
19. Gerstenblith AT, Thorne JE, Sobrin L, et al. (2007), *Punctate inner choroidopathy: a survey analysis of 77 persons.* Ophthalmology, **114**(6): p. 1201–4.
20. Grover S, Fishman GA, Anderson RJ, et al. (1997), *Rate of visual field loss in retinitis pigmentosa.* Ophthalmology, **104**(3): p. 460–5.
21. Grover S, Fishman GA, Brown J (1998), *Patterns of visual field progression in patients with retinitis pigmentosa.* Ophthalmology, **105**(6): p. 1069–75.
22. Hartong DT, Berson EL, Dryja TP (2006), *Retinitis pigmentosa.* Lancet, **368**(9549): p. 1795–809.

23. Heckenlively JR, Ferreyra HA (2008), *Autoimmune retinopathy: a review and summary.* Semin Immunopathol, **30**(2): p. 127–34.
24. Heckenlively JR, Yoser SL, Friedman LH, Oversier JJ (1988), *Clinical findings and common symptoms in retinitis pigmentosa.* Am J Ophthalmol, **105**: p. 504–11.
25. Heinen SJ, Skavenski AA (1991), *Recovery of visual responses in foveal V1 neurons following bilateral foveal lesions in adult monkey.* Exp Brain Res, **83**: p. 670–4.
26. Herse P (2005), *Retinitis pigmentosa: visual function and multidisciplinary management.* Clin Exp Optom, **88**(5): p. 335–50.
27. Holopigian K, Greenstein V, Seiple W, Carr R (1996), *Rates of change differ among measures of visual function in patients with retinitis pigmentosa.* Ophthalmology, **103**: p. 398–405.
28. Horton J, Hoyt W (1991), *The representation of the visual field in human striate cortex.* Arch Ophthalmol, **109**: p. 816–24.
29. Jampol LM, Sieving PA, Pugh D, et al. (1984), *Multiple evanescent white dot syndrome: I. Clinical findings.* Arch Ophthalmol, **102**: p. 671–4.
30. Khan JC, Shahid H, Thurlby DA, et al. (2008), *Charles Bonnet syndrome in age-related macular degeneration: the nature and frequency of images in subjects with end-stage disease.* Ophthalmic Epidemiol, **15**(3): p. 202–8.
31. Kiser AK, Mladenovich D, Eshraghi F, et al. (2005), *Reliability and consistency of visual acuity and contrast sensitivity measures in advanced eye disease.* Optom Vis Sci, **82**(11): p. 946–54.
32. Kiser AK, Pronovost PJ (2009), *Management of diseases without current treatment options: something can be done.* JAMA, **301**(16): p. 1708–9.
33. McDonald WI, Barnes D (1992), *The ocular manifestations of multiple sclerosis. 1. Abnormalities of the afferent visual system.* J Neurol Neurosurg Psychiatry, **55**: p. 747–52.
34. Massof RW, Dagnelie G, Benzschawel T, et al. (1990), *First order dynamics of visual field loss in retinitis pigmentosa.* Clin Vis Sci, **5**: p. 1–26.
35. Massof RW, Finkelstein D (1981), *Two forms of autosomal dominant primary retinitis pigmentosa.* Doc Ophthalmol, **51**(4): p. 289–346.
36. Masuda Y, Dumoulin SO, Nakadomari S, Wandell BA (2008), *V1 projection zone signals in human macular degeneration depend on task, not stimulus.* Cereb Cortex, **18**(11): p. 2483–93.
37. Menon GJ, Rahman I, Menon SJ, Dutton GN (2003), *Complex visual hallucinations in the visually impaired: the Charles Bonnet syndrome.* Surv Ophthalmol, **48**(1): p. 58–72.
38. Oishi A, Miyamoto K, Kashii S, Yoshimura N (2006), *Photopsia as a manifestation of digitalis toxicity.* Can J Ophthalmol, **41**(5): p. 603–4.
39. Pagon RA (1988), *Retinitis pigmentosa.* Surv Ophthalmol, **33**(3): p. 137–77.
40. Plummer C, Kleinitz A, Vroomen P, Watts R (2007), *Of Roman chariots and goats in overcoats: the syndrome of Charles Bonnet.* J Clin Neurosci, **14**(8): p. 709–14.
41. Ramachandran VS, Gregory RL (1991), *Perceptual filling in of artificially induced scotomas in human vision.* Nature, **350**: p. 699–702.
42. Rosenfeld PJ, Brown DM, Heier JS, et al. (2006), *Ranibizumab for neovascular age-related macular degeneration.* N Engl J Med, **355**(14): p. 1419–31.
43. Ross J, Rahman I (2005), *Charles Bonnet syndrome following enucleation.* Eye, **19**(7): p. 811–2.
44. Rovner BW (2006), *The Charles Bonnet syndrome: a review of recent research.* Curr Opin Ophthalmol, **17**(3): p. 275–7.
45. Sarks JP, Sarks SH, Killingsworth MC (1988), *Evolution of geographic atrophy of the retinal pigment epithelium.* Eye, **2**: p. 552–77.
46. Schuchard RA (1993), *Validity and interpretation of Amsler grid reports.* Arch Ophthalmol, **111**: p. 776–80.
47. Schumacher EH, Jacko JA, Primo SA, et al. (2008), *Reorganization of visual processing is related to eccentric viewing in patients with macular degeneration.* Restor Neurol Neurosci, **26**(4–5): p. 391–402.
48. Slotnick S, Klein S, Carney T, et al. (2001), *Electrophysiological estimate of human cortical magnification.* Clin Neurophysiol, **112**: p. 1349–56.

49. Sunness JS (2008), *The use of face fields for determining fixation location in eyes with central scotomas from macular disease.* J Vis Impair Blind, **102**: p. 679–89.

50. Sunness JS, Applegate CA, Gonzalez-Baron J (2000), *Improvement of visual acuity over time in patients with bilateral geographic atrophy from age-related macular degeneration.* Retina, **20**: p. 162–9.

51. Sunness JS, Liu T, Yantis S (2004), *Retinotopic mapping of visual cortex using fMRI in a patient with central scotomas from atrophic macular degeneration.* Ophthalmology, **111**: p. 1595–8.

52. Sunness JS, Rubin GS, Applegate CA, et al. (1997), *Visual function abnormalities and prognosis in eyes with age-related geographic atrophy of the macula and good acuity.* Ophthalmology, **104**: p. 1677–91.

53. Sunness JS, Rubin GS, Broman A, et al. (2008), *Low luminance visual dysfunction as a predictor of subsequent visual acuity loss in age-related geographic atrophy of the macula.* Ophthalmology, **115**: p. 1480–8.

54. Teunisse RJ, Cruysberg JR, Hoefnagels WH, et al. (1996), *Visual hallucinations in psychologically normal people: Charles Bonnet's syndrome.* Lancet, **347**: p. 794–7.

55. Teunisse RJ, Cruysberg JR, Hoefnagels WH, et al. (1999), *Social and psychological characteristics of elderly visually handicapped patients with the Charles Bonnet syndrome.* Compr Psychiatry, **40**(4): p. 315–9.

56. Tsujikawa M, Wada Y, Sukegawa M, et al. (2008), *Age at onset curves of retinitis pigmentosa.* Arch Ophthalmol, **126**(3): p. 337–40.

57. Turano KA, Geruschat DR, Baker FH, et al. (2001), *Direction of gaze while walking a simple route: persons with normal vision and persons with retinitis pigmentosa.* Optom Vis Sci, **78**(9): p. 667–75.

58. Weiss NJ (1991), *Low vision management of retinitis pigmentosa.* J Am Optom Assoc, **62**(1): p. 42–52.

59. White JM, Bedell HE (1990), *The oculomotor reference in human with bilateral macular disease.* Invest Ophthalmol Vis Sci, **31**: p. 1149–61.

60. Young RS, Fishman GA (1980), *Color matches of patients with retinitis pigmentosa.* Invest Ophthalmol Vis Sci, **19**(8): p. 967–72.

61. Zibrandtsen N, Munch IC, Klemp K, et al. (2008), *Photoreceptor atrophy in acute zonal occult outer retinopathy.* Acta Ophthalmol, **86**(8): p. 913–6.

Part II
Neural Stimulation of the Visual System

Part II
Neural Stimulation of the Visual System

Chapter 6
Structures, Materials, and Processes at the Electrode-to-Tissue Interface

Aditi Ray and James D. Weiland

Abstract This chapter reviews the basic concepts of neural stimulation along with safety considerations for both the electrode and tissue. The section on electrode–electrolyte interface describes the basic mechanism of charge injection at the interface introducing the reader to the electrode double layer. The use of circuit models to represent the physical processes at the interface and in the bulk tissue is discussed. The next section provides a detailed description of the biopotential electrode along with measurement techniques used in electrode characterization. Following this, an overview of popular electrode materials for neural stimulation is provided for the reader. These include conventional materials such as platinum and iridium oxide, as well as newer materials like conducting polymers and carbon nanotubes. The next section reviews the concept of extracellular stimulation introducing the reader to Goldman Equation used to describe the membrane potential. Finally the section dedicated to safe stimulation of tissue describes the mechanisms of neural injury and parameters considered to ensure safe neural stimulation. Special emphasis is placed on safety studies of retinal stimulation.

Abbreviations

AIROF	Anodic iridium oxide films
CMOS	Complimentary metal-oxide semiconductor
CNTs	Carbon nanotubes
CPE	Constant phase element
CV	Cyclic voltammetry
EAD	Early axonal degeneration
EIS	Electrochemical impedance spectroscopy
ERGs	Electroretinograms

A. Ray (✉)
Department of Biomedical Engineering, 1042 Downey Way, Denney Research Building (DRB) 140, Los Angeles, CA 90089, USA
e-mail: Aditi.Ray@AlconLabs.com

G. Dagnelie (ed.), *Visual Prosthetics: Physiology, Bioengineering, Rehabilitation*, DOI 10.1007/978-1-4419-0754-7_6, © Springer Science+Business Media, LLC 2011

IHP	Inner helmholtz plane
OHP	Outer helmholtz plane
PBS	Phosphate buffered saline
PEDOT	Poly(3,4-ethylenedioxythiophene)
PSTHs	Post stimulus time histograms
SIDNE	Stimulation induced depression in neuronal excitability
SIROF	Sputtered iridium oxide film
TiN	Titanium nitride
TIROF	Thermal iridium oxide film

6.1 Introduction

Electrodes used for neural stimulation must operate under demanding conditions. Along with exhibiting biocompatibility, they have to be small enough to cause localized excitation of the target neurons and large enough to support safe delivery of charge for effective neuronal excitation. Also in most cases, the implanted electrodes are required to function for the lifetime of the implant recipient. Consequently, for any neuroprosthesis employing electrical stimulation to be successful, the implanted electrodes must function for decades without significant degradation or damage to either themselves or to the tissue. This warrants understanding the characteristics of the metal-tissue interface in an effort to optimize electrode material selection and design stimulation protocols. Seminal studies of the interface were performed as part of larger consortia developing neural prostheses for paralysis and for implantation in the visual cortex. While retinal prosthesis development has benefited from these findings, the unique structure of the retina and eye require special consideration. Hence, increasing efforts are being made to understand the safety requirements of such retinal prostheses.

Material presented here has been mainly derived from three sources: *Principles of Neural Science by Kandel, Schwartz and Jessell* [30], *Electrochemical Methods: Fundamental and Applications by Bard and Faulkner* [3] and *Electrical stimulation of excitable tissue: design of efficacious and safe protocols by Merrill* [39].

6.2 Electrode–Electrolyte Interface

Whenever a metal electrode is placed in an electrolyte, thermodynamic processes operate to bring the two phases in electrochemical equilibrium. This causes attraction between the charge carriers in the two phases leading to the formation of a net potential across the interface. This interface is popularly known as the *electrical double layer* with the principal charge carriers in the metal phase being the electrons and those in the electrolyte being the ions. The importance of this interface lies in the fact that for any neural excitation to take place, current has to flow through tissue. Hence, the key to understanding and controlling stimulation through metal electrodes lies in understanding the different electrochemical processes that take place at the electrode–electrolyte interface.

When a metal electrode is placed in an electrolyte, a finite separation of charge occurs leading to the formation of the electrical double layer. This charge separation has several manifestations. One reason for charge redistribution at the interface is ions in the electrolyte combining with the electrode. This leads to a net transfer of electrons between the two phases causing a plane of charge at the metal electrode that is opposed by a plane of charge in the electrolyte. Other reasons for the formation of the double layer include the specific adsorption of certain chemical species and preferential orientation of polar molecules such as water. The solution side of the double layer is composed of several layers. The inner layer called the Helmholtz or Stern layer consists of solvent molecules and some other species such as specifically adsorbed ions or molecules. The locus of electrical centres of the specifically adsorbed ions defines the inner Helmholtz plane (IHP) while the locus of centres of the nearest solvated ions defines the outer Helmholtz plane (OHP). The solvated ions are said to be non-specifically adsorbed as their interaction with the charged metal is independent of the chemical properties of the ions. These ions are distributed in the three dimensional region called the diffuse layer extending from the OHP into the bulk solution. The thickness of the diffuse layer is dependent upon the total ionic concentration of the solution.

The metal electrode-solution interface has been shown to behave like a capacitor with a finite amount of charge residing in a very thin layer on the metal surface (excess or deficiency of electrons). In the solution side, the charge is made up of excess anions or cations residing close to the electrode surface. At any given potential, the double layer is characterized by its double layer capacitance C_{dl} (10–40 $\mu C/cm^2$).

6.2.1 Basic Mechanisms of Charge-Injection Across the Electrode–Electrolyte Interface

Before proceeding into understanding the basics of neural stimulation and electrode characterization, it is worth noting the different terminologies assigned to the electrodes employed, which vary depending upon the experimental conditions. For electrochemical characterization, a three-electrode system is employed where the electrode of interest is referred to as the working electrode, while the other two are called the counter and reference electrodes. For neural stimulation, a two-electrode system is employed where current enters the tissue through the stimulating electrode and exits the tissue through the return electrode. Neural stimulation can be further subdivided into monopolar and bipolar stimulation. Monopolar stimulation uses a small stimulating electrode and a large return electrode while in bipolar stimulation, two small electrodes are used as the source and sink in an effort to focus the current to small regions such as nerve cuff electrodes and electrodes used in cochlear implants. Measurements may contain a third electrode termed the reference electrode required for measuring precisely controlled electrical potentials. At equilibrium (no current), the potential of the system remains constant and is typically referred to as the open-circuit potential. Net electrochemical processes begin to take place as soon as the potential is forced away from equilibrium and resulting current begins to flow through the system. Charge transfer across the interface takes place through two primary mechanisms viz., Faradaic and non-Faradaic reactions.

Faradaic and non-faradaic reactions: Non-Faradaic processes include redistribution of the charge at the electrode–electrolyte interface and do not involve any net transfer of charge species across the interface. If charge injection is achieved through only non-Faradaic reactions, i.e. charging and discharging the double-layer capacitance, then the electrode–electrolyte interface can be modeled as a simple capacitor, viz. the double-layer capacitor C_{dl}. If the total amount of charge transferred is small then the transferred charge can be recovered by simply reversing the polarity of the applied pulse or by discharging the capacitor. In addition to charging-discharging of the double layer capacitance, charge injection can also be achieved by Faradaic processes such as oxidation–reduction reactions. These reactions involve the transfer of electrons between the two phases of the reaction and unlike the capacitive mechanism, may or may not be completely reversible in nature. In case of reactions in which at least one of the chemical species is surface bound, the reaction is completely reversible under steady state conditions. Such reactions are limited by the available surface area of electrode and the amount of species adsorbed onto the interface. However, reactions that do not involve at least one surface bound species, have no mechanism to force the reaction to be reversible in the steady state. Charge balancing in the Faradaic regimen is most often achieved via multiple partially reversible reactions that result in the release of one or more possibly cytotoxic chemical substances in the surrounding tissue.

6.3   Electrode Material

In order to depolarize neurons or to record biological potentials, an interface is required between the body and the electronic apparatus. This interface is called the *biopotential electrode*. Biopotential electrodes deal with challenges different from electrodes used in other systems. First the electrode material has to be biocompatible, i.e. non-toxic to the body and second it has to have the ability to serve as a transducer. This is because as we saw in the preceding section, current in the electrode is carried by electrons while in the electrolyte it is carried by ions.

Electrode potential: When a metal is brought into contact with a solution, a net rearrangement of charge occurs at the interface leading to a loss of neutrality of charge at the interface. As a result, the electrolyte in the immediate vicinity of the electrode is at a potential different from the rest of the solution. This difference in potential is called the half-cell potential and is determined by many different parameters such as the type of metal, the type and concentration of ions in the solution, temperature, etc. This half-cell potential is also referred to as the electrode interfacial potential. It is not possible to measure this potential without utilizing a second electrode. However, the second electrode would then create an interface of its own with the electrolyte thus making it impossible to separate the two resulting potentials from each other. To overcome this, electrochemical cells are evaluated in their entirety, generally composed of a working electrode and a reference electrode separated by the electrolyte. Thus a cell's potential is defined as the potential of the working electrode vs. the reference electrode.

Consider the reaction between a metal electrode and a redox couple in the electrolyte:

$$O + ne^- \leftrightarrow R$$

The equilibrium potential for any electrochemical cell can be calculated using the Nernst equation:

$$E_x = \frac{RT}{nF} \ln \frac{[X]_o}{[X]_i} \tag{6.1}$$

where, $[X]_o$ and $[X]_i$ are the concentrations of the species, R is the gas constant, T is the absolute temperature (Kelvin), F is Faraday's constant and n is the number of electrons transferred. For the electrochemical cell above, if the concentration of both species in solution is equal then the potential of the cell will equilibrate to its formal potential E^0. For unequal concentrations, using the Nernst equation, the equilibrium potential for the electrochemical cell is:

$$E_{eq} = E^0 + \frac{RT}{nF} \ln \frac{[O]}{[R]} \tag{6.2}$$

In the absence of any net current, the measured cell potential is called the open-circuit potential, which again is the sum of the two interfacial potentials. Now if instead a current is present, then the observed potential is different from the equilibrium potential. This is due to the polarization of the electrode and the difference between the observed potential and the equilibrium potential is known as the *overpotential η*.

$$\eta = E - E_{eq} \tag{6.3}$$

Three basic mechanisms contribute to overpotential: ohmic, concentration and activation overpotentials. Ohmic overpotential is due to the electrolyte resistance which leads to a voltage drop across the solution during the passage of current between the electrodes. Concentration overpotential occurs due to changes in the distribution of ions at the electrode–electrolyte interface. Activation overpotential occurs due to charge transfer processes involved during oxidation–reduction reactions that are not completely reversible. The net overpotential is simply a sum of three mechanisms.

Polarizable and non-polarizable electrodes: For ideally *polarizable electrodes*, no actual charge crosses the electrode–electrolyte interface during current flow. Instead, during current flow, redistribution of ions occurs at the interface thus exhibiting capacitor like properties. As a result the overpotential is dominated by the concentration overpotential. One example is titanium nitride electrode where charge injection takes place through capacitive charging–discharging processes. Noble metals such as platinum also behave as polarizable electrodes but over a limited range of voltages. Ideally *non-polarizable electrodes* on the other hand are the ones in which current passes freely between the electrode–electrolyte interface and hence causes no overpotential. Electrodes such as silver–silver chloride and saturated calomel come closest to behaving as non-polarizable electrodes. These electrodes are best used as reference electrodes during measurement of electrode potential as there is no change in voltage across their interface during current flow. However, it is essential to note that in reality no electrode behaves either as ideally polarizable or ideally non-polarizable. Electrodes come closest to ideal characteristics only over a limited range of voltages.

6.3.1 Electrode Characterization

Measurement of impedance: Electrochemical impedance spectroscopy [17] has been used successfully to characterize the electrode–electrolyte interface. Specifically for neuroprostheses employing current stimulation, impedance measurement techniques have been employed to test the efficacy of neural stimulation. Studies in the past have shown that for all stimulation strategies to efficiently inject charge across the electrode-tissue interface, an optimal relationship exists between the threshold of excitation and the distance between the electrode and tissue. For the auditory brainstem implants [31], measurements of threshold of excitation as a function of the distance of the electrodes from the target neurons have shown a strong correlation between the two [42]. The reason behind this is that in order to cause neuronal excitation a minimum amount of current density is required. If the interface impedance were high, it would lead to a higher applied voltage, which could then become a limiting factor in the power capabilities of the device. As shall be discussed in Sect. 6.5, in some cases, this high voltage can also lead to undesirable electrochemical reactions to take place at the interface thereby causing tissue damage.

As it is not possible to control the tissue properties of the target system, efforts are made instead to control the electrode design in order to allow safe and effective stimulation. To achieve this, equivalent circuit models of the electrode–electrolyte interface have been developed which along with impedance measurements, provide an estimate of optimal parameters for the electrode. The first ever model was proposed by Warburg in 1899 who modeled the interface as a polarization resistance in series with a polarization capacitance. This would produce a straight vertical line on the complex plane plots (Z imaginary vs. Z real). However, for solid electrodes it was often observed that the straight vertical line had an angle less than 90°. Thus, the electrode impedance consisted of a polarization resistance in series with complex impedance exhibiting frequency dependency. The phenomenon of constant phase angle was first shown by Fricke and the impedance associated with it is termed as the constant phase element (CPE). CPE is thought to arise from surface inhomogeneities and slow reaction kinetics [5]. Mathematically, CPE is represented as:

$$Z_{CPE} = \frac{1}{T(j\omega)^{\phi}} \tag{6.4}$$

where T is a constant in $F\,cm^{-2}\,s^{-1}$ and ϕ is related to the angle of rotation of a purely capacitive line on the complex plane plots. The CPE is often used to represent a "leaky capacitor" and only when $\phi = 1$, $T = C_{dl}$ and a purely capacitive behavior is obtained [33]. Equation (6.4) can be used to model the Warburg element that accounts for diffusion delay in Faradaic currents by assigning $\phi = 0.5$. Finally (6.4) can be used to describe a pure resistor for $\phi = 0$ and a pure inductor for $\phi = -1$. Randles's work showed the importance of the impedance associated with the faradaic processes occurring at the electrode–electrolyte interface. The popular Randles model consists of an interface capacitance shunted by charge transfer resistance (R_{CTe}) in series with the solution resistance (R_{Ss}) (Fig. 6.1) [21]. Since then studies have been done to characterize different electrode materials and their surfaces based on different combinations of the Randles model, constant phase element and Warburg impedance [19, 25, 54]. As platinum is the most widely used electrode material for biomedical applications, groups have focused on extensively characterizing its properties. Frank et al. used EIS techniques to compare three electrode materials geared towards biomedical applications: platinum, platinum black and titanium nitride [19].

The electrochemical impedance theory describes the response of a system to an alternating current or voltage input as a function of frequency. The basic approach of EIS is to apply small amplitude perturbations (sinusoidal current or voltage signals) to the electrodes and measure the system's current or voltage response. For microelectrodes used for neural stimulation, usually a sinusoidal voltage signal is used as the excitation signal and the resulting current is measured as the response of the system (potentiostatic EIS). Typically the single-sine technique is used where in the excitation signal is applied at discrete frequencies and the resulting response signal is measured at each frequency to develop the impedance spectrum. In most

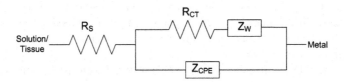

Fig. 6.1 Equivalent circuit model of electrode–electrolyte interface. R_s solution resistance; R_{CT} charge-transfer resistance; Z_W Warburg element; Z_{CPE} constant-phase element

experiments, measurement is started at the highest frequency and stepped down to progressively lower values until enough data has been collected to determine the impedance of the system as a function of frequency. This is done to ensure minimal sample perturbation, and to explore non-Faradaic before Faradaic charge transfer.

The impedance profiles of microelectrodes assist in developing electronic models analogous to the electrode–electrolyte interface such as the Randles model described in the preceding paragraphs. Values of the different circuit elements can be estimated from the impedance measurements. This aids in designing improved versions of the electrode in order to achieve optimal charge-injection situations. The profiles are viewed either through the 'Nyquist plot' or the 'Bode plot' and corresponding model parameters can be estimated. At high frequencies, the imped-ance of the Randles cell becomes almost entirely dominated by the solution resis-tance R_s while at low frequencies the resistance of the electrochemical reaction R_e also comes into play. The solution resistance has long been shown to have an inverse relationship with the radius of a disc electrode. However, recent work by Ahuja et al. suggests that this dependence may not hold for all frequencies [1]. Their work showed that the electrode impedance does scale with radius but only in the high frequency regime (~100 kHz), whereas at lower frequencies (~10 Hz) it scales with the area of the electrode. Thus, only the electrode edge contributes at higher frequencies due to the primary current distribution while at lower frequen-cies, a secondary current distribution comes into play that drives the current to the centre of the disk leading to an area dependence. They also showed that for micro-electrodes of radii less than 50 μm, the area dependence is exhibited even at rela-tively higher frequencies due to the decreased RC time constant and double layer charging of the electrodes at these frequencies.

Surface reactions and potential limits: Cyclic voltammetry (CV) falls under the class of voltammetric methods where the electrode potential is controlled and the resulting current is measured. In voltammetric methods, solutes in contact with the electrode undergo oxidation or reduction reactions producing current at the electrode surface that is measured. In case of cyclic voltammetry, the applied poten-tial is linearly varied with time (cycled) while the resulting current is measured. The applied potential has a triangular waveform with negative and positive turn-around potentials. Since, in a cyclic voltammogram, the range of applied potential is quite large, the measured current aids in understanding the reaction mechanisms available during stimulation. CV can characterize the potential at which the reaction proceeds maximally, the reaction kinetics, and the reversibility of the reaction, all of which are critical to determining if this reaction can be safely used to transfer charge to tissue. Platinum by far has been the most well studied electrode material (Fig. 6.2) but with increasing demands of neural stimulation treatment strategies, focus has shifted towards analyzing and characterizing other candidate electrode materials as reflected in the next section.

For microelectrode characterization in neural stimulation applications, CV plots are used to study a number of important parameters associated with the safe and effective charge-injection at the electrode-tissue interface.

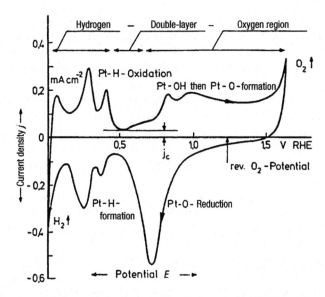

Fig. 6.2 Cyclic voltammogram of poly crystalline platinum in 1 M KOH at scan rate of 100 mV/s exhibits all the different processes involved during the cathodic and anodic direction. The potential scale is referred to a reversible hydrogen electrode (RHE) in the same solution. Reprinted from [28], with permission

1. *Voltage limits.* All electrodes must operate within the "water window," the term given for the potential range between hydrogen evolution potential (negative) and oxygen evolution potential (positive).

$$2H_2O + 2e^- \rightarrow H_2 \uparrow + 2OH^- \tag{6.5}$$

$$2H_2O \rightarrow O_2 \uparrow + 4H^+ + 4e^- \tag{6.6}$$

   Cyclic voltammetry is used to determine these voltage limits, which are material and solution dependent. As will be discussed in Sect. 6.5, during neural stimulation only reversible reactions are employed for charge injection, to avoid causing damage to either the electrode or tissue. For example, from cyclic voltammograms of IrOx and TiN done by Weiland et al., it is observed that the water window of TiN is in the voltage range of −0.6 to 0.8 V in phosphate-buffered saline solution (PBS). For IrOx, the water window has been estimated to be −0.7 to 0.8 V in PBS [55]. Note that the water window does not change in width, but can shift depending on the reference potential.
2. *Charge-Injection mechanism.* In a CV plot, the presence of peaks indicates electrochemical reactions occurring at the electrode–electrolyte interface along with the charging–discharging of the double layer capacitance. As an example,

the CV plot of platinum exhibits distinct peaks associated with the different surface reactions such as hydrogen-atom plating. Also, from the voltammograms of Weiland et al., IrOx CV traces exhibited distinct peaks indicating reduction–oxidation reactions involving transfer of electrons across the interface, along with current flow due to capacitive charging–discharging. On the other hand, the CV traces of TiN show no distinct peaks indicating that the current flow is dominated by the capacitive charging–discharging mechanism [55]. Also, from the nature of peaks, the type of reaction that is occurring can be determined. For example along with the oxidation–reduction peaks of water with the electrode metal, the presence of additional peaks indicates existence of other electro-active substances.

3. *Charge-storage capacity.* An important parameter for neural stimulation is the charge-storage capacity of the electrode. This is determined by integrating the area under either the cathodic or anodic sweep in the CV plot within the water window. The value obtained indicates the maximum charge that can be injected via reversible surface processes by an electrode. This is usually expressed in terms of charge density limit of the electrode. As an example, the charge storage capacity of activated iridium oxide has been reported to range from 10 to 240 mC/cm² depending upon the thickness of the film [52]. However, it should be noted that this is only capacity measured with cyclic voltammetry. The actual amount of charge injection that can be achieved during neural stimulation is usually only a fraction of the charge storage capacity and depends upon factors such as the thickness and morphology of the film, specific reactions of the redox material, pulse duration, etc.

4. *Reversibility of reaction.* Whether the electrochemical reaction occurring is reversible or irreversible in nature can be determined from the cyclic voltammogram of the electrode. All chemical reactions, including reactions occurring at the electrode–electrolyte interface, proceed at a finite rate. The reversibility of a reaction is thus governed by the rate of electron transfer and surface concentrations. In a reversible reaction, the cathodic peak height is equal to the anodic peak height and the reversible half-wave potential will lie exactly midway between the peaks. However, as the reaction becomes more and more irreversible, the cathodic peak height no longer remains equal to the anodic peak height and the separation between the peaks increase (Fig. 6.3). This situation can occur at high scan rates where due to slow reaction kinetics, the voltammogram changes from reversible to irreversible shape.

Voltage response: While charge storage estimates acquired from CVs give the maximum charge value that the electrode in question can store without causing hydrolysis, the actual amount that is injected during current stimulation is quite different. Hence, in order to get a comprehensive picture of how the electrode will behave during active stimulation, one must study the voltage response developed during stimulation. Whenever a current pulse is applied across the electrode–electrolyte or electrode–tissue interface, a resulting voltage response develops across the interface. This voltage waveform is characterized by an initial drop

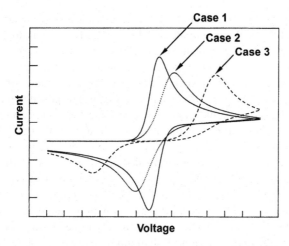

Fig. 6.3 Transition of cyclic voltammograms from reversible to irreversible domain. Case 1: reversible reaction with equal cathodic and anodic peak heights. Case 2: transition from reversible to irreversible reaction with increasing separation between cathodic and anodic peaks. Case 3: irreversible reactions indicated by large separation between the cathodic and anodic peaks. Modified from [18], with permission

called the iR drop or the access voltage, which results from the ohmic losses in the system due to resistance of the electrolyte or tissue. These iR losses do not contribute to the potential difference across the interface that drives the charge across the interface. Hence, before performing analysis of the potential transients, it is essential to subtract these losses from the total voltage response. Based on the net potential across the electrode for a given pulse amplitude and duration, estimates of the actual charge injection capacities can be made. Also, by monitoring the voltage drop across the electrode, the safe charge injection limits can be estimated for voltage drops that do not exceed the water window of the electrode.

6.4 Overview of Electrode Materials for Neural Stimulation

An ideal candidate for electrode material for neural stimulation is one which is biocompatible, mechanically stable to surgical implantation, maintains its electrical and mechanical properties for the entire duration of use, is able to support the charge-injection requirements without inducing damage to itself or to the target tissue. Parameters that govern the efficacy and safety of electrode materials have already been described in the preceding section while parameters that govern the safety of biological tissue will be described in Sect. 6.5. In this section, brief overview of materials that are most commonly used as electrodes for neural stimulation will be provided.

Platinum and its alloys with iridium is the most widely used electrode material for neural stimulation. Being a noble metal, it is highly resistant to corrosion and hence suitable for chronic implantations. The electrochemistry of platinum has been well studied along with its charge storage and injection capacities. Along with double layer charging, charge injection can occur through the reversible adsorption of hydrogen onto the platinum surface (H-atom plating) responsible for the pseudo-capacity of platinum. Brummer and Turner studied the underlying mechanisms during charge injection through platinum electrodes and its alloys [6–8] and found that these chemically reversible processes can provide charge injection up to 300–350 $\mu C/cm^2$ in simulated cerebrospinal fluid [8]. In practice, the safe charge injection limit of platinum depends upon a variety of factors such as the pulse duration, current density and geometry of the electrode surface. For square pulses 0.2 ms in duration, the safe charge injection limit with platinum was found to range from 50 to 150 $\mu C/cm^2$ [48]. Some studies have attempted to increase the electrochemical safe charge injection limit of platinum by increasing the real surface area of the electrode by roughening and have shown varying degrees of success [26, 57].

Iridium oxide belongs to the category of electrodes that are termed as valence change oxides. The oxide layer can be formed in three different ways. Anodic iridium oxide films (AIROF) are produced through repetitive potential cycling of the bulk metal between 0.0 and 1.5 V vs. a reversible hydrogen electrode in an acid or buffered neutral electrolyte [47]. The activated iridium is highly resistant to dissolution and corrosion and exhibit charge storage capacities ranging from 10 to 240 mC/cm^2 [52]. This charge storage capacity depends upon the thickness of the film and even moderate activation can lead to high values. However, during neural stimulation, only a fraction of this charge can actually be used. Weiland et al. found the reversible charge injection limits of AIROF to be about 4 mC/cm^2, which is greater than platinum and some other metals used for neural stimulation [55]. Beebe et al. showed charge injection limits of about 2 mC/cm^2 for biphasic pulses and 3.5 mC/cm^2 for monophasic pulses, 0.2 ms in duration with activated iridium wire electrodes [4]. Iridium oxide films can also be formed by thermal decomposition of layers of iridium salts (TIROF) or by reactively sputtering the oxide films onto a substrate from an iridium target (SIROF). Iridium oxide films on the whole have exhibited poor stability during chronic stimulation regimes however, recent work on SIROF shows improvement in in vitro stability during long-term pulsing [9]. In a separate study, Weiland et al. found the metal-tissue interface to be altered after chronic stimulation using thin film iridium oxide electrodes implanted in guinea pig cortex. They observed that current pulsing within safe limits increased the impedance at low frequencies (<100 Hz) after 1 or 2 days of stimulation and found the impedance change to correspond to a reduction in the charge storage capacity [54]. Other studies have also found iridium oxide electrodes to delaminate under high current pulsing with deposits in the surrounding tissue [10].

Capacitive electrodes are ideal for neural stimulation as they do not involve any reactions for charge injection and hence do not have to deal with problem of irreversible reactions. However, these electrodes still have to be operated within the water window in order to avoid hydrolysis. The metal is insulated from the solution by a thin layer of dielectric material that must be able to withstand the electric fields without any significant dc leakage. Materials in this group that have been found to be safe are anodized tantalum (Ta/Ta$_2$O$_5$), anodized titanium (Ti/TiO$_2$), thin films of barium titanate (BaTiO$_3$) and sputtered deposited titanium nitride (TiN). While anodized tantalum was found to have higher charge storage capacities than anodized titanium and thin films of barium titanate, titanium nitride was found to have charge storage capacities of 23 mC/cm^2 when combined with CMOS technology to develop microcolumnar structures [52]. However, the injectable charge limit of titanium nitride was found to be about 0.87 mC/cm^2 for microelectrodes while for Ta/Ta$_2$O$_5$ to be around 0.1–0.2 mC/cm^2 for large electrodes [47]. Hence, capacitive electrodes though safer than electrodes employing Faradaic reactions, have in general lower charge injection capabilities when operating within the water window.

Carbon nanotubes are also part of the capacitive electrode category exhibiting interesting electrochemical and mechanical properties. They are about five times stronger than steel and yet can be bent and twisted without breaking them. Recent work has shown them as potential electrode material for neural stimulation. Wang et al. developed vertically aligned multiwalled carbon nanotubes (CNTs) using catalytic thermal vapour deposition system [53]. They tested the properties of the CNTs and found that CNTs have a higher charge injection limit of 1–1.6 mC/cm^2 after some surface treatment had been performed. Also, continuous pulsing did not degrade the properties of the CNTs. They also found these carbon nanotubes to be capable of causing neuronal excitation in embryonic rat hippocampal neurons. With its precise control of size, geometry and location by lithographic patterning of the catalyst and high charge injection capabilities without any Faradaic reactions, carbon nanotubes may be an answer to the requirements of neuroprostheses employing localized chronic neural stimulation. However, CNTs generally are formed at very high temperatures, making them incompatible with most batch electrode processes.

Conductive polymers are one of the more recent members to the family of electrode materials for neural stimulation applications. Quite a few recent studies illustrate the feasibility of electrochemically polymerizing polypyrrole, polythiophene and their derivatives from aqueous solutions and depositing them on microelectrodes [12–15, 31, 45, 46, 58]. Some of these studies have also shown that these polymers can successfully be incorporated with cell adhesion molecules, growth factors, etc. to further enhance their properties. With its superior electrochemical stability and biocompatibility, poly (3,4-ethylenedioxythiophene) or more commonly known as PEDOT may be well suited for chronic neural interfaces. Recent work suggests that PEDOT coatings can be deposited over platinum electrodes and be used for chronic neural stimulation [16]. The impedance of PEDOT coated electrodes was found to be lower than the bare platinum electrodes with corresponding lower voltage excursion to applied current pulses in PBS. However, the stability of the PEDOT coated electrodes under chronic stimulation regimes was found to depend largely upon the thickness of the coating that can be controlled through deposition time. Physical degradation and changes in microstructure of the film have been suggested as possible modes of failure. Hence, more work needs to be done to make these polymers successful electrode materials for chronic neural stimulation.

6.5   Overview of Extracellular Stimulation

The bilipid layer membrane separates the intracellular region of the cell from the extracellular environment and acts as a barrier to the movement of ions between these two regions. It plays a crucial role in determining which ions are allowed to pass through and hence has the important properties of specificity and selectivity. The membrane also includes two specialized regions, the afferent region at which the neuron receives the signal and the efferent region at which the neuron sends the signal.

All cells have a resting transmembrane potential (from hereon referred to as membrane potential) with the interior of the cell negative with respect to the exterior of the cell. This membrane potential is dependent on the concentration of the ionic species such that the equilibrium potential of each ion differs from the membrane potential. In general, the ions of interest are K^+ (potassium), Na^+ (sodium) and Cl^- (chloride). At rest, concentration of K^+ ions is higher inside giving it a negative equilibrium potential compared to the membrane potential. This gradient tends to move the ions out of the cell. The concentration of Na^+ ions on the other hand is higher outside than inside the cell giving it a positive equilibrium potential, which causes them to move into the cell. At rest, the membrane acts as a barrier and is less permeable to Na^+ ions compared to K^+ ions. The concentration ratios of these ions are maintained by ionic pumps that force the movement of each of the ions in opposite directions thus maintaining a constant charge separation across the membrane and keeping the cell at its resting membrane potential. A typical value of the resting membrane potential is −60 mV measured inside the cell with reference to the outside. Using (6.1), the membrane potential associated with each of the ions is:

$$E_x = \frac{RT}{zF} \ln \frac{[X]_o}{[X]_i} \qquad (6.7)$$

where z is the valence of the ion. Although the membrane potential is dependent upon the ionic fluxes, it is not equal to either of their membrane potentials. Instead, the membrane potential of the cell is determined by the concentrations of the ions inside and outside the cell along with the ease with which each of ions can cross the membrane, i.e. on the conductivity and permeability of the membrane to the specific ions. The Goldman equation describes quantitatively the dependence of the membrane potential at steady state on ionic concentration and permeability (P):

$$V_m = \frac{RT}{F} \ln \frac{P_K[K^+]_o + P_{Na}[Na^+]_o + P_{Cl}[Cl^-]_o}{P_K[K^+]_i + P_{Na}[Na^+]_i + P_{Cl}[Cl^-]_i} \qquad (6.8)$$

As mentioned previously, the cell membrane is selectively permeable to certain ionic species. This is possible due to the presence of ion channels that are pore-like structures spanning across the membrane. As an example, the potassium channels remain open causing a leak of K^+ ions out of cell making the inside of the cell more negative. During neuronal signaling, the membrane potential rapidly changes in response to some stimulus. This is in part achieved by the reduction of membrane potential (depolarization) that leads to the opening of voltage-gated sodium ion channels causing a further reduction of membrane potential. Initially the cell's response is proportional to the stimulus strength, i.e. the cell responds as a graded potential. Once the membrane potential crosses threshold, the cell responds by generating an action potential that propagates down the cell's axon all the way to its axon terminals. The axon terminals in turn connect to other cells (through synapses) thereby activating them and thus initiating a signaling cascade. The action potential is described as an all-or-none phenomenon, i.e., once initiated it will actively propagate down the axon irrespective of the presence of the initial stimulus. Typically, action potentials last for about a millisecond after which the cells return to their resting state through the inactivation (closing) of voltage-gated sodium channels and activation of voltage-gated potassium channels. These two mechanisms have longer time constants compared to sodium activation but work together to bring the cell back to its resting membrane potential. This period of inactivation is called the refractory period.

Electrical stimulation of excitable tissue generates action potentials that in turn initiate neuronal signaling and enable partial restoration of lost functionality in sensory or motor systems. This process requires the extracellular region to be driven more negative by applying a rapid negative charge injection via an extracellular stimulating electrode. For the simplest case of stimulation, a single electrode is placed near the excitable tissue and the electrode is driven as a cathode causing the outside of the membrane to become more negative. This causes the membrane potential to become positive thus leading to a net reduction in the membrane potential (depolarizing the membrane). If on the other hand, the stimulating electrode is driven as an anode, then it will cause the outside of the cell to become more positive than the inside thus causing the membrane potential to become more negative. This will lead to a net increase in the membrane potential causing the membrane to hyperpolarize. Since a current generator must have a source and a sink, during extracellular stimulation, a second electrode is required for the current loop to be complete. This second electrode is usually called the return electrode and based upon its size and position can cause a number of different events to occur. If the return electrode is much larger than the stimulating electrode, then the current density is highest at the stimulating electrode causing excitation of neurons near it. However, if the return electrode is similar in size as the stimulating electrode, then the current density at both sites will be the same and hence neuronal excitation can occur at both sites. In this case, during cathodic stimulation, the neurons in close proximity to the stimulating electrode are depolarized while those underneath the return electrode are hyperpolarized. In some case this hyperpolarization may be large enough to suppress an action potential initiated near the electrode (anodic surround block) [41, 43]. On the other hand, if anodic stimulation is employed then the neurons near the stimulating electrode will be hyperpolarized while those near the return electrode will be depolarized. In this case, the action potentials are initiated in regions distant from the electrode known as virtual cathodes. The depolarization that occurs through anodic stimulation is about a seventh to a third of that accomplished through cathodic stimulation although this depends upon the electrode position [44]. Thus, cathodic stimulation requires less current to cause a cell to cross threshold and initiate action potentials.

The stimulation protocols described above may be effective at selectively activating one population of neurons without activating neighbouring neurons.

Activation thresholds are usually defined in terms of the amount of current needed to cause the excitation along with the duration that the current is applied. Another way to define excitation thresholds is in terms of the applied charge that is simply a multiplication of the amplitude of the applied pulse (current) with the duration of the pulse. Since in neural stimulation, currents applied are in the range of microamps and are applied typically for a few milliseconds, the charge delivered ranges from a few microcoulombs to a few nanocoulombs. By far, the best known law of stimulation is the one by Lapicque that relates the threshold current (I) required for stimulation to the duration (d) of the applied pulse [32]. He introduced the tissue specific excitability parameter called the chronaxie (c) and defined it as the pulse duration that required twice the rheobase current (b). Here, rheobase current is defined as the threshold current (I) for very long pulses. Mathematically, b is the limit of I, as pulse duration goes to infinity. The Lapicque law for stimulation is:

$$I = b(1 + c / d) \tag{6.9}$$

Based on the above equation, strength–duration curves can be plotted to graphically illustrate the relationship between the three parameters I, d and c, as shown in Fig. 6.4. The strength–duration curve is an essential tool in all types of studies where electrical stimulation of excitable tissue is employed. Studies have shown how different parameters can be calculated from these curves including charge and energy–duration relationships [24]. Although numerous studies illustrate chronaxie values of different excitable tissues, the accuracy of the measurements can be affected by factors such as the electrode characteristics, tissue inhomogeneity, stimulus waveform, etc. [22, 23]. Studies in motor nerves and different types of muscle have shown the dependence of chronaxie on different parameters such as temperature and location of electrodes [22].

Another way to define the relationship between stimulus strength and excitation is through amplitude-intensity function, as shown in Fig. 6.5. This is typically used where the response is an evoked potential and generates a plot of the stimulus strength at fixed pulse duration against the amplitude of the evoked response. It helps in determination of true threshold by simply extrapolating the curve to intersect the x-axis. Amplitude-intensity functions are useful because neural prostheses typically operate above threshold to provide a range of sensation or activation. Finally for the case of single units, analysis methods such as post-stimulus time histograms (PSTHs) are employed that sort the individual spikes based on their latencies. More sophisticated analyses of a mixture of action potentials produced by multiple cells involve grouping the individual spikes based on their individual waveform characteristics.

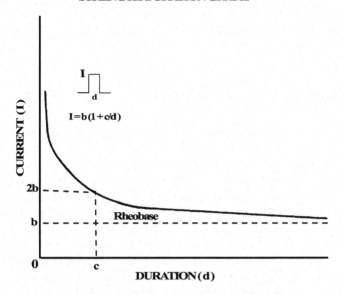

Fig. 6.4 Strength–duration graph illustrating threshold current required to elicit response at different pulse durations. Rheobase current=b; chronaxie=c. Modified from [23]

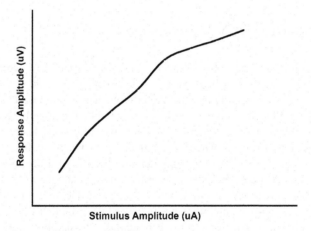

Fig. 6.5 Representative graph illustrating the gradual increase in response amplitude as the stimulus strength is increased. The amplitude of response is usually measured in microvolts (μV) while the applied stimulus amplitude is usually in microamps (μA)

6.6 Safe Stimulation of Tissue

A neural stimulation system that is not properly designed can cause damage to the tissue or to the electrode itself. For any neural stimulation system to be successful, it must elicit the required neuronal excitation without causing any damage to the biological system. Electrode shape, size and material along with stimulus pulse parameters need to be carefully chosen to meet the requirements of the system. Extensive work has been done in defining the role of all the different parameters that determine the safety limit of the tissue and electrode.

6.6.1 Mechanisms of Neural Injury

There are several mechanisms that may cause neural injury; they are broadly categorized into two main classes. The first mechanism of damage is associated with the electrochemical processes through which the stimulus current is injected into the target tissue. Damage is induced due to formation of toxic electrochemical reaction products during stimulation at a rate greater than what can be tolerated by the physiological system. These damaging processes have been well characterized using electrochemical methods as discussed in Sects. 6.2 and 6.3. A second mechanism of neural injury is associated with the flow of current through the target tissue [35]. This involves the metabolic stresses induced on the tissue causing a transient or permanent elevation of neurotransmitter release (excitotoxic effect). It may also include large depolarizations and hyperpolarizations induced by the voltage gradient (membrane electroporation). This second mechanism is multi-factorial and complex.

6.6.2 Parameters for Safe Stimulation

One of the well-established principles of neural stimulation is to achieve charge balancing during stimulation between the different phases of the stimulus pulse. This was first reported by Lilly in 1961 and ensures that the total net charge during stimulation at the electrode–tissue interface is zero [34]. If charge balancing is not accomplished, then a net accumulation of charge will ultimately lead to the rise of electrode potentials to levels where water hydrolysis will start. For monophasic stimulation, charge balancing is accomplished by the use of a blocking capacitor that slowly discharges after the application of the pulse. Although charge – balancing ensures that there is no net accumulation of charge, it does not guarantee safety. Such stimulus waveforms may momentarily exceed the established safety limits of total charge, charge density or electrode potential. Classically, safety limits for neural stimulation have been divided into two broad categories:

1. *Neural damage limits* dictated by the ability of biological tissue to withstand electric current without any degradation.
2. *Electrochemical limits* based on the ability of the electrode to store or dissipate electric charge without exceeding the water window, outside of which formation of harmful products start.

While neural injury limits are defined in terms of both charge density and charge per phase, electrochemical limits are defined in terms of charge density only. Charge density is simply the total charge per unit area of electrode and determines the magnitude of the depolarization or hyperpolarization induced in the neurons and axons close to the electrode. Charge per phase is the amount of charge injected during each phase of the stimulus pulse and determines the distance over which the applied stimulation can activate the neurons, i.e. the number of neurons activated. McCreery et al. [37] have shown that charge density and charge per phase act synergistically to determine the safe or unsafe levels of stimulation. They showed that neural damage is induced with low charge per phase but high charge density, as is often the case for microelectrodes. Based on these data delineating the boundary between safe and unsafe charge injection for different charge and charge density levels, Shannon et al. [51] developed the following empirical relationship:

$$\log(D) = k - \log(Q) \tag{6.10}$$

where, D is the charge density in $\mu C/cm^2/phase$ and Q is the charge per phase in $\mu C/phase$. The equation describes a family of lines for different values of k. The line for which $k = 1.5$ describes combinations of charge density and charge per phase values for which no damage was observed. Merrill et al. have graphically summarized the work of both studies and also included results of other studies assessing safety of neural stimulation (Fig. 6.6).

Along with charge density and charge per phase, other stimulus parameters such as frequency of stimulation, duration, etc. have been found to play an important role in determining the presence or absence of neural damage. McCreery et al. [38] demonstrated the effect of stimulus frequency as a parameter in causing injury during peripheral nerve stimulation. Their study showed that continuous stimulation of the cat sciatic nerve for 8 h over 3 days causes the myelin sheath to collapse into the axonal space leading to early axonal degeneration (EAD). The threshold of neural injury decreased with increasing stimulus pulse frequency (Fig. 6.7).

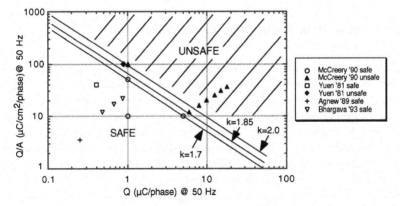

Fig. 6.6 Charge (Q) vs. charge density (Q/A) for safe stimulation. Different symbols indicate results of different studies. Reprinted from [39], with permission

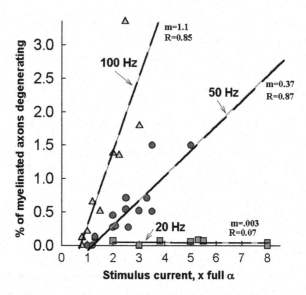

Fig. 6.7 Percentage of myelinated axons undergoing degeneration 7 days after undergoing 8 h of continuous stimulation. At higher stimulus frequency, the percentage of axons undergoing EAD can be substantial even at low stimulus current. Reprinted from [38], with permission

Most of the aforementioned studies have employed single electrode stimulation. However, a recent study [36] found that in the case of multi-electrode stimulation, both sequential and simultaneous stimulation at levels previously found to be safe create transient depression in the resulting neural response. One theory put forward by the authors is the creation of overlapping electric fields that cause certain neurons to be driven at rates higher than what is actually being delivered. The authors dubbed the observed effect "SIDNE" (stimulation induced depression in neuronal excitability).

6.6.3 Stimulation Induced Injury in the Retina

To date, most safety studies have been carried out in structures such as the cortex, muscle, etc. With increasing efforts towards developing retinal implants [20, 56, 59], extensive studies are being done to understand the response of the visual system to artificial stimuli [2, 29, 49, 50]. However, only a few studies so far have been dedicated towards understanding the consequences of long-term stimulation. Güven et al. [27] carried out chronic stimulation studies in dogs and found that the retina is able to tolerate chronic stimulation at $0.1 \, \text{mC/cm}^2$ without any histological detectable damage or change in the electroretinograms (ERGs). Another study investigated chronic stimulation effects through suprachoroidal-transretinal stimulation [40]. The results of the study showed that threshold for safe charge increased logarithmically or almost linearly with increasing stimulus duration but the threshold for safe current decreased logarithmically with increasing stimulus duration. There was severe damage in the inner layers when the applied current exceeded this threshold. Colodetti et al. [11] found that the retina is sensitive to pressure exerted by the electrode. They studied the type of damage due to pressure exerted by the electrode with and without accompanying high charge stimulation in the rodent retina. Although the type of damage exhibited in both cases were roughly similar, the extent of damaged area was significantly larger in the case of accompanying high charge stimulation. These studies although informative do not in any way give a complete picture of how the retina would respond to continuous stimulation. Also, as increasing efforts are being made to make these implants more sophisticated, the added requirement of a large number of closely spaced electrodes makes it imperative to study the possible consequences of high level stimulation on both the retina and associated cortical structures. Recent work in these areas is presented in Chaps. 7 (Loudin, Butterwick, Huie and Palanker) and 12 (Fried and Jensen).

References

1. Ahuja AK, Behrend MR, Whalen JJ, et al. (2008), *The dependence of spectral impedance on disc microelectrode radius*. IEEE Trans Biomed Eng, **55**(4): p. 1457–60.
2. Baig-Silva MS, Hathcock CD, Hetling JR (2005), *A preparation for studying electrical stimulation of the retina in vivo in rat*. J Neural Eng, **2**(1): p. S29–38.
3. Bard AJ, Faulkner LR (2004), *Electrocehmical Methods: Fundamentals and Applications*. Second ed, New York: Wiley.
4. Beebe X, Rose TL (1988), *Charge injection limits of activated iridium oxide electrodes with 0.2 ms pulses in bicarbonate buffered saline*. IEEE Trans Biomed Eng, **35**(6): p. 494–5.
5. Brug GJ, Van Den Eeden ALG, Sluythers-Rehbach M, Suythers JH (1984), *The analysis of electrode impedances complicated by the presence of a constant phase element*. J Electroanal Chem, **176**: p. 275–95.
6. Brummer SB, Turner MJ (1975), *Electrical stimulation of the nervous system: the principle of safe charge injection with noble metal electrodes*. Bioelectrochem Bioenerg, **2**: p. 13–25.
7. Brummer SB, Turner MJ (1977), *Electrical stimulation with Pt electrodes: I-a method for determination of "real" electrode areas*. IEEE Trans Biomed Eng, **24**(5): p. 436–9.
8. Brummer SB, Turner MJ (1977), *Electrical stimulation with Pt electrodes: II-estimation of maximum surface redox (theoretical non-gassing) limits*. IEEE Trans Biomed Eng, **24**(5): p. 440–3.
9. Cogan SF, Ehrlich J, Plante TD, et al. (2009), *Sputtered iridium oxide films for neural stimulation electrodes*. J Biomed Mater Res B Appl Biomater, **89**(2): p. 353–61.
10. Cogan SF, Guzelian AA, Agnew WF, et al. (2004), *Over-pulsing degrades activated iridium oxide films used for intracortical neural stimulation*. J Neurosci Methods, **137**(2): p. 141–50.
11. Colodetti L, Weiland JD, Colodetti S, et al. (2007), *Pathology of damaging electrical stimulation in the retina*. Exp Eye Res, **85**(1): p. 23–33.
12. Cui XY, Hetke J, F, Wiler JA, et al. (2001), *Electrochemical deposition and characterization of conducting ploymer polypyrrole/pss on multichannel neural probes*. Sens Actuators A Phys, **93**: p. 8–18.
13. Cui X, Lee VA, Raphael Y, et al. (2001), *Surface modification of neural recording electrodes with conducting polymer/biomolecule blends*. J Biomed Mater Res, **56**(2): p. 261–72.
14. Cui XY, Martin DC (2003), *Electrochemical deposition and characterization of poly (3,4-ethylenedioxythiophene) on neural microelectrode arrays*. Sens Actuators B Chem, **89**: p. 92–102.

15. Cui X, Wiler J, Dzaman M, et al. (2003), *In vivo studies of polypyrrole/peptide coated neural probes*. Biomaterials, **24**(5): p. 777–87.
16. Cui XT, Zhou DD (2007), *Poly (3,4-ethylenedioxythiophene) for chronic neural stimulation*. IEEE Trans Neural Syst Rehabil Eng, **15**(4): p. 502–8.
17. Eisenfeld AJ, Bunt-Milam AH, Sarthy PV (1984), *Muller cell expression of glial fibrillary acidic protein after genetic and experimental photoreceptor degeneration in the rat retina*. Invest Ophthalmol Vis Sci, **25**(11): p. 1321–8.
18. Evans DH (1991), *Review of voltammetric methods for the study of electrode reactions*, in Microelectrodes: Theory and Applications, Montenegro I, Queiros MA, Daschbach JL, eds., Dordrecht: Kluwer Academic.
19. Franks W, Schenker I, Hierlmann A (2005), *Impedance characterization and modeling of electrodes for biomedical applications*. IEEE Trans Biomed Eng, **52**(7): p. 1295–302.
20. Fujikado T, Morimoto T, Kanda H, et al. (2007), *Evaluation of phosphenes elicited by extraocular stimulation in normals and by suprachoroidal-transretinal stimulation in patients with retinitis pigmentosa*. Graefes Arch Clin Exp Ophthalmol, **245**(10): p. 1411–9.
21. Geddes LA (1997), *Historical evolution of circuit models for the electrode–electrolyte interface*. Ann Biomed Eng, **25**(1): p. 1–14.
22. Geddes LA (1999), *Chronaxie*. Australas Phys Eng Sci Med, **22**(1): p. 13–17.
23. Geddes LA (2004), *Accuracy limitations of chronaxie values*. IEEE Trans Biomed Eng, **51**(1): p. 176–81.
24. Geddes LA, Bourland JD (1985), *The strength-duration curve*. IEEE Trans Biomed Eng, **32**(6): p. 458–9.
25. Germain PS, Pell WG, Conway BE (2004), *Evaluation and origins of the differences between double-layer capacitance behaviour at Au-metal and oxidized Au surfaces*. Electrochim Acta, **49**: p. 1775–88.
26. Greenbaum E, Sanders C, Zhou D (2006), *Dynamic interactions at retinal prosthesis electrode interface*. Invest Ophthalmol Visual Sci 47, ARVO E-abstr#3200.
27. Guven D, Weiland JD, Fujii G, et al. (2005), *Long-term stimulation by active epiretinal implants in normal and RCD1 dogs*. J Neural Eng, **2**(1): p. S65–73.
28. Hamann CH, Hamnett A, Vielstich W (1998), *Methods for the study of electrode/electrolyte interface*, Chapter 5 in *Electrochemistry*, Weinheim, Germany: Wiley-VCH, p. 251–338.
29. Jensen RJ, Ziv OR, Rizzo JF, III (2005), *Thresholds for activation of rabbit retinal ganglion cells with relatively large, extracellular microelectrodes*. Invest Ophthalmol Vis Sci, **46**(4): p. 1486–96.
30. Kandel ER, Schwartz JH, Jessell TM (2000), *Principles of Neural Science*. Fourth ed, New York: McGraw-Hill.
31. Kim DH, Abidian M, Martin DC (2004), *Conducting polymers grown in hydrogel scaffolds coated on neural prosthetic devices*. J Biomed Mater Res A, **71**(4): p. 577–85.
32. Lapicque L (1909), *Definition experimentale de l'excitabilite*. C R Acad Sci, **67**(2): p. 280–3.
33. Lasia A (2002), *Electrochemical impedance spectroscopy and its application*, in *Modern Aspects of Electrochemistry*, Conway BE, Bockris JOM, White RE, eds., New York: Kluwer Academic.
34. Lilly JC (1961), *Injury and excitation by electric currents: the balanced pulse-pair waveform*, in *Electrical Stimulation of the Brain*, Sheer DE, ed., Austin, TX: Hogg Foundation for Mental Health.
35. McCreery DB (2004), *Tissue reaction to electrodes: the problem of safe and effective stimulation of neural tissue*, in *Neuroprosthetics: Theory and Practice*, Horch KW, Dhillon GS, eds., Singapore: World Scientific Publishing Co. Pte. Ltd.
36. McCreery DB, Agnew WF, Bullara LA (2002), *The effects of prolonged intracortical microstimulation on the excitability of pyramidal tract neurons in the cat*. Ann Biomed Eng, **30**(1): p. 107–19.
37. McCreery DB, Agnew WF, Yuen TG, Bullara L (1990), *Charge density and charge per phase as cofactors in neural injury induced by electrical stimulation*. IEEE Trans Biomed Eng, **37**(10): p. 996–1001.

38. McCreery DB, Agnew WF, Yuen TG, Bullara LA (1995), *Relationship between stimulus amplitude, stimulus frequency and neural damage during electrical stimulation of sciatic nerve of cat.* Med Biol Eng Comput, **33**(3 Spec No): p. 426–9.

39. Merrill DR, Bikson M, Jefferys JG (2005), *Electrical stimulation of excitable tissue: design of efficacious and safe protocols.* J Neurosci Methods, **141**(2): p. 171–98.

40. Nakauchi K, Fujikado T, Kanda H, et al. (2007), *Threshold suprachoroidal-transretinal stimulation current resulting in retinal damage in rabbits.* J Neural Eng, **4**(1): p. S50–7.

41. Plonsey R, Barr RC (1991), *Functional neuromuscular stimulation*, Chapter 12 in *Bioelectricity: A Quantitative Approach*, New York: Plenum, p. 271–299.

42. Prokhorov E, Llamas F, Morales-Sanchez E, et al. (2002), *In vivo impedance measurements on nerves and surrounding skeletal muscles in rats and human body.* Med Biol Eng Comput, **40**(3): p. 323–6.

43. Rattay F (1989), *Analysis of models for extracellular fiber stimulation.* IEEE Trans Biomed Eng, **36**(7): p. 676–82.

44. Rattay F (1999), *The basic mechanism for the electrical stimulation of the nervous system.* Neuroscience, **89**(2): p. 335–46.

45. Richardson-Burns SM, Hendricks JL, Foster B, et al. (2007), *Polymerization of the conducting polymer poly(3,4-ethylenedioxythiophene) (PEDOT) around living neural cells.* Biomaterials, **28**(8): p. 1539–52.

46. Richardson-Burns SM, Hendricks JL, Martin DC (2007), *Electrochemical polymerization of conducting polymers in living neural tissue.* J Neural Eng, **4**(2): p. L6–13.

47. Robblee LS, Rose TL (1990), *Electrochemical guidelines for selection of protocols and electrode materials for neural stimulation*, in *Neural Prostheses: Fundamental Studies*, Agnew WF, McCreery DB, eds., Englewood Cliffs, NJ: Prentice Hall.

48. Rose TL, Robblee LS (1990), *Electrical stimulation with Pt electrodes. VIII. Electrochemically safe charge injection limits with 0.2 ms pulses.* IEEE Trans Biomed Eng, **37**(11): p. 1118–20

49. Sachs HG, Gekeler F, Schwahn H, et al. (2005), *Implantation of stimulation electrodes in the subretinal space to demonstrate cortical responses in Yucatan minipig in the course of visual prosthesis development.* Eur J Ophthalmol, **15**(4): p. 493–9.

50. Sekirnjak C, Hottowy P, Sher A, et al. (2006), *Electrical stimulation of mammalian retinal ganglion cells with multielectrode arrays.* J Neurophysiol, **95**(6): p. 3311–27.

51. Shannon RV (1992), *A model of safe levels for electrical stimulation.* IEEE Trans Biomed Eng, **39**(4): p. 424–6.

52. Stieglitz T (2004), *Electrode materials for recording and stimulation*, in *Neuroprosthetics: Theory and Practice*, Horch KW, Dhillon GS, eds., Vol. 2, Singapore: World Scientific.

53. Wang K, Fishman HA, Dai H, Harris JS (2006), *Neural stimulation with a carbon nanotube microelectrode array.* Nano Lett, **6**(9): p. 2043–8.

54. Weiland JD, Anderson DJ (2000), *Chronic neural stimulation with thin-film, iridium oxide electrodes.* IEEE Trans Biomed Eng, **47**(7): p. 911–8.

55. Weiland JD, Anderson DJ, Humayun MS (2002), *In vitro electrical properties for iridium oxide versus titanium nitride stimulating electrodes.* IEEE Trans Biomed Eng, **49**(12 Pt 2): p. 1574–9.

56. Weiland JD, Liu W, Humayun MS (2005), *Retinal prosthesis.* Annu Rev Biomed Eng, **7**: p. 361–401.

57. Whalen JJ, Weiland JW, Searson P (2005), *Electrochemical deposition of platinum from aqueous ammonium hexachloroplatinate solution.* J Electrochem Soc, **152**(11): p. C738–43.

58. Yang JY, Martin DC (2004), *Microporous conducting polymers on neural microelectrode arrays II. Physical characterization.* Sens Actuators A Phys, **113A**: p. 204–11.

59. Zrenner E (2002), *The subretinal implant: can microphotodiode arrays replace degenerated retinal photoreceptors to restore vision?* Ophthalmologica, **216**(Suppl 1): p. 8–20; discussion 52–3.

Chapter 7
Delivery of Information and Power to the Implant, Integration of the Electrode Array with the Retina, and Safety of Chronic Stimulation

James Loudin, Alexander Butterwick, Philip Huie, and Daniel Palanker

Abstract The fundamental function of a visual prosthesis is to deliver information about a patient's surroundings to his/her neurons, usually via patterned electronic stimulation. In addition to transmitting visual information from the outside world to the implanted stimulating array, visual prostheses must also pass the electrical power necessary for such stimulation from the external world to the intraocular electrode array. The first section of this chapter reviews three common methods for achieving this data and power transfer: direct wireline connections (suitable for research studies), inductively coupled coils, and photodiode-based optical systems which utilize the natural optics of the eye.

Once the data and power has been received, retinal prostheses must effectively deliver stimulation currents to surviving retinal neurons. This necessitates an understanding of the electrode/retina interface. The second section of this chapter is a histological description of this interface for the case of subretinal implants, investigating the tissue response to flat implants coated with different materials. Several three-dimensional geometries are also described and evaluated to decrease the implant–neuron distance.

Finally, stimulation currents must not damage the stimulated neurons. The third section of this chapter describes measurements and scaling laws associated with tissue damage from electric currents. Damage thresholds are found to be approximately 50–100 times stimulation thresholds.

Abbreviations

AC	Alternating current
ASR	Artificial silicon retina, a retinal prosthesis fabricated by Optobionics
CMOS	Combined metal on silicon
CMP	Computational molecular phenotyping

J. Loudin (✉)
Department of Applied Physics, Stanford University, 450 Serra Mall,
Stanford, CA 94305, USA
e-mail: loudin@stanford.edu

G. Dagnelie (ed.), *Visual Prosthetics: Physiology, Bioengineering, Rehabilitation*, 137
DOI 10.1007/978-1-4419-0754-7_7, © Springer Science+Business Media, LLC 2011

DC   Direct current
EU   European Union
IMI   Intelligent medical implants, a company fabricating a retinal prosthesis
INL   Inner nuclear layer
IR   Infrared
LCD   Liquid crystal display
MPDA   Microphotodiode array, retinal prosthesis fabricated by retina implant
            AG
ONL   Outer nuclear layer
P45   45 days after birth
PI   Propidium iodide
RCS rat   Royal College of Surgeons rat, a common animal model of retinal
            degeneration
RF   Radio frequency
RPE   Retinal pigmented epithelium
SIROF   Sputtered iridium oxide film
SU-8   A photo-curable epoxy
USC   University of Southern California

7.1 Introduction

One of the fundamental challenges for a visual prosthesis is to efficiently deliver visual stimuli from the external world to target neurons in the retina, optic nerve, or visual cortex. Power and visual information must be transmitted and subsequently distributed over an electrode array while ideally not interfering with residual vision, and keeping the natural association between visual information and eye movements. Four basic methods have been used to achieve this: direct wireline connection to implanted stimulators, radio frequency (RF) telemetry, serial optical telemetry, and parallel optical telemetry. In the first part of this chapter we review these techniques in their various incarnations.

After the data is received, providing the appropriate stimulus to the retina presents a new set of challenges: high-resolution prostheses require that nearby neurons are stimulated with high selectivity and broad dynamic range. While the electric field created by the electrode array and the constraints on cellular proximity have been characterized [35], the process of *maintaining* this proximity between electrodes and cells is less understood. Chronically preserving apposition between an *epiretinal* prosthesis and neurons requires only mechanical stabilization of the implant in the vitreous cavity. However, doing so with a *subretinal* prosthesis requires controlling the response of the retina to an implant. In the second part of this chapter we describe techniques used to mechanically stabilize implants, and the response of the retina to various implant geometries and coatings.

One of the critically important issues in development of retinal prosthesis is understanding the safe limits of electrical stimulation for prolonged periods of time.

In the third part of this chapter we describe the dependence of the damage threshold on pulse duration, electrode size and its separation from the cells, as well as on the number of pulses. We also compare damage and stimulation thresholds to assess the safe therapeutic range at various pulse durations.

7.2   Power and Data Transmission

7.2.1   Wireline Connection

William Dobelle led one of the earliest attempts at constructing a visual prosthesis. In a series of studies begun in 1968, he used direct wireline connections to link electrodes placed in the visual cortex with a stimulator worn externally to the body [16]. Subsequent electrical stimulation successfully evoked visual responses in 19 blind patients, offering hope that future prostheses would one day restore some degree of useful sight.

Direct percutaneous connections are far from ideal, as they can provide pathogens with a direct pathway through the skin and are prone to severe scarring [30]. Despite this, transdermal cables have often been used in short-term human trials of various visual prostheses [51, 53, 65, 76], because of the unrivalled electrical versatility which they offer. For example, a group at the Naval Research Lab has developed a 3,200 electrode epiretinal prosthesis which is driven with a cable containing ten wires [58]. This prosthesis is intended for acute experiments; a future version under development is wireless. In at least one case, percutaneous cables driving a retinal prosthesis have been left in place for a period exceeding 1 year [76]. Though direct connections will likely continue to be used in research settings for years to come, any future commercial prosthesis will be wireless.

7.2.2   Inductive Coils

Inductively coupled coils are used for wireless data and power transmission in a wide variety of applications, including medical implants such as cardiac pacemakers [4] and cochlear prostheses [74]. More recently, the unique power and data requirements of visual prostheses have spurred much research in the field, with inductive coil systems currently developed for epiretinal [33, 69], subretinal [35, 63], visual cortex [64], and optic nerve stimulators [65].

In all of these designs, an AC current driven through an external transmitting coil induces an AC voltage on an implanted coil, which is converted to DC power by implanted circuitry. Sometimes the transmitter encodes data onto this signal, which is also recovered by the implanted circuitry. Since the coils are only weakly coupled to each other (typical values for coupling coefficient k are in the range 0.08–0.24 [68], compared to ~0.9 for standard transformers), great care must be taken to optimize the receiving circuitry. With this in mind, a capacitor is added in series with the receiving coil to create a tuned resonance at the transmitter frequency, f_t. The resulting circuit amplifies the received voltage by the quality factor Q, typically in the range 10–100. High Q values yield more efficient power transfer thus helping to decrease the body's exposure to radiation. The optimization of coil geometry and receiving circuitry to maximize Q has been the subject of numerous studies [19, 20, 27, 28, 31, 62, 66, 68, 72]. Since Q is proportional to the transmission frequency, high frequency operation yields higher Q values; however, tissue's RF absorption increases exponentially beyond a few MHz [43] limiting transmission frequency f_t to 1–10 MHz.

Inductive coils have been used to deliver data to visual prostheses for over half a century. In the 1960s, a team led by Giles Brindley of the Medical Research Council in London implanted an array of 80 coils beneath the pericranium of a blind patient [8]. The 80 coils were connected through separate rectifying circuits [7] to 80 platinum electrodes placed onto the surface of the patient's visual cortex. Individual electrodes were activated by placing a transmitting coil on the scalp directly above the electrode's receiver. Interference was minimized by tuning adjacent receivers to different frequencies. Though this scheme was rather successful (of the 80 electrode placements, 39 elicited phosphenes), it is hardly scalable. With the goal of scaling visual prostheses to hundreds and eventually thousands of pixels, higher data rates must be extracted from fewer coils.

Ironically, while high-Q coils are efficient power receivers, they are rather poor data receivers. According to the Shannon–Hartley theorem [59], the data capacity C of a coil may be expressed as

$$C = B \cdot \log_2 (1 + SNR) = \frac{f_t}{Q} \log_2 (1 + SNR) \qquad (7.1)$$

where C is in bits per second, B is the bandwidth of the receiving circuit, f_t is the transmission frequency in Hz, and SNR is the signal to noise power ratio. Thus, while received power is directly proportional to coil Q, the attainable data rate is *inversely* proportional to Q. For this reason, many visual prosthesis designs use two coil pairs: one for power, and one for data, where data transmission is accomplished at a higher frequency [27, 68] or with a lower-Q coil [35]. In addition, complex single-coil systems capable of delivering both power and data over one coil pair have also been developed [18, 19, 67], in one case achieving a data rate in excess of 1 Mb/s [33].

Ignoring the time involved in implant monitoring feedback signals, transmitting control signals, and other housekeeping functions, the maximum number of pixels N that can be individually driven at the refresh rate, R, is determined by the data rate, C, and the number of the stimulation strength levels S, as

$$N = \frac{C}{R \cdot \log_2 (S)} \qquad (7.2)$$

For example, the system presented in [33] with a data rate of 1 Mb/s, refresh rate of 60 Hz, and 16 different stimulation levels can adequately support more than 4,096 pixels (a 64 by 64 array).

Alternatively, James Weiland's group in USC is developing an implantable intraocular video camera [69]. Such a design would do away with the need for inductive data transmission altogether, and will require only power and (low bandwidth) control signals.

Power and data transmission efficiency are inextricably intertwined with the physical coil placements. As such, the ideal surgical placement for receiving coils is a matter of ongoing debate. In Second Sight's first generation prosthesis, two coils implanted subcutaneously behind the ear were coupled to a second pair, attached to the frame of eyeglasses worn by the patient [23]. This location was chosen for two reasons: first, because surgeons have years of experience implanting coils behind the ear, due to the success of cochlear implants. Secondly, the outer coil may be placed very close to the implanted one, thereby maximizing coupling efficiency. The disadvantages are also twofold: first, any operation implanting both a retinal stimulator and a posterior auricular coil-set necessitates both a retinal surgeon and an otolaryngologist. This increases both the cost and the length of the operation. Secondly, a trans-scleral cable must connect the coil to the intraocular stimulator. This wire must be thin and flexible enough to allow normal eye movement, while also robust enough to withstand years of bending without failure. Fabricating such a wire is a challenge, though not an insurmountable one. For these reasons Second Sight has changed to a periocular design for their second generation prosthesis (Argus II), with receiving coils mounted on the front of the eye globe. The transmitting coils are mounted on the front of the patient's eyeglasses [2].

The Boston Retinal Implant Project has designed a pair of coils and receiving circuitry which is sutured to the side of the eye under the conjunctiva, facing towards a transmitting coil mounted on the side arm of eyeglasses worn by the patient [63]. The German EPIRET consortium attempted a different approach, removing the lens and placing the receiving coil in the lens capsule [40]. This technique is appealing due to its similarity to cataract surgery. However, the relatively small size of the lens capsule puts tight constraints on the coil size (approximately 12 mm).

7.2.3 Serial Optical Telemetry

Photodiodes are excellent data receivers. Standard CMOS integrated photodiodes can have bandwidths in excess of 1 GHz [49], over two orders of magnitude higher than those attainable with inductive coils, whose bandwidths cannot exceed ~10 MHz due to transmission frequency and Q-factor constraints (see (7.1)). The German company IMI Technologies is developing an epiretinal prosthesis with a subconjunctively placed inductive coil and a photodiode placed inside the vitreous cavity [69]. The coil receives power inductively, while the photodiode receives stimulation data optically. As in the case of inductive coil systems, the data is delivered serially over one channel, so it must be decoded and distributed over the electrode array using a data processing chip connected to the retinal array by intraocular cable.

7.2.4 Photodiode Array-Based Prostheses

The Chow brothers were the first to propose and investigate the use of photodiodes in a retinal prosthesis, in the early 1990s [12]. Optobionics, a company they founded, developed a two dimensional array of 5,000 photodiodes on a 2 mm silicon disk. These so-called artificial silicon retinas (ASRs) were fabricated such that all 5,000 photodiode anodes were connected together, while their cathodes were electrically isolated from each other [48]. Each cathode contained its own iridium oxide electrode, resulting in a 2 mm device with 5,000 light-controllable electrical sources. Since each photodiode in the array collected light simultaneously, visual data could be delivered to all 5,000 pixels in parallel. This is in contrast to the serial RF and optical telemetry systems described above.

The Optobionics design assumed that ambient light would be directly converted by the photodiodes to currents strong enough to stimulate surviving retinal neurons. Indeed, initial human trials did result in some vision improvement [13]; however, this improvement was not due to electrically-elicited action potentials, but from neurotrophic effects resulting from ASR implantation [46]. Unfortunately, ambient light intensities provide insufficient current to directly stimulate nerve tissue, by at least three orders of magnitude [45]. In addition, since the electrode–electrolyte interface is capacitive when driven in a biologically compatible way, a photodiode array should be driven by pulsed, rather than continuous illumination. Indeed, later in vivo experiments on RCS rats with subretinally implanted ASRs successfully demonstrated neural activity in the superior colliculus in response to intense pulsed infrared illumination of the retina [15].

Eberhart Zrenner and his group from the University of Tuebingen, Germany have constructed a photosensitive array equipped with built-in differential amplifiers, called the microphotodiode array prosthesis (MPDA) [77]. Each pixel contains a photodiode and active circuitry which measures the difference between local and global brightness, and then drives a current corresponding to this difference through an associated microelectrode. The system is capable of utilizing almost all the electrochemical capacity of the electrodes; however, it requires separate power delivery to drive the active circuitry. So far, short-term human trials have relied on percutaneous wireline connections to deliver this power; future, wireless implants will use an inductive coil system for this purpose [6]. Recent result with a human patient confirmed the ability of the 1,500 pixel array to provide visual acuity on the order of 20/1,000, allowing a patient to read large fonts [53].

Daniel Palanker's group at Stanford University has taken a different approach to an active photodiode array system by using video goggles to project pulsed

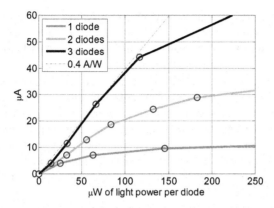

Fig. 7.1 Average stimulation current produced by one, two, and three photodiodes connected in series. The diodes were oriented such that the biphasic stimulation pulses were anodal-first. Data taken for 25 Hz, 500 μs pulses with a 50 μm SIROF active electrode coupled to a much larger return electrode

infrared (905 nm) images onto the subretinal array. A single, photovoltaically-driven photodiode can only produce up to 0.6 V at physiologically safe light intensities [35], a fraction of the 1.4 V electrochemically-safe "water window." By providing a pulsed bias voltage and utilizing the diodes in a photoconductive rather than photovoltaic manner, they can produce bi-phasic currents sufficient for neural stimulation, and limited only by the electrode charge injection capacity [35]. The common photodiode bias is provided by a periocular coil-based system. Recently Palanker's group proposed the use of series photodiodes to receive sufficient current photovoltaically [34]. The voltage increase afforded by series photodiodes, combined with the nonlinear electrochemical capacitance of iridium oxide electrodes [14], greatly increases the attainable current, as shown in Fig. 7.1. Since pulsed infrared illumination is directly converted into electric currents sufficient for stimulation, there is no need for a wired connection to a separate power-receiving module. The pixels do not even need to be physically connected to each other. The arrays may be separately placed into the subretinal space, greatly simplifying surgery.

The information transfer rate C from goggles with N pixels operating at S levels of gray at frame rate R can be estimated in a manner similar to (7.2). With an XGA LCD display ($N = 1,024 \times 768$) operating at 25 Hz and 128 levels of gray, the data rate is $C = 138$ Mb/s. The limit in this approach is clearly on the receiving end of the system – the photodiode array and its interface with the retina.

7.2.5   Thermal Safety Considerations

Power losses due to tissue absorption and intrinsic imperfections in the receiving circuitry lead to heating. For coil systems this includes absorption of RF radiation in tissue between the transmitting and receiving coils, resistive losses in the coils themselves, and losses in the rectifying circuitry. In photodiode systems this includes light absorption in ocular pigments such as melanin and in the implant itself. In both system types, the resulting tissue heating must be understood and controlled to within acceptable safety limits.

Tissue RF-absorption has a strong frequency dependence, increasing exponentially beyond a few MHz [43]. However, power transfer efficiency also increases with frequency due to the linear increase of the quality factor Q. There is an optimal frequency region balancing these counteracting effects where RF tissue exposure is minimized. Most coil designs operate at a frequency between 1 and 10 MHz [19, 33, 67]. Once a frequency is chosen, there exist design methodologies to maximize receiving circuit efficiency [27, 28]. Such systems can have power-transfer efficiencies exceeding 65% [28].

Photodiodes are rather inefficient at converting light into electrical power. A photodiode's maximum conversion efficiency (ratio of current output to incident light) typically does not exceed 0.6 A/W. Since photodiodes produce a photovoltage of at most 0.5 V at physiologically safe light intensities [35], 1 W of incident light power cannot produce more than 0.3 W of electrical power – an efficiency of at most 30%. Thus, photovoltaic retinal stimulation is a rather energy intensive task, rendering it imperative to examine safety limits for intense retinal illumination.

According to established ocular safety standards [3, 60], the maximum permissible retinal irradiance for prolonged exposure to near-IR light is 2.8 mW/mm^2.[1] Similar thermal considerations apply to heating of the iris. Peak irradiance can significantly exceed the average value during short pulses if the duty cycle is decreased. For example, in a goggles-based system with 1 ms pulses delivered at 25 Hz, the duty cycle is 1/40. The peak irradiance during the pulse can then be increased by a factor of 40 – to 112 mW/mm^2. Assuming a light-to-current conversion efficiency of 0.4 A/W, the maximum current that can be produced by photodiodes with this irradiance is 45 mA/mm^2, corresponding to a charge density of 45 µC/mm^2. This value exceeds the retinal stimulation threshold on large electrodes by at least three orders of magnitude [24].

Most retinal-heating studies to date have been acute; little data is yet available on the effects of chronic retinal heating. However, it has been observed that chronic lens heating by 2–3°C can lead to cataract formation [57], in which case chronic heating due to electronic implants could also cause cataracts. A "less than 1°C" criterion for implantable devices is codified in EU safety regulation [1] since this level is comparable to natural variations of the body temperature [50]. For a disk-shaped heater of diameter D which dissipates power P the maximum temperature rise ΔT in the adjacent medium is [52]

$$\Delta T = \frac{P}{4\lambda D} \tag{7.3}$$

---

[1] The ED50 level for producing a minimally-visible lesion with near-IR light ($\lambda = 810$–950 nm) for spot sizes larger than 1.7 mm on the retina and exposure times exceeding 1000s is 56 mW/mm^2 [42, 43]. With a safety factor of 20, the maximum permissible exposure is then 2.8 mW/mm^2.

where λ is the heat conductivity of the medium. For example, keeping temperature rise under 1°C for a 1 cm disk in water ($\lambda = 0.58\,W\,K^{-1}\,m^{-1}$) requires dissipating less than 23 mW of power. Blood perfusion in the eye helps to cool the tissue more efficiently, especially at temperature rises exceeding 3°C [55, 56].

7.2.6 Conclusions: Comparing the Different Approaches

Though the drawbacks of percutaneous connections are many, they will continue to be used in research environments through the foreseeable future. The electrical access afforded by percutaneous cables is invaluable when studying stimulation thresholds or reading the changing electrical properties of tissue. In addition, the development of a wireless system can be quite challenging – direct connections can greatly simplify experiments. However, any commercial prosthesis will need to incorporate a wireless power and data delivery system of some sort.

Of the three wireless system types described (RF, serial optical, and parallel optical telemetry), wireless cortical and optic nerve prostheses exclusively use radio links [8, 64, 65]. For retinal prostheses, there is as of yet no clear answer as to which system is superior. The circuit complexity of inductive systems is typically much higher than for photodiode-array based ones. Data signals must be separated from the carrier frequency, decoded, stored, and routed from the coil to individual electrodes. Photodiode array pixels receive all data simultaneously, with no need for complex wiring schemes. In addition, photodiode systems can produce DC voltage directly, whereas inductive coils produce AC current which must then be converted to DC. However, in terms of power transfer efficiency, inductive coil systems have a clear edge over photodiode systems, achieving efficiencies of greater than 65%, vs. ~30% achievable with photodiodes.

Photodiode-array based prosthetics keep the natural association between eye movements and visual perception, since they use the eye's natural optics. A shift of the patient's gaze directly changes what part of the visual field falls on the photodiode array. In contrast, current coil designs, as well as a serial optical telemetry with one receiving photodiode deliver visual information based solely on the orientation of a head-mounted camera; shifts in gaze do not change the visual stimuli. This could be fixed in the future by adding an eye-tracking system which controls what part of the visual field is transmitted to the stimulator. The Weiland group's proposed eye-implanted camera [69] would also produce images which shift naturally with eye movements, although it would add significant complexity to implanted electronics.

7.3 Tissue Response to a Subretinal Implant

The sophistication of the visual system requires a prosthetic much more complex than previous electrical stimulation devices including pacemakers, cochlear implants and deep brain stimulators. Ensuring that electrodes maintain sufficiently close proximity to the target cells is an important aspect of interfacing a prosthetic with the retina.

Visual acuity of 20/200 (the threshold of legal blindness in the United States) geometrically corresponds to a spatial frequency of three cycles per degree or ten lines per millimeter on the retina [29]. Since at least 2 pixels per cycle are required for appropriate sampling (Nyquist–Shannon sampling theorem), achieving this resolution constrains the maximum pixel size to 50 μm, or a pixel density of 400 pixels/mm². If the electrode is to be no larger than half the size of the pixel, then the electrode diameter should not exceed approximately 25 μm. The divergence of the electric field from the electrode requires greater currents to stimulate cells at increasing distances. Since increasing separation also causes the divergent electric field to influence larger regions of the retina, this effectively reduces the specificity of neural stimulation, resolution and contrast [45]. Charge and current injection limits, determined by the electrochemical capacitance of the electrode material, also determine the maximum distance at which neurons can be stimulated. In addition, significant variation in the distance between electrodes and neurons across an implant will lead to position-dependent differences in stimulation thresholds and responses. These factors dictate that the distance between the electrodes and target cells should not exceed the electrode size [45]. In the case of 50 μm pixels with 25 μm electrodes, the separation of electrodes from cells should ideally not exceed 25 μm.

Retinal neurons can be stimulated electrically using arrays of electrodes positioned either epiretinally [22, 37, 38] or subretinally [54, 61, 75]. Although surgically more challenging, a subretinal array placement to stimulate bipolar cells has the potential advantage that the electrical stimuli can be simpler. Since the cells in the inner nuclear layer (INL) have a graded response (they do not spike), they may not require as precise stimulus timing as the spiking ganglion cells. Subretinal electrical stimulation can produce a graded response which is then converted to spiking ganglion cell output via natural signal transduction pathways [61]. Addressing the visual system earlier in the signal processing cascade may also utilize some preserved natural signal processing mechanisms between the inner retina and ganglion cells. In contrast, direct ganglion cell excitation with epiretinal electrodes bypasses inner retinal circuitry and requires significantly more complex signal processing by external elements.

In the subretinal approach, the proximity between the stimulating array and the bipolar cells can be limited by subretinal gliosis and fibrosis, whereas in the epiretinal approach, proximity is limited by the inner limiting membrane and the nerve fiber layer. Mounting the implant epiretinally presents an additional challenge: attachment via a single retinal tack often results in tens or even hundreds of microns of separation between the retina and peripheral parts of the implant [36]. In contrast, the subretinal approach appears to provide more consistent proximity to the retina along the implant [47], although the thickness of the degenerating retina may be uneven.

In the following section we present results of investigations on the effect of an implant's material and shape on its integration with the retina.

7.3.1 Flat Implants

SU-8, a photo-curable epoxy, was used to manufacture devices to investigate tissue response to subretinal implants. SU-8 polymers are ideal for this purpose because they can be coated with materials commonly used in retinal prostheses, can form high aspect-ratio structures and are soft enough to be easily sectioned, in situ, on a conventional microtome to histologically evaluate the implant–retina interface. Three different coating materials have been compared: silicon oxide, parylene-C, and iridium oxide.

The Royal College of Surgeons (RCS) rat is a commonly studied model of retinal degeneration and is ideal for subretinal implantations of these devices because of its sufficiently large eyes and vascularized retina [32]. All implantations were performed at 45–60 days of age; at this stage the photoreceptor cells have largely degenerated. The implants were placed in the subretinal space using a custom implantation tool that protects the implant from mechanical damage during insertion [9]. The implant is delivered trans-sclerally through a small incision behind the pars plana; the retina is detached from the RPE by injecting BSS with a 30-gauge cannula. Implant placement was evaluated after each surgery by fundus examination.

Figure 7.2a illustrates the normal wild type rat retina, and Fig. 7.2b shows RCS retina at 45 days of age (P45). Figure 7.2c demonstrates a flat SU-8 implant in the subretinal space of a P45 RCS rat 6 weeks after surgery. Changes in the degenerating retina are easily seen comparing 7.2b to 7.2a; in Fig. 7.2b, the outer segments have disintegrated and there is significant thinning of the outer nuclear layer (ONL) though the inner nuclear, inner plexiform, and ganglion cell layers are generally well preserved.

As shown in Fig. 7.2c, two types of tissue reaction to the subretinal implant are evident: gliosis and fibrosis. Differentiation of retinal pigment epithelial (RPE) cells into long fibrotic membranes that encapsulate the foreign body is called fibrosis – this

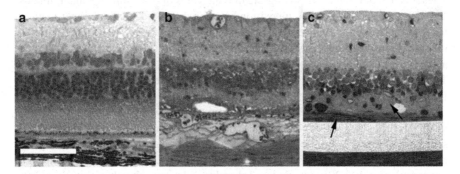

Fig. 7.2 Histological sections depicting (**a**) wild type rat retina, (**b**) RCS rat retina 45 days post natal (P45), and (**c**) RCS rat retina with a flat SU-8 implant in the subretinal space 6 weeks post-op. A fibrotic seal running along the length of the implant is denoted with the *left arrow*. A region of gliosis separating the implant from the INL by 40 μm is shown by the *right arrow*. Scale bar is 50 μm. Figure reprinted from [9], with permission

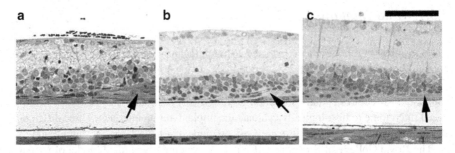

Fig. 7.3 Comparison of typical tissue responses to flat implants with different coatings 6 weeks post-op. (**a**) SiO$_2$ coating appears to induce significant fibrosis over the implant. (**b**) IrOx causes a mild gliosis above to the implant, pointed by an *arrow*. (**c**) Parylene-c coating allows the INL to settle down very close to the implant. The INL is separated from the upper surface of the implant by only 15–30 μm. Scale bar is 50 μm. Figure reprinted from [9], with permission

dark stained layer apposed to the implant is denoted by an arrow in Fig. 7.2c. The lightly stained reaction in the inner nuclear layer (INL), denoted by an arrow in Fig. 7.2c, is the hypertrophy of glial cells processes and is called gliosis. Fibrosis and gliosis can separate the INL from the surface of the implant by approximately 40 μm.

The silicon oxide coating (Fig. 7.3a) induced significant fibrosis around the implant. Generally, the iridium oxide and parylene coatings (Fig. 7.3b, c) were well tolerated with only a mild gliotic response, resulting in 15–30 μm separation from INL somata.

7.3.2 Chamber Implants

The tendency of retinal cells to migrate into the voids of three-dimensional implants can be used to provide closer apposition between the implant and the inner nuclear layer [44]. The vertical movement of the retina is hypothesized to be largely due to the movement of glial processes into the voids. Retinal neurons appear to be pulled along with retinal glia – this movement occurs within 72 h after implantation [17]. The ability of these structures to maintain a stable interface and suppress severe fibrosis is discussed below.

Chamber structures were fabricated with an array of wells (40 × 40 × 20 μm tall) in an SU-8 substrate (Fig. 7.4a, b). These devices were implanted into the subretinal space of RCS rats. A representative histological section of the retina, 6 weeks post-implantation, is shown in Fig. 7.5. In general, INL cell bodies migrate through apertures larger than 20 μm (right three chambers). In some chambers with apertures greater than 20 μm a retinal microvasculature developed [9].

Computational molecular phenotyping (CMP) analysis was used to distinguish superclasses and classes of neurons and glia, determine their status, examine morphological circuitry changes in response to retinal degeneration, and document any influence the implant may have upon these processes [25, 26]. Within the time

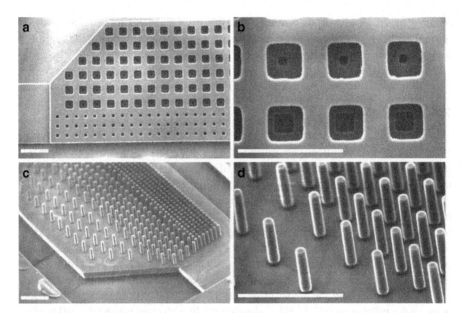

Fig. 7.4 SEM of three-dimensional implant structures. (**a**) Two microfabricated layers of the SU-8 chamber structures prior to adhesion to the basal membrane. Chamber sizes are 40 and 20 μm, and aperture sizes are 20 and 10 μm. (**b**) High magnification view of the chamber array. The 10 and 20 μm apertures can be seen clearly in the center of the 40 μm chambers. (**c**) Implant with an array of SU-8 pillars at three densities, with center-to-center distances of 60, 40 and 20 μm. (**d**) High magnification SEM of the pillar array. Pillars are 10 μm in diameter and 65 μm in height. All scale bars in this figure are 100 μm. Figure reprinted from [9], with permission

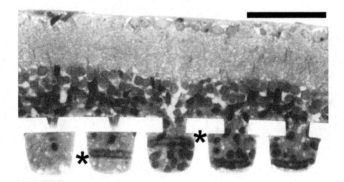

Fig. 7.5 The chamber structure implanted into P45 RCS rat subretinally for 6 weeks. The three chambers on the *right* have 20 μm apertures, the two on the *left* are 10 μm. This is an example of a typical section where cell bodies have migrated through the wider apertures while only processes migrated through 10 μm apertures. Artifactual folds are marked with a *. Scale bar is 50 μm. Figure reprinted from [9], with permission

window of this study, up to 6 weeks postoperatively, the neurons maintain their normal narrowly defined small molecular signatures in the presence of the implant, indicating normal metabolic status [9].

7.3.3 Pillar Arrays

While chamber arrays effectively improved neuron-implant proximity, the migrated tissue was somewhat isolated from the rest of the retina. Pillar arrays were designed to provide proximity to cells by utilizing retinal migration, while avoiding this isolation. The pillars used in this study were made from uncoated SU-8, approximately 10 μm in diameter, spaced 20, 40 and 60 μm center-to-center and 65 μm in height (Fig. 7.4c, d). The devices were implanted into P45 RCS rats for 6 weeks.

Figure 7.6 shows retinal histology 6 weeks after implantation. In the area with lower pillar density (Fig. 7.6a, 40 μm spacing) some cell bodies appear to be pulled down past the tops of the pillars, but many cell bodies remain apposed to the electrode surface at the top of the pillars. At the higher pillar density (Fig. 7.6b, 20 μm spacing) the space between the pillars seems to be filled almost entirely with neuropil and the cell bodies remain stable near the tops of the pillars. The inner retina appears well ordered and healthy for RCS retina. CMP results show that the neurons retain their phenotype and

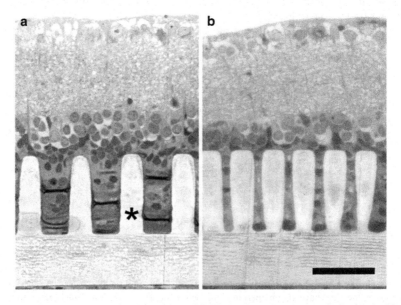

Fig. 7.6 Pillar implant in subretinal space of RCS rat 6 weeks post-op. Scale bar is 50 μm. (**a**) Area with 40 μm pillar spacing. (**b**) Area with 20 μm pillar spacing. Figure reprinted from [9], with permission

function through the migration process, for 6 weeks after implantation, and that there is excellent apposition between neuronal cell bodies and electrode tips [9].

In summary, the best proximity between electrodes and target cells in the inner retina is achieved using three-dimensional implants that utilize retinal plasticity for intimate integration of the inner retina with the implant. An implant with a multitude of voids allows retinal cell bodies and cell processes to migrate into the voids within 72 h after implantation, and appears to be stable at least up to 6 weeks post-op.

7.4   Damage to Retinal Tissue from Electrical Stimulation

Understanding the safe limits of electrical stimulation of neural tissue is critically important for maintaining a stable interface between the retina and the prosthesis. Some studies along these lines have been performed with cortical stimulation in cats, using charge-balanced 400 μs pulses at 50 Hz over the course of 7 h, and analyzed by histology [39]. It was determined that charge per phase and charge density were cofactors that determined cellular damage in different regimes. However, no detailed understanding has been achieved regarding dependence of the damage threshold on pulse duration, electrode size, distance from the electrode and number of pulses. This section explores these dependences using chick retina, validated with a limited number of experiments on mammalian retina in vitro.

Electrical stimulation was biphasic, with the same duration in both phases, leading with the cathodal phase. All durations mentioned below refer to time per phase. Cellular damage was assessed using propidium iodide (PI), a normally cell impermeable molecule that becomes fluorescent upon binding to nucleic acids [70]. PI was added to the medium prior to the treatment and dye fluorescence was assessed 15 min after the electrical pulse. Causes of cellular damage may include the direct effect of electric field, thermal damage from the applied current, or toxic products from the electrochemical reactions at the electrode–electrolyte interface. Our estimations of Joule heating within the pulse durations and currents used in our experiments indicated that this effect is negligible – temperature rise did not exceed 0.02°C, significantly lower than thermal damage thresholds [10]. Glass pipettes pulled to various tip diameters were used as stimulating electrodes. This design allowed large platinum wire bundles inside the pipette to have low current density on the metal surface to avoid generation of gas inside the capillary, while having high current densities at the electrode tip.

Damage thresholds were established for pulse durations in the range of 6 μs/phase to 6 ms/phase, and for electrode diameters of 0.1–1 mm. All plots include two points for each setting: a maximum safe value and a minimum damaging level, evaluated 5–15 min after the insult.

As shown in Fig. 7.7, the damage threshold decreased with the number of pulses, stabilizing after 100 exposures at approximately 15% of the single pulse value. This level remained stable up to the maximum number of pulses tested – 7,500 exposures at 25 Hz. The pulse duration was 600 μs in these measurements.

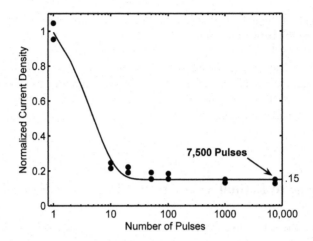

Fig. 7.7 Retinal damage threshold current density as a function of the number of pulses, applied during 5 min, normalized to a single exposure damage threshold. After approximately 50 pulses, the damage threshold reaches a constant level. A pipette of 1 mm in diameter was used in these measurements with a 600 µs pulse duration. Figure reprinted from [10], with permission; © 2009 IEEE

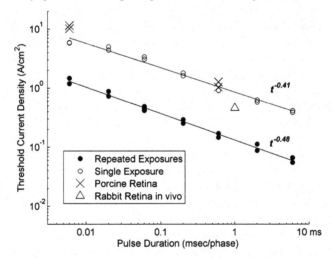

Fig. 7.8 Strength-duration dependence of the damage thresholds on the retina. (○) Measured on the chick retina with single shots (*open symbols*) and with sustained repetitive exposures (● *solid symbols*). Current density relates to pulse duration t roughly as $t^{-0.5}$, which is characteristic of electroporation [69, 70]. For comparison, (×) represents the damage thresholds of the porcine retina by single pulses in vitro and (△) presents chronic damage thresholds on the rabbit retina measured in vivo [13]. Figure reprinted from [10], with permission; © 2009 IEEE

7.4.1   Effect of Pulse Duration

Pulse duration was varied between 6 µs and 6 ms using a large electrode (1 mm) for single and repeated exposures. As shown in Fig. 7.8, the damage thresholds scale with pulse length as approximately $1/\sqrt{t}$, or more exactly $t^{-0.48}$ for chronic stimulation (7,500 pulses), and $t^{-0.41}$ for single exposures. Two additional measurements have been performed with single exposures on porcine retina to validate the chick model; they are presented with the X. For comparison we also plot the in vivo result of chronic retinal stimulation in rabbits (Δ) [71].

The approximate scaling of the strength–duration curves as $t^{-0.5}$ is characteristic of electroporation [41, 42], indicating that cellular damage is produced by the opening of pores in the cell membrane. Though cells can recover from the transient occurrence of these pores, it is unlikely that the cell would be able to sustain this abnormal state chronically. The scaling also indicates that neither charge, nor charge density, $q=j{\cdot}t$, are conserved along the strength–duration curve. It used to be believed that charge and charge density per phase were the two determinants of damage threshold [39].

7.4.2 Electrode Size

The dependence of damage threshold on electrode size was investigated using 600 μs biphasic pulses on chick retina with electrode size ranging from 0.1 to 1 mm. As shown in Fig. 7.9, damage threshold current density is nearly constant with large electrodes (diameter greater than 300 μm). With smaller electrodes the current increases, asymptotically approaching a $1/d^2$ dependence, indicating a constant current regime characteristic of a point source. This asymptotic constant current value was about 140 μA for a 600 μs pulse duration.

The strength–duration relationship for large and small pipettes (0.115 and 1.0 mm) are compared in Fig. 7.10 with the retinal stimulation thresholds published by Jensen et al. [24]. The inset plot depicts the safe dynamic range of retinal

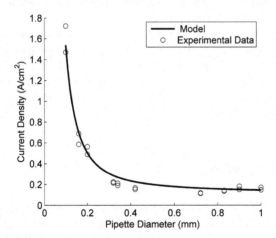

Fig. 7.9 Dependence of the threshold current density on pipette diameter for sustained exposures on chick retina with pulse duration of 0.6 ms/phase. The *solid line* represents the current density at the tissue, calculated using the model of a disk electrode separated from the retina by 125 μm. On electrodes smaller than 200 μm, the current density scales as $1/d^2$, corresponding to a constant current of 139 μA. Figure reprinted from [10], with permission; © 2009 IEEE

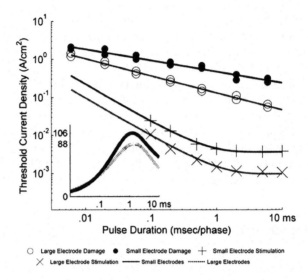

Fig. 7.10 Dependence of the chronic retinal damage threshold on pulse duration measured with pipettes of 0.12 (●) and 1.0 mm (○) in diameter. For comparison, we plot stimulation thresholds of the retinal ganglion cells measured by [44] using disk electrodes of similar sizes: 0.12 (+) and 0.5 (×) mm in diameter. Ratios of the damage thresholds to the stimulation thresholds are shown in the insert for both electrodes. Figure reprinted from [10], with permission; © 2009 IEEE

stimulation (the ratio of the damage threshold to stimulation threshold) as a function of pulse duration for both electrodes. The maximum (on the order of 100) of these curves occur near chronaxie for both electrode sizes. It is important to note that although the damage and the stimulation thresholds are dependent on electrode size, their ratio, which determines the dynamic range of safe stimulation, appears to be practically size independent.

Comparison of the recent measurements of the stimulation threshold in humans (electrode size 0.4 mm, 1 ms, 0.01 A/cm²) [36] and the in vivo damage threshold in rabbits (electrode size 0.4 mm, 1 ms, 0.46 A/cm²) [71] results in a slightly lower ratio, 46. A safe dynamic range of 50–100 is sufficiently broad to cover the linear response range of neural cells (typically 10–30 [5, 73]), and is therefore adequate for the purpose of prosthetic vision.

7.5 Concluding Remarks

The development and testing of retinal prostheses by multiple groups throughout the world is rapidly advancing. The delivery of a vast amount of information and sufficient power to the retinal neurons has proved to be technically challenging, and has required the development of new technologies in many disparate fields. Sophisticated coil systems have been developed to transmit and receive power and data; others have developed novel optical approaches for serial and parallel data delivery. In both approaches care has been taken to avoid thermal damage to surrounding tissues in the process of power transmission.

Once received by the implanted prosthesis, power and data must be delivered to target neurons, a task which requires close neuron-electrode proximity. Many materials have been tested to characterize tissue response to the implanted devices. In addition, three-dimensional subretinal arrays have been developed to utilize retinal plasticity to achieve intimate proximity between neurons and stimulation sites. Finally, electrical damage thresholds have been carefully measured to characterize the safe dynamic range of stimulation.

Despite the incredible advancements made in recent decades, there is much left to be done. This includes implementation of already proposed ideas, and improvements to the currently used approaches. Higher resolution implants will allow for more sophisticated evaluation of prosthetic vision and will most probably generate a need for development of more advanced signal processing algorithms. The past two decades of research have been very fruitful – several prosthetic technologies are currently being tested in human trials [6, 11, 21, 53]. The results from the current trials are eagerly awaited by researchers around the world, as they will likely dictate the direction of technological development for the next decade.

References

1. (2004), *Active Implantable Medical Devices*, in *Directive 90/285/EEC*.
2. (2007), *Second Sight Medical Retinal Prosthesis Receives FDA Approval for Clinical Trials*, in *medGadget*.
3. [Anon], Radiati ICN (Anon) (2000), *ICNIRP statement on light-emitting diodes (LEDs) and laser diodes: Implications for hazard assessment*. Health Phys, **78**: p. 744–52.
4. Abrams LD, Hudson WA, Lightwood R (1960), *A surgical approach to the management of heart-block using an inductive coupled artificial cardiac pacemaker*. Lancet, **1**: p. 1372–4.
5. Berntson A, Taylor WR (2000), *Response characteristics and receptive field widths of on-bipolar cells in the mouse retina*. J Physiol, **524**(Pt 3): p. 879–89.
6. Besch D, Sachs H, Szurman P, et al. (2008), *Extraocular surgery for implantation of an active subretinal visual prosthesis with external connections: feasibility and outcome in seven patients*. Br J Ophthalmol, **92**(10): p. 1361–8.
7. Brindley G (1964), *Transmission of electrical stimuli along many independent channels through a fairly small area of intact skin*. J Physiol, **177**: p. 44–6.
8. Brindley G, Lewin W (1968), *The sensations produced by electrical stimulation of the visual cortex*. J Physiol, **196**: p. 479–93.
9. Butterwick A, Huie P, Jones BW, et al. (2009), *Effect of shape and coating of a subretinal prosthesis on its integration with the retina*. Exp Eye Res, **88**: p. 22–9.
10. Butterwick A, Vankov A, Huie P, et al. (2007), *Tissue damage by pulsed electrical stimulation*. IEEE Trans Biomed Eng, **54**(12): p. 2261–7.
11. Caspi A, Dorn JD, McClure KH, et al. (2009), *Feasibility study of a retinal prosthesis: spatial vision with a 16-electrode implant*. Arch Ophthalmol, **127**(4): p. 398–401.
12. Chow A (1993), *Electrical stimulation of the rabbit retina with sub-retinal electrodes and high density microphotodiode array implants*. ARVO abstracts. Invest Ophthalmol Vis Sci, **34**: p. 835.
13. Chow A, Chow V, Packo K, et al. (2004), *The artificial silicon retina microchip for the treatment of vision loss from retinitis pigmentosa*. Arch Ophthalmol, **122**(4): p. 460–9.

14. Cogan S, Troyk P, Ehrlich J, et al. (2006), *Potential-biased, asymmetric waveforms for charge-injection with activated iridium oxide (AIROF) neural stimulation electrodes*. IEEE Trans Biomed Eng, **53**(2): p. 327–32.

15. DeMarco P, Yarbrough G, Yee C, et al. (2007), *Stimulation via a subretinally placed prosthetic elicits central activity and induces a trophic effect on visual responses*. Invest Ophthalmol Vis Sci, **48**(2): p. 916–26.

16. Dobelle WH, Mladejovsky MG, Girvin JP (1974), *Artifical vision for the blind: electrical stimulation of visual cortex offers hope for a functional prosthesis*. Science, **183**(123): p. 440–4.

17. Fisher SK, Erickson PA, Lewis GP, Anderson DH (1991), *Intraretinal proliferation induced by retinal detachment*. Invest Ophthalmol Vis Sci, **32**(6): p. 1739–48.

18. Ghovanloo M, Najafi K (2002). *Fully integrated power supply design for wireless biomedical implants*, in *Microtechnologies in Medicine & Biology 2nd Annual International IEEE-EMB Special Topic Conference*. Madison, WI.

19. Hamici Z, Itti R, Champier J (1996), *A high-efficiency power and data transmission system for biomedical implanted electronic devices*. Meas Sci Technol, **7**: p. 192–201.

20. Heetderks W (1988), *RF powering of millimeter and submillimeter sized neural prosthetic implants*. IEEE Trans Biomed Eng, **35**: p. 323–6.

21. Humayun M (2009), *Preliminary results from Argus II feasibility study: a 60 electrode epiretinal prosthesis*. Invest Ophthalmol Vis Sci, **50**: E-Abstr# 4744.

22. Humayun MS, de Juan E, Jr., Weiland JD, et al. (1999), *Pattern electrical stimulation of the human retina*. Vision Res, **39**(15): p. 2569–76.

23. Humayun M, Weiland J, Fujii G, et al. (2003), *Visual perception in a blind subject with a chronic microelectronic retinal prosthesis*. Vision Res, **43**: p. 2573–81.

24. Jensen RJ, Rizzo JF, Ziv OR, et al. (2003), *Thresholds for activation of rabbit retinal, ganglion cells with an ultrafine, extracellular microelectrode*. Invest Ophthalmol Vis Sci, **44**(8): p. 3533–43.

25. Jones BW, Marc RE (2005), *Retinal remodeling during retinal degeneration*. Exp Eye Res, **81**(2): p. 123–37.

26. Jones BW, Watt CB, Frederick JM, et al. (2003), *Retinal remodeling triggered by photoreceptor degenerations*. J Comp Neurol, **464**(1): p. 1–16.

27. Kelly S (2003). *A system for efficient neural stimulation with energy recovery*. Thesis, Electrical Engineering and Computer Science, Massachusetts Institute of Technology, Cambridge.

28. Kendir G, Liu W, Wang G, et al. (2005), *An optimal design methodology for inductive power link with Class-E amplifier*. IEEE Trans Circ Syst, **52**(5): p. 857–65.

29. Khanani AM, Brown SM, Xu KT (2004), *Normal values for a clinical test of letter-recognition contrast thresholds*. J Cataract Refract Surg, **30**(11): p. 2377–82.

30. Knutson J, Naples G, Peckham P, Keith M (2002), *Electrode fracture rates and occurences of infection and granuloma associated with percutaneous intramuscular electrodes in upper-limb functional electrical stimulation applications*. J Rehabil Res Dev, **39**(6): p. 671–84.

31. Ko W, Liang S, Fung C (1977), *Design of radio-frequency powered coils for implant instruments*. Med Biol Eng Comput, **15**(6): p. 634–40.

32. Li L, Sheedlo HJ, Turner JE (1993), *Muller cell expression of glial fibrillary acidic protein (GFAP) in RPE-cell transplanted retinas of RCS dystrophic rats*. Curr Eye Res, **12**(9): p. 841–9.

33. Liu W, Vichienchom K, Clements M, et al. (2000), *A neuro-stimulus chip with telemetry unit for retinal prosthesis device*. IEEE Solid-State Circuits, **35**: p. 1487–97.

34. Loudin JD, Palanker D (2008), *Photovoltaic retinal prosthesis*. Invest Ophthalmol Vis Sci, **49**: E-Abstr# 3014.

35. Loudin JD, Simanovskii DM, Vijayraghavan K, et al. (2007), *Optoelectronic retinal prosthesis: system design and performance*. J Neural Eng, **4**(1): p. S72–84.

36. Mahadevappa M, Weiland JD, Yanai D, et al. (2005), *Perceptual thresholds and electrode impedance in three retinal prosthesis subjects*. IEEE Trans Neural Syst Rehabil Eng, **13**(2): p. 201–6.

37. Margalit E, Maia M, Weiland J, et al. (2002), *Retinal prothesis for the blind*. Surv Ophthalmol, **47**(4): p. 335–56.

38. Margalit E, Weiland JD, Clatterbuck RE, et al. (2003), *Visual and electrical evoked response recorded from subdural electrodes implanted above the visual cortex in normal dogs under two methods of anesthesia.* J Neurosci Methods, **123**(2): p. 129–37.
39. McCreery DB, Agnew WF, Yuen TG, Bullara L (1990), *Charge density and charge per phase as cofactors in neural injury induced by electrical stimulation.* IEEE Trans Biomed Eng, **37**(10): p. 996–1001.
40. Mokaw W (2004), *MEMS technologies for epiretinal stimulation of the retina.* J Micromech Microeng, **14**: p. S12–6.
41. Neumann E (1992), *Membrane electroporation and direct gene-transfer.* Bioelectrochem Bioenerg, **28**: p. 247–67.
42. Neumann E, Toensing K, Kakorin S, et al. (1998), *Mechanism of electroporative dye uptake by mouse B cells.* Biophys J, **74**(1): p. 98–108.
43. Osepchuck JM (1983), *Biological Effects of Electromagnetic Radiation.* IEEE Press Selected Reprint Series. New York: IEEE.
44. Palanker D, Huie P, Vankov A, et al. (2004), *Migration of retinal cells through a perforated membrane: implications for a high-resolution prosthesis.* Invest Ophthalmol Vis Sci, **45**(9): p. 3266–70.
45. Palanker D, Vankov A, Huie P, Baccus S (2005), *Design of a high-resolution optoelectronic retinal prosthesis.* J Neural Eng, **2**(1): p. S105–20.
46. Pardue M, Phillips M, Yin H, et al. (2005), *Possible sources of neuroprotection following subretinal silicon chip implantation in RCS rats.* J Neural Eng, **2**: p. S39–47.
47. Pardue MT, Stubbs EB, Jr., Perlman JI, et al. (2001), *Immunohistochemical studies of the retina following long-term implantation with subretinal microphotodiode arrays.* Exp Eye Res, **73**(3): p. 333–43.
48. Peachey N, Chow A (1999), *Subretinal implantation of semiconductor-based photodiodes: progress and challenges.* J Rehabil Res Dev, **36**(4): p. 371–6.
49. Radovanovic S, Annema A, Nauta B (2004), *Bandwidth of integrated photodiodes in standard CMOS for CD/DVD applications.* Microelectron Reliab, **45**: p. 705–10.
50. Refinetti R, Menaker M (1992), *The circadian rhythm of body temperature.* Physiol Behav, **51**: p. 613–37.
51. Rizzo J, Wyatt J, Loewenstein J, et al. (2003), *Methods and perceptual thresholds for short-term electrical stimulation of human retina with microelectrode arrays.* Invest Ophthalmol Vis Sci, **44**: p. 5355–61.
52. Rohsenow W, Hartnett J, Gani E (1985), *Handbook of Heat Transfer Fundamentals.* New York: McGraw-Hill, p. 164.
53. Sachs HG, Bartz-Schmidt U, Gekeler F, et al. (2009), *The transchoroidal implantation of subretinal active micro-photodiode arrays in blind patients: long term surgical results in the first 11 implanted patients demonstrating the potential and safety of this new complex surgical procedure that allows restoration of useful visual percepts.* Invest Ophthalmol Vis Sci, **50**: E-Abstr# 4742.
54. Sachs HG, Gekeler F, Schwahn H, et al. (2005), *Implantation of stimulation electrodes in the subretinal space to demonstrate cortical responses in Yucatan minipig in the course of visual prosthesis development.* Eur J Ophthalmol, **15**(4): p. 493–9.
55. Sailer H, Shinoda K, Blatsios G, et al. (2007), *Investigation of thermal effects of infrared lasers on the rabbit retina: a study in the course of development of an active subretinal prosthesis.* Graefes Arch Clin Exp Ophthalmol, **245**(8): p. 1169–78.
56. Schule G, Huttmann G, Framme C, et al. (2004), *Noninvasive optoacoustic temperature determination at the fundus of the eye during laser irradiation.* J Biomed Opt, **9**(1): p. 173–9.
57. Scott J (1988), *The computation of temperature rises in the human eye induced by infrared radiation.* Phys Med Biol, **33**(2): p. 243–57.
58. Scribner D, Johnson L, Skeath P, et al. (2005), *Microelectronic array for stimulation of retinal tissue,* in *NRL Review.* Naval Research Lab, p. 53–61.
59. Shannon C (1998), *Communication in the presence of noise.* Proc IEEE, **86**(2): p. 447–57.

60. Sliney D, Aron-Rosa D, DeLori F, et al. (2005), *Adjustment of guidelines for exposure of the eye to optical radiation from ocular instruments: statement from a task group of the International Commission on Non-Ionizing Radiation Protection (ICNIRP).* Appl Opt, **44**(11): p. 2162–76.
61. Stett A, Barth W, Weiss S, et al. (2000), *Electrical multisite stimulation of the isolated chicken retina.* Vision Res, **40**(13): p. 1785–95.
62. Sullivan C (1999), *Optimal choice for number of strands in a litz-wire transformer winding.* IEEE Trans Power Electron, **14**(2): p. 283–91.
63. Theogarajan L, Wyatt J, Rizzo J, et al. (2006). *Minimally invasive retinal prosthesis*, in *IEEE International Solid-State Circuits Conference.*
64. Troyk P, Bradley D, Bak M, et al. (2005). *Intracortical visual prosthesis research – approach and progress*, in *IEEE Engineering in Medicine and Biology 27th Annual Conference.* Shanghai: IEEE.
65. Veraart C, Wanet-Defalque M, Gerard B, et al. (2003), *Pattern recognition with the optic nerve visual prosthesis.* Artif Organs, **27**(11): p. 996–1004.
66. Von Arx J (1998). *A single chip, fully integrated, telemetry powered system for peripheral nerve stimulation.* Thesis, Electrical Engineering, University of Michigan, Ann Arbor.
67. Von Arx J, Najafi K (1997). *On-chip coils with integrated cores for remote inductive powering of integrated microsystems*, in *1997 International Conference on Solid-State Sensors and Actuators.* Chicago.
68. Wang G, Liu W, Sivaprakasam M, Kendir G (2005), *Design and analysis of an adaptive transcutaneous power telemetry for biomedical implants.* IEEE Trans Circ Syst, **52**(10): p. 2109–17.
69. Wickelgren I (2006), *A vision for the blind.* Science, **312**: p. 1124–6.
70. Wilde GJ, Sundstrom LE, Iannotti F (1994), *Propidium iodide in vivo: an early marker of neuronal damage in rat hippocampus.* Neurosci Lett, **180**(2): p. 223–6.
71. Yamauchi Y, Enzmann V, Franco M, et al. (2005). *Subretinal placement of the microelectrode array is associated with a low threshold for electrical stimulation,* in *Annual Meeting of the Association for Research in Vision and Opthalmology.* Fort Lauderdale, FL.
72. Yang Z, Liu W, Basham E (2007), *Inductor modeling in wireless links for implantable electronics.* IEEE Trans Magn, **43**(10): p. 3851–60.
73. Yang XL, Wu SM (1997), *Response sensitivity and voltage gain of the rod- and cone-bipolar cell synapses in dark-adapted tiger salamander retina.* J Neurophysiol, **78**(5): p. 2662–73.
74. Zierhofer C, Hochmair-Desoyer I, Hochmair E (1995), *Electronic design of a cochlear implant for multichannel high-rate pulsatile stimulation strategies.* IEEE Trans Rehab Eng, **3**(1): p. 112–6.
75. Zrenner E (2002), *The subretinal implant: can microphotodiode arrays replace degenerated retinal photoreceptors to restore vision?* Ophthalmologica, **216**(Suppl 1): p. 8–20; discussion 52–3.
76. Zrenner E (2007), *Restoring neuroretinal function: new potentials.* Doc Ophthalmol, **115**: p. 56–9.
77. Zrenner E, Gabel V, Gekeler F, et al. (2004). *From passive to active subretinal implants, serving as adapting electronic substitution of degenerated photoreceptors*, in *IEEE International Joint Conference.*

Chapter 8
Retinal Cell Excitation Modeling

Carlos J. Cela and Gianluca Lazzi

Abstract As the electrode density of implantable retinal prosthesis increases, simulation becomes a valuable tool to characterize excitation performance, evaluate implant electrical safety, determine optimal geometry and placement of implant current return, and understand charge distribution due to stimulation. To gain an insight into the effectiveness of a retina stimulator, quasi-static numerical electromagnetic methods can help estimate current densities, potentials, and their gradients in retinal layers and neural cells. Detailed discrete three-dimensional models of the retina, implant and surrounding tissue can be developed to account for the anatomical complexity of the human eye and appropriate dielectric properties. This chapter will cover the basics of quasi-static methods that can be used for this purpose. Specifically, authors will focus on the admittance method, the output it produces, and possibilities it offers to determine the potential effectiveness of a retinal stimulator, ranging from evaluating the current density magnitude in the ganglion cell layer, to calculating local activation function in the areas targeted by the electrical stimulation.

Abbreviations

GCL   Ganglion cell layer
NFL   Nerve fiber layer
SAR   Specific absorption rate

C.J. Cela (✉)
Department of Electrical and Computer Engineering, University of Utah,
50 S. Central Campus Drive, Room 3280, Salt Lake City, UT 84112-9206, USA
e-mail: carlos.cela@utah.edu

G. Dagnelie (ed.), *Visual Prosthetics: Physiology, Bioengineering, Rehabilitation*,
DOI 10.1007/978-1-4419-0754-7_8, © Springer Science+Business Media, LLC 2011

8.1 Introduction

Retinal implants can help partially restoring vision to patients suffering from degenerative diseases of the retina. Age-related macular degeneration and retinitis pigmentosa by replacing the functionality of no longer working photoreceptors with systematic electrical stimulation to neural cells further down the optical neural path [8, 12].

Clinical trials show that electrical stimulation using epiretinally implanted electrodes causes the appearance of localized white or yellow round phosphenes [12]. These percepts must correspond to excitation of cells in the ganglion cell layer (GCL) or deeper in the retina, as only these cells map to a location under the stimulating electrode; the more superficial nerve fiber layer (NFL) is formed by axons of GCL neurons going towards the optic nerve that belong to ganglion cells away from the stimulating point (Fig. 8.1). Epiretinal electrode arrays having 16 (4×4) electrodes have already been successfully implanted in clinical trials, and efforts are ongoing to increase the resolution of the implant; versions with 60 electrodes are currently undergoing the FDA approval process and 240 and more electrodes are being worked on.

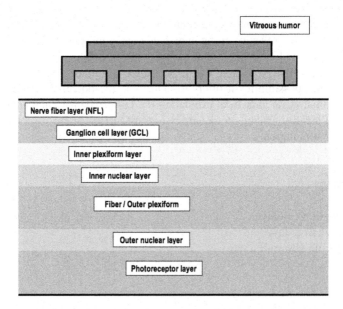

Fig. 8.1 Diagram of a transverse cut of retinal model with epiretinal implant close to its surface (not to scale). The retina geometry has been approximated as flat in this model. The epiretinal implant electrically stimulates the ganglion cell layer (GCL) across the nerve fiber layer (NFL) by injecting electrical current using charge-balanced biphasic pulses. The NFL is formed by the axons of cells in the GCL, which curve and bundle together, eventually shaping into the optic nerve, which relays the visual signal to the brain. In all layers but the NFL, the neural pathway is predominantly vertical. In this configuration the 25 electrodes are arranged in a regular 5×5 matrix and partially embedded in a dielectric substrate. A three-dimensional view of the implant model is shown in Fig. 8.2

Fig. 8.2 Geometry of the 5×5 electrode array used in the model (not to scale). The electrode array is positioned inside the ocular globe, in close contact with the retina. In the model simulated the electrodes are $10 \mu m$ away from the retinal surface, and the gap in between is filled with vitreous humor. The electrodes are encased in a block of a dielectric. The current return is positioned on the back of the assembly, and exposed to the vitreous humor, which is a relatively good electrical conductor

As arrays become denser, electrodes are packed closer and have smaller size. There is a compromise between electrode size, stimulation rate, and charge injected. A smaller electrode must use higher current densities to inject the same amount of charge in the same time compared to a larger electrode. Current densities must be limited so non-reversible electrochemical effects in the electrode-tissue interface are minimized, and there is no permanent damage to the living tissue [2]. Because of inherent difficulties in performing in vivo experiments, modeling and simulation are valuable aids in designing retinal implants [6, 13, 15, 17].

The admittance method is attractive for bioelectromagnetic problems because it can solve complex heterogeneous models using a wide variety of electromagnetic stimulation types in a computationally efficient way.

8.2 Quasistatic Numerical Methods: The Admittance Method

Quasistatic electromagnetic methods have been successfully used over the last 25 years to model bioelectromagnetic interactions; in particular, variations of the finite difference method – the admittance method – and its complement, the imped-ance method have proven useful for diverse bioelectromagnetic problems, including calculation of specific absorption rate (SAR) and novel ways to induce hyper-thermia in patients [1, 5]. Armitage and Ghandi were the first to study realistic models with generic lumped circuit element quasistatic electromagnetic numerical methods. Before them, other authors have used analytic formulations of gross geometric body tissues approximations such as cylinders and multilayered spheres (see references of [1]). Our research group extended the admittance method formulation introducing a general two-dimensional multiresolution meshing algorithm, in which cells close to tissue boundaries are discretized with fine resolution while cells surrounded by large areas of homogeneous space are progressively larger [3]; more recently, a three-dimensional method based on the same principle has also been developed [18].

In the impedance method, the electrical properties of the model are described using an impedance network; similarly, an admittance network is used in the admittance method; otherwise the methods are equivalent. For both methods there are several possible formulations suited to different electromagnetic stimulation mechanisms, be it an external electromagnetic field, a capacitive electrode, a metallic electrode, an implanted coil, etc. For the rest of this chapter we will focus on the admittance method.

The quasistatic constraint assumes that the highest significant frequency component of the electromagnetic fields involved in the simulation has a wavelength much larger than the size of the model simulated. Another way of looking at the validity of the quasistatic assumption is to consider if at the single highest frequency component used for excitation it is reasonable to assume that the phase differences across the model are negligible. For the model sizes used in this application, the quasistatic condition is met up to frequency components in the range of tens of megahertz.

The general idea of the admittance method is to obtain an equivalent circuit model starting from the geometry of the implant and surrounding tissue and the bulk electrical properties of the substances involved, and then apply electrical or magnetic stimulation using ideal current and voltage sources. The resulting equivalent circuit is solved for the node voltages using circuit theory and a numerical linear solver. Branch currents can then be calculated from the node voltages and circuit components values. Noting that each node in the equivalent circuit will correspond to the location of a spatial point in the model, the numerical solution of the simulation using the admittance method is the current vector field and the matching scalar potential field at each significant point in the model. Electric field, current densities, equivalent impedances, etc., can then be derived from these results.

8.2.1   Layered Retinal Model

For retinal implants, detailed three-dimensional models of retinal tissues and surrounding areas can be constructed using a layered retinal model [17]. In the multi-layer retinal models each layer represents a different tissue type. The values for conductivity (σ) used for the simulation results presented have been obtained from experimental measurements performed on a frog's retina [11]. The thickness values for each layer have been measured from electron microscopy images of a transverse cut of a mammalian retina [16]. The retinal layers considered and their thickness and conductivity are shown in Table 8.1.

In addition to the retina, models for retinal stimulation often must include the implant, electrodes, current return, and surrounding tissue, including choroid

Table 8.1 Properties of retinal layers used for layered retina model

Retinal layer	σ (s m^{-1})	Thickness (μm)
Photoreceptors	0.0198	60
Outer nuclear	0.0166	30
Fiber/outer plexiform	0.0143	60
Inner nuclear	0.0153	30
Inner plexiform	0.0555	30
Ganglion cell	0.0143	30
Optic fiber	0.0143	30

The conductivity values used for this set of simulations correspond to frog retinal tissue [11]

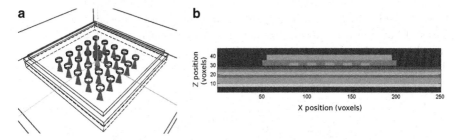

Fig. 8.3 (**a**) Simulation model is obtained from a geometric description of the retinal layered substrate, electrode array and current return. The 2.5 mm × 2.5 mm × 0.48 mm model was discretized to a resolution of 10 μm, resulting in a mesh having 250 × 250 × 48 voxels. In this view, current sources are noted as triangles under the electrodes, and the position of the current return marked by the central triangle. (**b**) Transverse slice of the model at 45Y = 125, showing detail of the retinal layers and the implanted electrode array, dielectric backing and current return (topmost layer of assembly)

($\sigma = 0.92$ s m^{-1}), sclera ($\sigma = 0.50$ s m^{-1}) and vitreous humor ($\sigma = 1.5$ s m^{-1}) [4]. The model used is depicted in Fig. 8.3. For the example presented in this chapter, a 25 electrode array having cylindrical electrodes of 200 μm diameter spaced 500 μm between centers, and with the current return exposed to the vitreous humor on the back of the electrode array dielectric substrate has been included. Each electrode is constantly injecting – 50 μA for cathodic stimulation, and the simulation considers the model to be static, i.e. purely resistive.

A discrete model is then obtained by spatially sampling a three-dimensional geometric description of the tissue and implant at regular intervals on the three Cartesian coordinate axes. This sampled model can then be interpreted as a collection of voxels, each surrounding a sampled point and considered made of a single material. Since the thickness of individual retinal layers are in the order of tens of micrometers, to calculate electrical activity inside a layer the model must be resolved with voxels of size 15 μm or less.

In order to reduce the voxel count and thus the size of the resulting linear system, the spatial sampling can optionally be made to create an expanding grid in one of more spatial dimensions and additional multiresolution techniques can be applied to the discrete model. For purposes of simplifying the discussion we will consider a discrete model having uniform voxel sizes.

8.2.2 Equivalent Electric Circuit

An equivalent electrical circuit is then constructed from the discrete model, by placing circuit nodes at the vertices of each voxel, and calculating lumped resistors from the dimensions and conductivity of the material or tissue the voxel is made of (Fig. 8.4). Each voxel is considered subdivided in 12 sub-volumes, four along each of the coordinate axes, and the electrical resistivity for each sub-volume is used to calculate the equivalent lumped resistor value, as shown in (8.1), where W, H, and L are the width, height, and length of the sub-volumes being considered. For purposes of building the equivalent circuit, each equivalent resistor is placed at the edge between the accessible node for the sub-volume it models. The process is illustrated in Fig. 8.3.

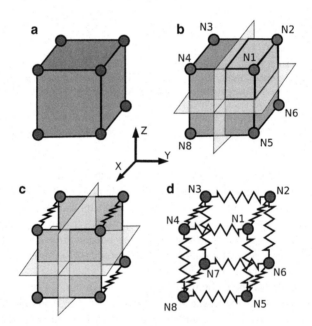

Fig. 8.4 Calculation of equivalent circuit for individual voxel. (**a**) Circuit nodes are places at each vertex of the voxel. (**b**) The voxel is considered subdivided in four sub-volumes along the X coordinate axis. (**c**) An equivalent resistor is used to model each of the sub-volumes in that direction, and placed in the edge in between the accessible nodes for that sub-volume. Equivalent resistors are calculated using (8.1). (**d**) The same process is then repeated for the Y and Z coordinates axis, resulting in the final circuit equivalent model for the voxel. After all voxels are processed, the circuit for the model is formed considering each shared vertex as a shared node. Finally, the current sources for excitation and the ground are placed

$$R = \frac{L}{W \cdot H} \cdot \frac{1}{\sigma}. \tag{8.1}$$

An equivalent electrical circuit is created by considering all the resistors for every voxel, and knowing that contiguous voxels share two nodes at each shared edge. The ground node is assigned, and the stimulation is added using current sources. In the example case considered in this chapter we have used one current source per electrode; since we considered cathodic stimulation, for each current source the negative terminal is connected to a node belonging to an electrode and the positive terminal to ground. Additional circuit elements can also be added to the equivalent circuit to model other electrical behaviors of the tissue or implant.

8.2.3   Electric Potentials and Current Density Magnitude Calculation

This equivalent electric circuit can be solved using standard circuit theory tools. Using Kirchoff current law, the model's equations can then expressed as a linear combination, in terms of an admittance matrix G, an unknown voltage vector V, and a current vector I, as shown in (8.2).

$$G \cdot V = I. \tag{8.2}$$

On a circuit having n nodes including ground, G is a symmetric, sparse, $(n-1) \times (n-1)$ admittance matrix, V is the unknown voltage vector, and I is the vector of independent current sources modeling the stimulating electrodes. While any method can be used for small systems, the size of the matrix G grows with the square of the number of nodes. Because of this, direct methods such as Gaussian elimination may not be the most suitable for this application; instead, iterative approaches including Krylov sub-space methods present advantages. In particular, the biconjugate gradient method with appropriate preconditioning has proven to be efficient in the considered cases (Figs. 8.5 and 8.6).

The solution of the system provides the value for the potentials at each node of the circuit, which are equivalent to the potentials at the vertexes of each voxel. The components of the current density vector field at each point in the model can then be determined from the cross-sectional areas of each voxel, the equivalent resistors, and the branch currents flowing through each of them:

$$J_X = \sum_{n=1}^{4} \frac{I_n}{Y_n \cdot Z_n}. \tag{8.3}$$

For each voxel, the branch currents parallel to one coordinate axis are calculated by considering the voltage at the nodes and the four resistors along that direction. For the X axis, for example, the current density component is determined by taking

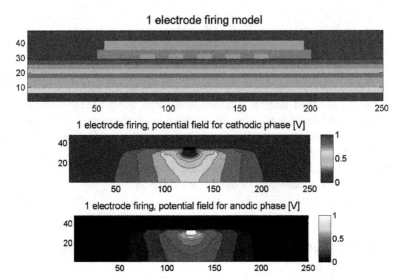

Fig. 8.5 *Top*: Model slice (Y = 125) and contour plot of the electric potential resulting from admittance method simulation of the retinal prosthesis applying (*center*) cathodic and (*bottom*) anodic stimulation through the central electrode of the array. Units on the X and Y axes are voxels

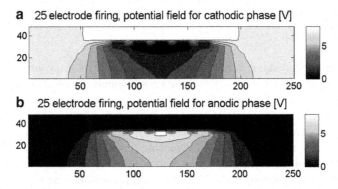

Fig. 8.6 Contour plot of the electric potential resulting from admittance method simulation of the retinal prosthesis applying (**a**) cathodic and (**b**) anodic stimulation through all 25 electrodes. Values correspond to Y = 125 slice. Units on the X and Y axes are voxels

into account the contribution of each sub-voxel in the X direction (8.3), where J_x is the current density component for direction X, I_n is the branch current going through each of the resistors parallel to the X axis, and $Y_n \cdot Z_n$ is the transverse area for each sub-voxel in the X direction. The current densities in the remaining axes are calculated in a similar way.

Figures 8.7–8.9 show the resulting current density magnitudes for the model slice considered in Fig. 8.5, and the values on a line at the ganglion cell layer, under the middle row of electrodes, for both the case of a single electrode and all 25 electrodes excited.

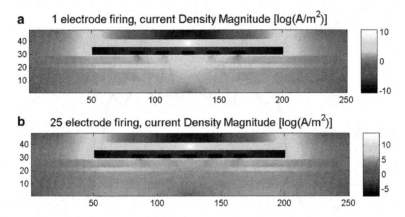

Fig. 8.7 Colorplot of current density magnitude for (**a**) single stimulating electrode and (**b**) all electrodes stimulating simultaneously. The values plotted are from a transverse slice of the three-dimensional model cutting through the center of the central electrode, at the plane Y = 125

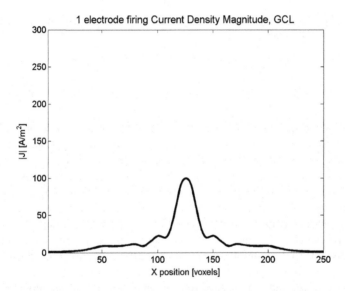

Fig. 8.8 Lineplot of current density magnitude at the GCL, centered under the central electrode, with a single electrode stimulating

8.3 Three-Dimensional Activation Function Calculation

In his early work, when analyzing a discretized formulation of Hodgkin and Huxley equations [7], and assuming initial conditions of transmembrane potential at rest, Rattay noted that the only dependency of the cellular transmembrane potential with respect to the extracellular potential V_E was given by the relation described in (8.4), and called it the activation function [13,15]:

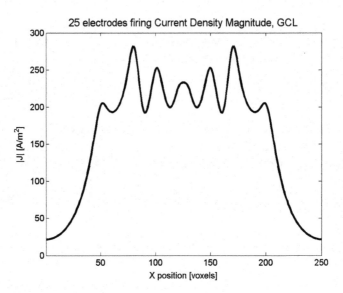

Fig. 8.9 Lineplot of current density magnitude at the GCL, centered under the central electrode, with all 25 electrodes stimulating

$$f = \frac{d^2 V_E}{dx^2}. \tag{8.4}$$

In (8.4), dx is the length differential along the axon of the neuron we seek to stimulate, taken in the direction of action potential propagation. Taking into account that the intracellular field of sensory neurons is very weak compared to the fields created by an implanted stimulator, in a retinal prosthesis the activation function can be approximated as an exclusive function of V_E. If calculated close to the cellular wall, the activation function is positive for zones that tend to be depolarized by the influence of the extracellular potential and negative for zones that are hyperpolarized, allowing to characterize a stimulation configuration in terms of its potential capability to trigger or not trigger action potentials in neural cells starting from the extracellular fields it creates (Fig. 8.10).

When using layered retinal three-dimensional models and numerical methods, the activation function can be calculated if the scalar electric potentials are known for points along the axons. Note that the activation function describes a necessary but not sufficient condition for neural stimulation.

In the case of the GCL, the axons we look to stimulate are assumed perpendicular to the surface of the retina, so the activation function has been calculated along that direction. Note that while in general cathodic stimulation is desirable to anodic stimulation [14], in this particular case, because we assume that activation is taking place at the small portion of axon coming from the ganglion cell towards the electrode, anodic stimulation takes place. Had we positioned the stimulator array so excitation will happen in a segment of axon going away from the electrode instead of coming towards it, or had we considered that activation happened in a segment

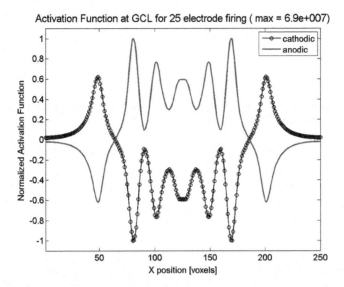

Fig. 8.10 Plot of normalized activation function corresponding to GCL from slice Y = 125. In this case, all 25 electrodes of the array are injecting current. The peaks in the figure align under a row of electrodes. The five central peaks correspond to one electrode each, and the lateral peaks correspond to the place where the dielectric backing of the electrode ends, allowing current flow to go up towards the vitreous and the current return at the back of the implant assembly. Traces for the anodic and cathodic stimulation are shown. From this plot we can conclude that for this particular stimulator configuration, since the activation function is greater than zero under the electrodes in the anodic case, stimulation happens during the anodic phase of the pulse. In addition, during the cathodic phase the side peaks have a magnitude comparable to the electrode peaks for the anodic case – this is a reason of concern, hinting that perhaps a wider dielectric should be used for this array, to allow currents to spread through a larger volume and obtain a lower value for the activation function at that place

of axon going away from the stimulating electrode, cathodic stimulation would have taken place instead.

One way to think about why anodic stimulation takes place in this configuration is to consider that in anodic stimulation the electric potential at the stimulating electrode is positive with respect to all other points in the model; that implies that for an axon having its action potential propagation direction coming towards the stimulating electrode, the potential will be lower at points farther away from the stimulating electrode. The first and second derivatives of the voltage along the axon taken in the direction of action potential propagation will then be positive, making the activation function positive (Fig. 8.11).

8.4   Safety of Implant

The standards concerning electrical exposure safety [9, 10] describe different types of electromagnetic interactions with living tissue and report thresholds beyond which the body will react to the stimulus. Since the intention of the implant is to

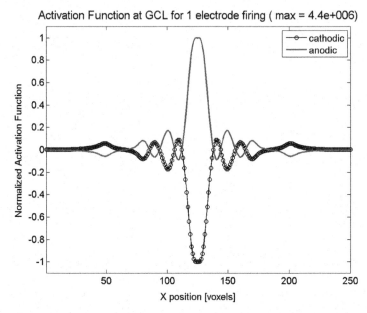

Fig. 8.11 Plot of normalized activation function corresponding to GCL from slice Y = 125, for the case of only the central electrode injecting current. As the previous case, the activation function takes positive values during the anodic phase of the pulse, indicating possible stimulation. Note that during the cathodic phase the side peaks are small compared with the side peaks in the simulation firing all 25 electrodes

reach those threshold values, the safety standards must be interpreted in this case as a general guideline of the order of reasonable values for stimulation and not as a set limit for exposure.

With reference to low-frequency electrical stimulation, the coupling mechanisms include conduction electric current, polarization of bond charges, and reorientation of existing dipoles [10]. In the case of retinal implants, because of the very low stimulation frequencies, conduction currents will be more noticeable than displacement currents.

8.5 Conclusion

Bioelectromagnetic modeling applied to simulating retinal implants is a complex and multidisciplinary topic. Variations of the admittance and impedance methods are suitable to solve this type of problems, but can only provide results as good as the underlying model. Understanding if a particular implant and excitation configuration will be successful at triggering action potentials in the target neurons requires taking into account the excitation patterns, surrounding tissue anatomy, and the mechanics of triggering action potentials in the target neural cells.

Part of what makes solving these systems very challenging is the fact that the size of the model tends to be large in comparison with the minimum feature size, resulting in extremely large linear systems, which can only be solved by using iterative algebraic methods. In addition, the material properties and consequently the current and voltage magnitudes involved can vary multiple orders of magnitude, making the convergence of the system harder for iterative solvers. Some of the models we are currently working with involve matrices having over 50 million rows and columns. These systems are being solved using multi-resolution techniques and sparse iterative linear solvers [17, 18].

The configurations analyzed in this article considered an intraocular current return. Characterizing performance for different current return configurations in epiretinal implants is complex. Part of the issue is that biological tissue in the area is arranged in layers, and the range of conductivities involved varies by several orders of magnitude. Further, if the current return electrode is implanted extraocularly and the eye retain movement after surgery, current densities will vary with the position of the eye as well. In general, as the electrode array is pressed into the retina, the current injected tend to penetrate the retinal surface under each of the active electrodes regardless of the current return configurations. The currents injected for each electrode will then seek a path towards the current return, and any asymmetry in the conductive path from the electrode through the tissue to the current return will result in different current density patterns; areas having shadows of lower current densities will appear. This shadow effect is more pronounced with larger and denser electrode arrays, and it is hard to characterize. Some of the current return related factors that affect performance include current return shape, size, material, surrounding tissue structures, and distance.

References

1. Armitage DW, LeVeen HH, Pethig R (1983), *Radiofrequency-induced hyperthermia: computer simulation of specific absorption rate distributions using realistic anatomical models.* Phys Med Biol, **28**(1): p. 31–42.
2. Brummer SB, Roblee LS (1983), *Criteria for selecting electrodes for electrical stimulation: Theoretical and practical considerations.* Ann NY Acad Sci, **405**: p. 159–171.
3. Ebert M, Brown PK, Lazzi G (2003), *Two-dimensional SPICE-linked multiresolution impedance method for low frequency electromagnetic interactions.* IEEE Trans Biomed Eng, **50**(7): p. 881–889.
4. Gabriel S, Lau RW, Gabriel C (1996), *The dielectric properties of biological tissues: II Measurements in the frequency range 10 Hz to 20 GHz.* Phys Med Biol, **41**(11): p. 2251–2269.
5. Gandhi OP, DeFord JF, Kanai H (1984), *Impedance method for calculation of power deposition patterns in magnetically induced hyperthermia.* IEEE Trans Biomed Eng, **BME-31**: p. 644–651.
6. Greenberg RJ, Velte TJ, Humayun MS, et al. (1999), *A computational model of electrical stimulation of the retinal ganglion cell.* IEEE Trans Biomed Eng, **46**(5): p. 505–514.
7. Hodgkin AL, Huxley AF (1952), *A quantitative description of membrane current and its application to conduction and excitation in nerve.* J Physiol, **117**(4): p. 500–544.

8. Humayun MS, De Juan Jr E, Weiland JD, et al. (1999), *Pattern electrical stimulation of the human retina.* Vision Res, **39**: p. 2569–2576.

9. IEEE International Committee on Electromagnetic Safety (2005), *IEEE standard for safety levels with respect to human exposure to radio frequency electromagnetic fields, 3 kHz to 300 GHz.* IEEE Std C95.1.

10. International Commission for Non-Ionizing Radiation Protection (1998), *Guidelines for limiting exposure to time-varying electric, magnetic, and electromagnetic fields (up to 300 GHz).* Health Phys **74**: p. 494–522.

11. Karwoski CJ, Frambach DA, Proenza LM (1985), *Laminar profile of resistivity in frog retina.* J Neurophysiol **54**(6): p. 1607–1619.

12. Mahadevappa M, Weiland JD, Yanai D, et al. (2005), *Perceptual thresholds and electrode impedance in three retinal prostheses subjects.* IEEE Trans Neural Syst Rehabil Eng, **13**(2): p. 201–206.

13. Rattay F (1986), *Analysis of models for external stimulation of axons.* IEEE Trans Biomed Eng, **BME-33**: p. 974–978.

14. Rattay F (1989), *Analysis of models for extracellular fiber stimulation.* IEEE Trans Biomed Eng, **BME-36**(7): p. 676–682.

15. Rattay F (1991), *Electrical Nerve Stimulation: Theory, Experiments and Applications.* Springer, New York.

16. Rodieck RW (1988), *The primate retina.* Comp Primate Biol, **4**: p. 203–278.

17. Schmidt S, Cela CJ, Singh V, et al. (2007), *Computational modeling of electromagnetic and thermal effects for a dual-unit retinal prosthesis: inductive telemetry, temperature increase, and current densities in the retina.* Artificial Sight, eds. Humayun MS, Weiland JD, Chader G, et al.: Springer, Berlin.

18. Cela CJ (2010), A Multiresolution Admittance Method for Large-Scale Bioelectromagnetic Interactions. PhD dissertation, North Carolina State University, Raleigh, NC.

Chapter 9
Neurotransmitter Stimulation for Retinal Prosthesis: The Artificial Synapse Chip

Raymond Iezzi and Paul G. Finlayson

Abstract Retinal prostheses may one day improve the lives of hundreds of thousands of patients with retinitis pigmentosa (RP) or millions of blind patients with advanced age-related macular degeneration (ARMD), depending on their effectiveness. While considerable progress has been made in electrical stimulation of the retina, herein we explore some possible alternatives to electrical stimulation for retinal prosthesis. Since neurotransmitters normally shape visual responses, some groups have been developing visual prostheses based upon the spatially and temporally controlled delivery of neurotransmitters to the retina. This chapter examines the possibilities for utilizing these chemical messengers, as a means to effectively stimulate retinal ganglion cells and produce vision along established visual information channels.

Abbreviations

5HT	5-Hydoxytryphan, serotonin
AGB	1-Amino-4-guanidobutane
AMPA	α-Amino-3-hydroxyl-5-methyl-4-isoxazole-propionate
EAAT	Excitatory amino acid transporters
GABA	Gamma-aminobutyrate
iGluR	Ionotropic glutamate receptor (GluR1, GluR2, GluR3, GluR4)
INL	Inner nuclear layer
IPL	Inner plexiform layer
mGlur	Metabotropic glutamate receptor
NMDA	N-methyl-D-aspartate
OPL	Outer plexiform layer
P#	Postnatal day
PR	Photoreceptors

R. Iezzi (✉)
Department of Ophthalmology, Mayo Clinic, 200 First Street, SW,
Rochester, MN 55905, USA
e-mail: iezzi.raymond@mayo.edu

G. Dagnelie (ed.), *Visual Prosthetics: Physiology, Bioengineering, Rehabilitation*,
DOI 10.1007/978-1-4419-0754-7_9, © Springer Science+Business Media, LLC 2011

RCS   Royal College of Surgeons
RD1   Retinal degeneration type 1 mouse
RGC   Retinal ganglion cell
RP   Retinitis pigmentosa
S334ter   Opsin gene bearing a termination codon at residue 334

9.1 Pathophysiology of Retinal Degeneration

Two major classes of retinal disorders, retinitis pigmentosa and age-related macular degeneration, result in the loss of vision, due to progressive loss of photoreceptors (PR). Retinitis pigmentosa is a term used to designate diverse genetic disorders [15, 34, 38, 64] that vary in their hereditary linkage – autosomal recessive, autosomal dominant, sex linked, mitochondrial or digenic, and in the underlying genetic mutations (see Chap. 3). Although, the onset, rate and type of PR loss vary between these genetic deficits, they all result in a progressive loss of photoreceptors. It also appears that several different factors, including genetic mutations play a role in PR degeneration in ARMD [14, 33, 56, 67] (see also Chap. 3). The diverse etiologies of RP and ARMD suggest that a single treatment will likely not be possible. Animal models of retinitis pigmentosa indicate that although further neurodegeneration and reorganization in the remaining neural retina occurs (see below), much of the rich network within the retina remains intact for extended periods of time. This presents the opportunity to produce visual sensations through the artificial stimulation of the degenerated retina.

9.2 Modes of Interneuronal Communication Within the Normal Retina

Although, excitatory (glutamate) and inhibitory (GABA and glycine) amino acids are the major neurotransmitter systems in the retina, other transmitters, including acetylcholine, serotonin, dopamine and a variety of neuropeptides shape the visual response (Table 9.1). There is a large diversity of receptors on retinal cell somata and dendrites in the inner and outer plexiform layers (IPL and OPL) and retinal ganglion cell layer (Table 9.2). The outer and inner plexiform layers are near enough to the subretinal and epiretinal surfaces, respectively, for effective activation by application of exogenous agents. In addition, the diversity and location of receptors may allow for differential stimulation of pathways, such as OFF and ON.

Table 9.1 Major transmitters released by retinal cells

Cell type	Transmitters
Photoreceptors	Glutamate
Horizontal	GABA
Bipolar cells	Glutamate
Amacrine	
AII	GABA, glycine
A4	Glycine
A8	Glycine
A10	GABA
A17, A18, A20	GABA, serotonin
A18	GABA, dopamine
A22	GABA, substance P
Starburst	GABA, acetylcholine
Retinal ganglion cells (RGC)	Glutamate

9.2.1   Outer Plexiform Layer

In the OPL, glutamate release from photoreceptors directly modulates horizontal and bipolar cell responses. Horizontal cells exhibit AMPA type receptors, depolarize to glutamate release from PR, and reciprocally modulate photoreceptor responses through GABAergic neurotransmission. Horizontal cell transmission plays a major role in the surround inhibition, but not by GABAergic inhibition [36, 65, 104]. GABA released from horizontal cell has a depolarizing effect on photoreceptors via $GABA_A$ receptors [54, 55], and modulates the temporal properties of light responses [43, 110]. Bipolar cell responses are dependent on the glutamate receptor types they express: OFF-center bipolar (human: flat midget bipolar (FMB)) express ionotropic glutamate receptors (AMPA) and hyperpolarize to light (like photoreceptors); whereas rod and ON-center cone bipolar cells (invaginating midget bipolar [IMB]), depolarize in response to light, due to decreased activation of g-protein coupled metabotropic glutamate, mGluR6 receptors [70, 72, 94, 95], which invert the signal from photoreceptors. Therefore, glutamate in the OPL differentially activates ON and OFF pathways.

9.2.2   Inner Plexiform Layer

Transmitter and interneuronal signaling in the IPL is more complex, but a simplification of the major interactions can be used to examine possible pathways for artificially stimulating the retina.

Table 9.2 Simplified transmitter/receptor basis of retinal interactions

Postsynaptic cell receptors		Presynaptic cells					
		Photoreceptors	Horizontal	Bipolar cells			Amacrine
				Cone-ON	Cone-OFF	Rod	AII, A4, A8, A10, A17, A18, A20, A22, starburst
Photoreceptors			GABA				
Horizontal		AMPA					
Bipolar cells	Cone-ON	mGlur6					Gap junctions
	Cone-OFF	AMPA					
	Rod	mGlur6					
Amacrine	AII, A4, A8, A10, A17, A18, A20, A22, starburst			AMPA?			Gap junctions; GABA; glycine; mGluR; dopamine; other receptors
RGCs	ON			AMPA, NMDA			
	OFF				AMPA, NMDA		
	ON–OFF			AMPA, NMDA	AMPA, NMDA		

Blank cells no known/normal synaptic contacts from cell type above to cell type at left

9.2.2.1   Bipolar Cell Excitation of Retinal Ganglion Cells

Glutamate is released from cone bipolar cell ribbon synapses in the IPL, where it directly activates retinal ganglion cells (RGCs), through AMPA, NMDA and kainate receptors [10, 12, 42, 58, 59, 109], and also excites amacrine cells [3]. Rod bipolar cells do not directly contact RGCs, but contact amacrine cells (review: [2, 93]). Amacrine (AII) cells contact cone bipolar cells and other AII cells through gap junctions. Therefore, depolarization of AII amacrine cells by ionotropic (AMPA and kainate) glutamate receptors leads to depolarization of cone bipolar cells and excitation of RGCs.

9.2.2.2   Amacrine Cell Modulation of Signal Processing

The many different types of amacrine cells in the IPL perform different functional roles, including lateral inhibition, and contributing to spatial tuning, direction selectivity and center surround receptive fields [13, 24, 88, 101]. For example, direction selectivity of RGCs involve GABA, likely originating in starburst amacrine cells [48]. Amacrine cells modulate bipolar, ganglion and other amacrine cells by releasing glycine, GABA, biogenic amines (5HT, dopamine and acetylcholine) and neuropeptides in the IPL. In addition to ionotropic glutamate excitation, a subset of amacrine cells are likely modulated by metabotropic (mGlur1, mGlur2/3, mGlur5) receptors [9, 39, 49].

9.2.2.3   Inhibitory Transmitters

Inhibitory transmitters differentially affect ON, OFF and rod pathways, based on cell type, pre- or post-synaptic action and current duration of $GABA_A$, $GABA_C$ and glycine receptors [19–21, 68, 106]. $GABA_B$ receptor subunits R1a are found in the INL and RGC layer, while R1b is only found in the RGC layer [111, 112]. RGCs exhibit spontaneous and light evoked GABA and glycine responses [102], and are inhibited by GABA and glycine released by amacrine cells [24, 57]. In addition, OFF-bipolar cells are predominantly inhibited by glycine. Rod bipolar cells also receive glycinergic inputs but their major inhibitory response is through GABA acting on $GABA_C$ receptors, which have slow kinetics, while ON bipolar cells are inhibited through $GABA_A$ receptors.

9.2.2.4   Acetylcholine and Dopamine

Acetylcholine is also released from amacrine cells [40]. Both muscarinic [32, 33, 110] and nicotinic cholinergic receptors are found in retina (chick IPL and RGC layer [46]. Nicotinic alpha$_7$ receptors are also localized on bipolar, amacrine and ganglion cells in rabbit retina [17]. Nicotinic receptors are also expressed by a subset of the starburst cells [47]. Acetylcholine (Ach) excites RGCs, particularly Y cells [18, 44, 46, 53, 63, 91], and a role for Ach has been implicated in direction-sensitivity in the retina.

Dopamine released by amacrine cells regulates the spread of activity through gap junctions in the retina. Dopamine D1 receptors decrease the conductance of gap junctions between amacrine cells and bipolar cells [35, 107]. Therefore connections are dynamically regulated in photopic and scotopic light conditions by dopamine [4, 5].

9.2.2.5 Neuropeptides

Amacrine cells also produce a number of neuropeptides, including substance P, somatostatin, vasoactive intestinal peptide (VIP), neuropeptide Y (NPY), corticotropin releasing factor (CRF) and opiates. The roles of peptides in retinal processing are less well understood, and due to the long-term instability of proteins, and complications in exogenous application of peptides, they are not likely to be useful in a neurotransmitter-based prosthesis.

9.2.2.6 Putative neurotransmitters for retinal prosthesis

The neurotransmitter and gap–junction interactions in the IPL and ganglion cell layer (GCL) provide a variety of means to stimulate the retina, possibly in a more naturalistic way. Glutamate application to the retina can directly excite RGCs, and indirectly activate RGCs through the amacrine-bipolar-RGC pathway. In addition, glutamate stimulation may activate amacrine pathways which are used for feature detection. Activation of amacrine cells can modulate many retinal processing pathways. Acetylcholine may also be effective in selectively activating large ganglion cells such as the Y or type A RGC. In addition, GABA or glycine application could reduce activity and may also evoke rebound activity at the offset of application [30, 96].

9.3   Neurophysiological Changes in Retinal Degeneration

An important consideration for any retinal prosthesis is how retinal function is affected beyond photoreceptor loss due to neurodegeneration and reorganization. Degenerative changes in biophysical and morphological cell properties, reorganization of connections, endogenous transmitter release, and transmitter receptor alterations have been observed in animal models of retinitis pigmentosa [60–62]. Such changes may affect the excitability of RGCs to exogenously applied neurotransmitters. Late stages of retinal degeneration have been shown to severely limit RGC stimulation via electrical charge, as thresholds for eliciting electrically evoked cortical potentials increase and will likely impact the efficacy of neurotransmitter stimulation [45, 76].

Various animal models of RP express many similarities, but differ in time course of degenerative and physiological changes. Photoreceptor loss in the pink-eyed RCS rat (rdy+/rdy+) is apparent by postnatal day 20 (P20), progresses rapidly to only a few nuclear layers by P40, and is nearly complete by P100 [6, 50, 69]. The S344-ter rat has a true rhodopsin gene mutation and therefore is an important model for studying human RP. Different lines of S344-ter rats exhibit different rates of progressive photoreceptor loss. In the rd1 mouse model, which has a mutation in phosphodiesterase [7], PRs exhibit a rapid loss of in the first 2–3 postnatal weeks [22]. This early loss of PR is associated with abnormal development of bipolar mGLUR6 receptors, and an early remodeling both in horizontal cells, which exhibit atrophy of terminal dendrites, and in rod bipolar cells, where photoreceptor directed dendrites do not develop [99]. However, amacrine cells do not appear to be affected [100], and in recent work from the same group the many types of RGCs also exhibit normal morphology in rd10 mice [99]. Bipolar and other cell remodeling occurs in stage 3, with onset varying with molecular deficit. In the RCS, s334ter, and P23H rat models, remodeling is relatively late in the disease with the onset on or after P270 [60, 62].

Visual function in RCS rats based on electroretinogram (ERG) recordings [8, 28, 75, 79, 90] shows a progressive loss of rod function to near total loss by P100. Cone function, although declining, can be measured up to P200 [85]. Visual receptive fields in pigmented RCS rats are recorded in the superior colliculus up to P180, albeit with expected increases in threshold [90]. Thus, even after substantial loss of visual function due to photoreceptor loss, RGCs are relaying information to the central nervous system.

Studies on degenerated retinas have in part focused on the changes in neurotransmitter levels and glutamate receptors. Glutamate and aspartate are reduced by approximately 50% in RCS rats at 23 weeks of age [77], and this is likely to be a consequence of photoreceptor loss. GABA is reduced to a lower extent, while glycine levels increase in 23 week RCS rats [77]. However, other studies found that both GABA and glycine levels increase in degenerating retinas [23, 78, 92]. In addition, of the transmitters used by amacrine cells, dopamine is reduced by approximately 50%, but acetylcholine levels are not affected [77]. The reduced dopamine levels correspond with a loss of dopaminergic amacrine cells associated with retinal degeneration [16, 23]. A reduction or loss of many subunits of NMDA receptors (NR1, NR2A-D) has been found in RCS rat by P120 [29]. However, decreased expression of NMDA NR1 subunits in IPL was also observed in congenic non-pigmented rats compared to brown Norway [29]. Kainate binding sites also decrease by P180 in the IPL and OPL of RCS rats [98]. Excitation of RGCs can be shown in response to activation of AMPA, kainate and NMDA receptors [10, 58, 59, 109]. AMPA receptors subunit mRNA for GluR2, GluR3 and GluR4 increase in degenerating retinas of rd1 mice by P40, but the flop:flip ratio (the ratio of the two AMPA receptor splice variants which affect binding and currents evoked by glutamate) is unchanged [71]. The levels of GluR1 mRNA do not change, but the flop:flip ratio of GluR1 (flip responses have slower desensitization and a greater steady-state component) does not exhibit the normal increase between p10 and p40 [71].

The activation of RGCs by exogenous glutamate may also be affected by excitation of bipolar and amacrine cells. Bipolar cells express either mGluR or kainite glutamate receptors. Kainate receptor expression in the IPL and OPL is high at early stages of development (P17) and decreases by postnatal 180 days in pink-eyed

RCS rats [98]. Messenger RNA for metabotropic glutamate receptors (mGlur6), which are likely expressed by ON bipolar cells, increase in the INL of pink-eyed RCS rats at the longest (P60 and P120) periods examined [1], suggesting an up-regulation of these signal-inverting receptors. Physiological studies of RGCs in vitro have found many changes between p20 and p100, although there are conflicting results. Extracellular recordings in whole mount RCS rat retinas [86] demonstrated an increase in spontaneous activity up to P100, which coupled with decreased responses to light, resulted in significantly lower signal-to-noise levels. These investigators noted a predominance of cells with "OFF" responses by P47, and a decrease in receptive field size by P36. Intracellular recordings of RGCs in dystrophic RCS rat retinal slices, however, demonstrated a decrease in the number of cells with sustained responses. Action potentials could not be evoked in 62% of RGCs from 9 to 12-week old animals [11].

Recent studies on functional glutamate receptors in retinal degeneration [61] based on cellular uptake of organic cations (AGB) found significant and differential changes in retinal cell glutamate responses. In two models of retinal degeneration due to loss of photoreceptors, rodless/coneless mice (*rd/rd cl*) and rhodopsin knock-in mutation model mice (*brboG*), a severe loss of glutamate (kainate) sensitivity of bipolar cells was found in the late, stage 3 of degeneration and remodeling. Glutamate still activates amacrine and ganglion cells, although reduced in this late stage of degeneration. However, in small islands where apparently non-functional cones survive, bipolar cells exhibit ionotropic glutamate responses. A high number of bipolar cells activated by kainate suggest that rod bipolar cells begin to express iGluR, in comparison to normally expressing mGluR [61]. In addition, AGB uptake suggests that some amacrine and ganglion cells exhibit increased activity. In a single retinal sample from the posterior pole of a human male RP patient with 90–100% rod PR loss and remodeled cone PRs, all inner retinal cell types exhibited a robust glutamatergic response [61].

Overall, evidence from numerous studies indicates that despite decreased number and possibly excitability, surviving RGCs in retinas undergoing photoreceptor degeneration can transmit information to the brain. Glutamate receptor changes may reduce the efficacy of exogenous glutamate application, but this needs to be examined experimentally at specific points during the degenerative process.

9.4   Rationale for a Neurotransmitter-Based Retinal Prosthesis

Transmitter application may be a more effective and naturalistic means of conveying visual information to the brain, than other methods such as electrical stimulation. The effect of stimulation will depend on the location of application. In addition all types of retinal prosthesis and vision restoration strategies must be designed to stimulate the retina according to the type and stage of retinal degeneration.

Subretinal application of neurotransmitters, such as glutamate, in the normal retina would inhibit ON and activate OFF ganglion cells, but these physiological effects would be superimposed upon the effects of continuous glutamate release from photoreceptors. In eyes with retinal degeneration, however, exogenous glutamate application could replace endogenous glutamate release lost due to PR cell death. Pulsatile glutamate application would activate OFF ganglion cells and inhibit ON cells. Following glutamate application, the disinhibition of ON bipolar cells, may elicit a rebound response. Therefore, differential stimulation of OFF and ON pathways could be achieved, but the signals would be reversed – i.e. OFF cells respond first during stimulation, and ON-cells respond at the offset. Continuous release of glutamate with reductions to mimic light responses could mimic normal photoreceptor releases. However, this would likely produce too high of a glutamate load on cellular systems which clear glutamate from the extracellular space, such as excitatory amino acid transmitter pumps (EAATs). The distance from the sub-retinal surface and OPL in normal retina is over $100\,\mu m$, whereas after PR loss this distance can be less than $50\,\mu m$.

Neurotransmitter application at the epiretinal surface can stimulate the retina by activating receptors in the ganglion cell layer (40–$60\,\mu m$ from surface) and in the IPL (60–$75\,\mu m$ from the surface). Epiretinal glutamate or acetylcholine application could directly activate RGCs through receptors on their somata, which are within $50\,\mu m$ of the surface. Glutamate could also stimulate receptors in synapses within the IPL, including RGC dendritic fields, bipolar-ON cell synapses in IPL b, bipolar-OFF cell in IPL a, rod bipolar-amacrine in IPL, and amacrine cells. Epiretinal application of GABA or glycine could be used to inhibit RGCs and amacrine cells. This may be useful if for example RGCs become highly active in degenerated retinas, as those observed in rd1 mice [97]. Inhibition of amacrine cells could result in disinhibition of other cells, including RGCs in the surround area, due to decreased inhibitory transmitter release. Our preliminary results indicate that ganglion cells exhibit robust excitatory responses to exogenously applied glutamate in 180 day RCS and s334ter line 4 rats. We also have observed that spontaneous firing rates of RGCs in these animals range from absent to high in degenerating retinas.

9.4.1 Limitations of Electrical Stimulation

Prostheses based on electrical stimulation of the retina have been under development over the past two decades. Testing in acute humans studies have had limited success in providing useful vision. Chronic human experiments have been limited to low-resolution devices, since large electrodes are required to handle the high currents required to stimulate degenerated retinal tissues. Small-diameter electrodes, required for a high-resolution prosthesis, are prone to failure due to high charge-densities that erode metals and stimulation voltages that often exceed those required to dissociate water. These facts make small-diameter electrodes more capable of inducing retinal tissue damage from free radicals that are toxic to the lipid membranes of neurons and glia. Further limiting the efficacy of current stimulation methods is the fact that electricity cannot selectively stimulate specific types of visual pathways (e.g. ON and OFF channels) within the visual system. Thus, at this point, electricity cannot encode important sensory features used in normal central visual processing.

9.4.2   Requirements and Benefits of Neurotransmitter Stimulation

Many of these limitations could be circumvented by using more naturalistic means of stimulating retinal ganglion cells (RGCs) for a retinal prosthesis. Natural vision is encoded as neurotransmitter signals. Neurotransmitter-based retinal prosthesis designs will enable us to design and build a device based upon the physiological requirements for RGC stimulation by exogenous neurotransmitters in retinal degeneration. Our preliminary results show that glutamate is effective in stimulating retinal ganglion cells. RGC responses to exogenously applied glutamate are brief, since excitatory amino acid transmitter transporter systems (EAATs) rapidly remove glutamate from the extracellular space. A neurotransmitter-based retinal prosthesis could also take advantage of other transmitters, simultaneously. For example, OFF responses may be mimicked by applying inhibitory transmitters such as glycine or GABA. By applying these inhibitory neurotransmitters adjacent to areas of glutamate stimulation, we may be able to simulate visual contrast. This approach is not possible with electrical stimulation alone. Finally, effective prostheses may use both transmitter and electrical stimulation, synergistically. However, the parameters for stimulating RGCs using glutamate and other neurotransmitters in diseased retinas have not been established.

9.5   Technical Considerations and Design Approaches

9.5.1   Operating Principles for a Neurotransmitter-Based Retinal Prosthesis

Retinal prosthetic devices produce artificial vision by replacing the function of photoreceptors lost due to retinal degeneration. Ideally, these devices pattern afferent stimulation to the remaining retinal neurons, in spatially and temporally naturalistic patterns. Electrically based retinal prosthetic devices initiate neuronal stimulation by inducing local electrical fields that activate voltage-gated ion channels. Neurotransmitter-based devices are capable of directly stimulating or inhibiting neurons by selectively activating ligand-gated ion channels. This requires stimulation hardware capable of accurately modulating the localized delivery of neurotransmitters in space and time. While microelectronic circuitry for the control of release has evolved considerably over many decades, methods for delivering neurotransmitters with these devices are still within relatively early stages of development.

9.5.2   Establishing a Retinal Prosthesis/Synaptic Interface

9.5.2.1   The Proximity Requirement

Prior to the fabrication of microfluidic devices for retinal prosthesis, the general requirements for retinal stimulation via neurotransmitters must the considered. It should be noted that inter-neuronal communication occurs primarily at the synapse. Thus, neurotransmitter-based retinal prosthesis devices must localize their delivery to retinal layers that contain synapses for the target cells of interest. Proximity between target dendrites and sites of neurotransmitter delivery is critical for two primary reasons. First, diffusion is a relatively slow process that will increase the latency between stimulation and response, significantly reducing the effective stimulus update rate. Taking into account the tissue tortuosity factor, the coefficient for diffusion of L-glutamate, the primary excitatory retinal neurotransmitter, at 37°C is approximately $10 \times 10^{-6} \, \text{cm}^2 \text{s}^{-1}$ [37, 74]. This translates to a linear diffusion rate of approximately 33 µm/s. Thus, if the site of neurotransmitter release is 33 µm away from the target dendrites, the response latency will be 1 s. Limited to diffusional delivery, neurotransmitter-based retinal prostheses would be constrained to very low frame rates. Proximity is also critical for efficient delivery of neurotransmitter to target synapses. The concentrations required to elicit neuronal responses to the exogenous application of L-glutamate are relatively high (see discussion below). Thus, diffusional dilution over longer distances would necessitate higher total doses of L-glutamate. In addition, excitatory amino acid transmitter pumps actively remove L-glutamate from the extracellular space. This is desirable in that these pumps rapidly dampen neuronal responses to the exogenous application of L-glutamate, improving the dynamic range, spatial and temporal resolution of response. However, if there is poor proximity between stimulation sites and target dendrite populations, these pumps may increase the threshold quantity of L-glutamate release required to achieve neuronal stimulation.

The proximity requirement for neurotransmitter-based retinal prostheses may necessitate that these devices penetrate into dendritic retinal sublaminae of the inner or outer plexiform layers. The concept of chemically inducing neurons to extend synaptic contacts to a retinal prosthesis has been proposed [51, 52, 66]. Epiretinal or sub-retinal neurotransmitter-based retinal prostheses or versions of these devices that penetrate into the retina could, incorporate drug-delivery methods to release chemo-attractant molecules that induce the migration of dendrites toward stimulation sites. The loss of afferent input to bipolar cells due to photoreceptor cell loss in retinal degeneration does induce bipolar cells to re-direct their dendrites toward the inner retina where they have been reported to create self-stimulation loop circuits [60–62]. Thus, there may be a period of time during which these de-afferented bipolar cells may be induced to synapse upon a sensory substitution implant. This may occur as a consequence of the sensory substitution, itself. Or, perhaps the controlled release of growth factors from a retinal prosthetic device could provide a signal to dendrites that would promote the extension and maintenance of synapses to the device. Retinal ganglion cells maintain their synaptic contacts with their afferent bipolar and amacrine cells and do not become de-afferented as a consequence of the retinal degeneration. Thus, it may be more difficult to induce these cells to alter their well established dendritic organization.

9.5.2.2 Convective Delivery of Neurotransmitters Via Microfluidics

To overcome the temporal constraints of neurotransmitter diffusion some retinal prosthesis designs employ microfluidic technology capable of convective delivery. Two groups have worked on the development of microfluidic devices, capable of the controlling the release of neurotransmitter in space and time. Iezzi and colleagues at Wayne State University first introduced the concept of a microfluidic neurotransmitter-based retinal prosthetic device [41]. Devoid of valves, the design employs the use of phototriggered neurotransmitters. These neurotransmitters do not activate ligand-gated ion channels prior to their flash photolysis. The "uncage and release" device employs microfluidic channels that incorporate an optical subsystem for the spatially and temporally controlled activation of phototriggered neurotransmitters. An electrical current is then used to iontophoretically and/or electro-osmotically eject the charged, uncaged neurotransmitter from a microfluidic aperture or microneedle into close proximity to the target dendrites. This design involves storing a reservoir of caged L-glutamate prodrug and involves optical and electrical means for controlled release. This potentially minimizes the possibility of a dose-related L-glutamate induced excitoxicity. Finlayson and Iezzi [80] have shown that the localized convective delivery of L-glutamate via pneumatic ejection results in linear RGC dose-response firing with response latencies of 200 ms. These preliminary results validate the utility of convective neurotransmitter delivery for retinal prosthesis.

Another group at Stanford University has also developed microfluidic circuits that employ electroosmotic flow for the controlled delivery of neurotransmitters in space and time. They have demonstrated that electric field-driven fluid ejection of bradykinin was effective in stimulating PC-12 cells cultured on the stimulation system [81–84].

9.5.2.3 Functionalized Surfaces for Neurotransmitter Stimulation

Pepperberg and associates have been developing functionalized surfaces coated with tethered neurotransmitters for neuronal stimulation [73, 89, 105, 108]. According to the design concept, an electrical or other control signal will modulate the capacity of tethered molecules to bind to synaptic or extra-synaptic neurotransmitter receptors. Neurotransmitter analogs such as the muscimol, bound to biotin for the future purpose of adsorption to surfaces, rendering them "functionalized" have been shown to activate GABA receptors in an oocyte model. Since the neurotransmitter–biotin conjugates will ultimately be adsorbed to the surface of the implant, solid posts could be used to assure that stimulation occurs within the desired retinal layers.

9.5.2.4 Synaptic Requirements for L-Glutamate Mediated Neuronal Stimulation

Any system for delivering neurotransmitters to the retina for the purpose of retinal prosthesis will be required to match doses of L-glutamate required by target neurons. Consequently, an analysis of the anatomy and physiology of the synapse may be useful in establishing operating parameters for neurotransmitter-based retinal stimulators.

The requirements for neurotransmitter stimulation of the retina differ according to the target cells for stimulation. ON and OFF pathways are first established at the bipolar cell level. Thus, stimulation at this level may permit selective ON and OFF stimulation selectivity. Depending upon whether the retinal prosthesis is placed epiretinal or subretinal, microneedles may be necessary to deliver neurotransmitter to target neuronal cell dendrites. In degenerating retina, bipolar cells that have lost their photoreceptor input redirect their afferent dendrites toward the inner plexiform layer (IPL). Within the IPL RGC afferents synapse. Neurotransmitter stimulation directed toward ganglion cells must reach this region. Within the IPL, it may be possible to stimulate bipolar cell dendrites and/or RGCs directly.

The rate of quantal excitation to RGCs in response to visual stimulation has been examined. Any neurotransmitter-based retinal prosthesis will need to mimic patterns of quantal excitation induced by visual stimulation. Freed determined that the just-maximal sustained RGC response to visual stimulation was induced by 3,700 quanta of L-glutamate per second, among all synapses [25, 26]. Studies of the number of L-glutamate molecules per synaptic vesicle report a range between 500 and 10,000 [87]. Thus, between 1.85 and 37×10^6 L-glutamate molecules per second would be required to induce a sustained RCG response. Freed and Sterling reported that there are approximately 550 bipolar synapses upon an ON alpha-RGC in the area centralis [27]. At 10° eccentricity, the larger membrane surface area of ON alpha-RGCs causes them to have approximately 2,200 bipolar cell synapses, since the density of bipolar cell synapses on the membrane is constant [25, 26]. Based upon a synapse diameter of 200 nm² and a synaptic cleft of 20 nm, the volume of each synapse is approximately 2.5 al [103]. Thus, the total synaptic volume for a single ON alpha-RGC ranges between 1.38 fl near the area centralis and 5.5 fl at 10° eccentricity. Using the lowest molar quantity of L-glutamate needed for sustained RGC stimulation, combined with the largest total synaptic volume for an ON alpha RGC we arrive at a predicted minimum molar concentration of L-glutamate necessary for stimulation by a neurotransmitter-based retinal prosthesis of 0.55 mM L-glutamate. By taking the higher molar quantity of L-glutamate from the above computations, divided by the smallest total synaptic volume for an ON alpha-RGC, we predict that the upper concentration for L-glutamate required for sustained stimulation is 11.1 mM. This range is consistent with our unpublished experimental findings for RGC stimulation via exogenous application of L-glutamate in normal Sprague–Dawley, RCS and S334-ter-4 rats.

9.6 Summary

A neurotransmitter-based retina prosthesis is a feasible option for restoration of visual function in humans with retinal degeneration. The diverse and differential actions of glutamate, GABA, glycine and acetylcholine on surviving retinal cells allow for both excitatory and inhibitory stimulation of the retina. These neurotransmitters differentially influence a number cell types that underlie feature detection processed in the retina. Although, pathophysiological changes in retinal degeneration may reduce the effectiveness of stimulation, neurotransmitter based prostheses offer the ability to activate retinal circuits, or suppress hyperactive ones. Technical considerations such as diffusion and EAATs ensure that exogenous local application of transmitters will affect a small restricted area of retina with responses that are temporally dampened. In addition, since passing axons are not stimulated, a neurotransmitter-based retinal prosthesis can maintain a high spatial specificity, even at suprathreshold stimulation. Technological advances in electrical prosthesis will aid in the development of a neurotransmitter based prosthesis, since existing circuits may be used to control electro-osmotic ejection of transmitters. In addition, there are new advances in other drug delivery technologies, such as caging and tethering molecules, which may be adapted for a neurotransmitter based retinal prosthesis. In conclusion, neurotransmitters offer a promising new approach to stimulation for retinal prosthesis, which may also supplement and piggy-back upon existing technology.

References

1. Armata IA, Giompres P, Smith A, et al. (2006), *Genetically induced retinal degeneration leads to changes in metabotropic glutamate receptor expression.* Neurosci Lett, **393**(1): p. 12–7.
2. Bloomfield SA, Dacheux RF (2001), *Rod vision: pathways and processing in the mammalian retina.* Prog Retin Eye Res, **20**(3): p. 351–84.
3. Bloomfield SA, Dowling JE (1985), *Roles of aspartate and glutamate in synaptic transmission in rabbit retina. II. Inner plexiform layer.* J Neurophysiol, **53**(3): p. 714–25.
4. Bloomfield SA, Xin D (1997), *A comparison of receptive-field and tracer-coupling size of amacrine and ganglion cells in the rabbit retina.* Vis Neurosci, **14**(6): p. 1153–65.
5. Bloomfield SA, Xin D, Osborne T (1997), *Light-induced modulation of coupling between aII amacrine cells in the rabbit retina.* Vis Neurosci, **14**(3): p. 565–76.
6. Bok D, Hall MO (1971), *The role of the pigment epithelium in the etiology of inherited retinal dystrophy in the rat.* J Cell Biol, **49**(3): p. 664–82.
7. Bowes C, Li T, Danciger M, et al. (1990), *Retinal degeneration in the rd mouse is caused by a defect in the beta subunit of rod cGMP-phosphodiesterase.* Nature, **347**(6294): p. 677–80.
8. Bush RA, Hawks KW, Sieving PA (1995), *Preservation of inner retinal responses in the aged Royal College of Surgeons rat. Evidence against glutamate excitotoxicity in photoreceptor degeneration.* Invest Ophthalmol Vis Sci, **36**(10): p. 2054–62.
9. Cai W, Pourcho RG (1999), *Localization of metabotropic glutamate receptors mGluR1alpha and mGluR2/3 in the cat retina.* J Comp Neurol, **407**(3): p. 427–37.
10. Chen S, Diamond JS (2002), *Synaptically released glutamate activates extrasynaptic NMDA receptors on cells in the ganglion cell layer of rat retina.* J Neurosci, **22**(6): p. 2165–73.

11. Chen ZS, Yin ZQ, Chen S, Wang SJ (2005), *Electrophysiological changes of retinal ganglion cells in Royal College of Surgeons rats during retinal degeneration.* Neuroreport, **16**(9): p. 971–5.

12. Cohen ED, Miller RF (1994), *The role of NMDA and non-NMDA excitatory amino acid receptors in the functional organization of primate retinal ganglion cells.* Vis Neurosci, **11**(2): p. 317–32.

13. Cook PB, McReynolds JS (1998), *Lateral inhibition in the inner retina is important for spatial tuning of ganglion cells.* Nat Neurosci, **1**(8): p. 714–9.

14. de Jong PT (2006), *Age-related macular degeneration.* N Engl J Med, **355**(14): p. 1474–85.

15. Delyfer MN, Leveillard T, Mohand-Said S, et al. (2004), *Inherited retinal degenerations: therapeutic prospects.* Biol Cell, **96**(4): p. 261–9.

16. Djamgoz MB, Hankins MW, Hirano J, Archer SN (1997), *Neurobiology of retinal dopamine in relation to degenerative states of the tissue.* Vision Res, **37**(24): p. 3509–29.

17. Dmitrieva NA, Strang CE, Keyser KT (2007), *Expression of alpha 7 nicotinic acetylcholine receptors by bipolar, amacrine, and ganglion cells of the rabbit retina.* J Histochem Cytochem, **55**(5): p. 461–76.

18. Downing JE, Kaneko A (1992), *Cat retinal ganglion cells show transient responses to acetylcholine and sustained responses to l-glutamate.* Neurosci Lett, **137**(1): p. 114–8.

19. Eggers ED, Lukasiewicz PD (2006), *GABA(A), GABA(C) and glycine receptor-mediated inhibition differentially affects light-evoked signalling from mouse retinal rod bipolar cells.* J Physiol, **572**(Pt 1): p. 215–25.

20. Eggers ED, Lukasiewicz PD (2006), *Receptor and transmitter release properties set the time course of retinal inhibition.* J Neurosci, **26**(37): p. 9413–25.

21. Eggers ED, McCall MA, Lukasiewicz PD (2007), *Presynaptic inhibition differentially shapes transmission in distinct circuits in the mouse retina.* J Physiol, **582**(Pt 2): p. 569–82.

22. Farber DB, Flannery JG, Bowes-Rickman C (1994), *The rd mouse story: Seventy years of research on an animal model of inherited retinal degeneration.* Vis Neurosci, **13**: p. 31–64.

23. Fletcher EL (2000), *Alterations in neurochemistry during retinal degeneration.* Microsc Res Tech, **50**(2): p. 89–102.

24. Flores-Herr N, Protti DA, Wassle H (2001), *Synaptic currents generating the inhibitory surround of ganglion cells in the mammalian retina.* J Neurosci, **21**(13): p. 4852–63.

25. Freed MA (2000), *Parallel cone bipolar pathways to a ganglion cell use different rates and amplitudes of quantal excitation.* J Neurosci, **20**(11): p. 3956–63.

26. Freed MA (2000), *Rate of quantal excitation to a retinal ganglion cell evoked by sensory input.* J Neurophysiol, **83**(5): p. 2956–66.

27. Freed MA, Sterling P (1988), *The ON-alpha ganglion cell of the cat retina and its presynaptic cell types.* J Neurosci, **8**(7): p. 2303–20.

28. Fulton AB (1983), *Background adaptation in RCS rats.* Invest Ophthalmol Vis Sci, **24**(1): p. 72–6.

29. Grunder T, Kohler K, Guenther E (2001), *Alterations in NMDA receptor expression during retinal degeneration in the RCS rat.* Vis Neurosci, **18**(5): p. 781–7.

30. Grunfeld ED, Spitzer H (1995), *Spatio-temporal model for subjective colours based on colour coded ganglion cells.* Vision Res, **35**(2): p. 275–83.

31. Gupta N, Drance SM, McAllister R, et al. (1994), *Localization of M3 muscarinic receptor subtype and mRNA in the human eye.* Ophthalmic Res, **26**(4): p. 207–13.

32. Gupta N, McAllister R, Drance SM, et al. (1994), *Muscarinic receptor M1 and M2 subtypes in the human eye: QNB, pirenzipine, oxotremorine, and AFDX-116 in vitro autoradiography.* Br J Ophthalmol, **78**(7): p. 555–9.

33. Haddad S, Chen CA, Santangelo SL, Seddon JM (2006), *The genetics of age-related macular degeneration: a review of progress to date.* Surv Ophthalmol, **51**(4): p. 316–63.

34. Hamel C (2006), *Retinitis pigmentosa.* Orphanet J Rare Dis, **1**: p. 40.

35. Hampson EC, Vaney DI, Weiler R (1992), *Dopaminergic modulation of gap junction permeability between amacrine cells in mammalian retina.* J Neurosci, **12**(12): p. 4911–22.

36. Hare WA, Owen WG (1996), *Receptive field of the retinal bipolar cell: a pharmacological study in the tiger salamander.* J Neurophysiol, **76**(3): p. 2005–19.

37. Hille B (1992), *Ionic Channels of Excitable Membranes.* Sunderland, MA: Sinauer.

38. Hims MM, Diager SP, Inglehearn CF (2003), *Retinitis pigmentosa: genes, proteins and prospects.* Dev Ophthalmol, **37**: p. 109–25.

39. Hoffpauir BK, Gleason EL (2002), *Activation of mGluR5 modulates GABA(A) receptor function in retinal amacrine cells.* J Neurophysiol, **88**(4): p. 1766–76.

40. Hutchins JB, Hollyfield JG (1987), *Cholinergic neurons in the human retina.* Exp Eye Res, **44**, 363–376.

41. Iezzi R, Auner G, McAllister P, Abrams GW (2003), *Method and apparatus for activating molecules to stimulate neurological tissue,* United States patent US 6, 668, 190.

42. Ikeda H, Kay CD, Robbins J (1989), *Properties of excitatory amino acid receptors on sustained ganglion cells in the cat retina.* Neuroscience, **32**(1): p. 27–38.

43. Kamermans M, Werblin F (1992), *GABA-mediated positive autofeedback loop controls horizontal cell kinetics in tiger salamander retina.* J Neurosci, **12**(7): p. 2451–63.

44. Kaneda M, Hashimoto M, Kaneko A (1995), *Neuronal nicotinic acetylcholine receptors of ganglion cells in the cat retina.* Jpn J Physiol, **45**(3): p. 491–508.

45. Kent TL, Glybina IV, Abrams GW, Iezzi R (2008), *Chronic intravitreous infusion of ciliary neurotrophic factor modulates electrical retinal stimulation thresholds in the RCS rat.* Invest Ophthalmol Vis Sci, **49**(1): p. 372–9.

46. Keyser KT, Britto LR, Schoepfer R, et al. (1993), *Three subtypes of alpha-bungarotoxin-sensitive nicotinic acetylcholine receptors are expressed in chick retina.* J Neurosci, **13**(2): p. 442–54.

47. Keyser KT, MacNeil MA, Dmitrieva N, et al. (2000), *Amacrine, ganglion, and displaced amacrine cells in the rabbit retina express nicotinic acetylcholine receptors.* Vis Neurosci, **17**(5): p. 743–52.

48. Kittila CA, Massey SC (1997), *Pharmacology of directionally selective ganglion cells in the rabbit retina.* J Neurophysiol, **77**(2): p. 675–89.

49. Koulen P, Kuhn R, Wassle H, Brandstatter JH (1997), *Group I metabotropic glutamate receptors mGluR1alpha and mGluR5a: localization in both synaptic layers of the rat retina.* J Neurosci, **17**(6): p. 2200–11.

50. LaVail MM, Mullen RJ (1976), *Role of the pigment epithelium in inherited retinal degeneration analyzed with experimental mouse chimeras.* Exp Eye Res, **23**(2): p. 227–45.

51. Lee CJ, Blumenkranz MS, Fishman HA, Bent SF (2004), *Controlling cell adhesion on human tissue by soft lithography.* Langmuir, **20**(10): p. 4155–61.

52. Leng T, Wu P, Mehenti NZ, et al. (2004), *Directed retinal nerve cell growth for use in a retinal prosthesis interface.* Invest Ophthalmol Vis Sci, **45**(11): p. 4132–7.

53. Lipton SA, Aizenman E, Loring RH (1987), *Neural nicotinic acetylcholine responses in solitary mammalian retinal ganglion cells.* Pflugers Arch, **410**(1–2): p. 37–43.

54. Lukasiewicz PD, Shields CR (1998), *Different combinations of GABAA and GABAC receptors confer distinct temporal properties to retinal synaptic responses.* J Neurophysiol, **79**(6): p. 3157–67.

55. Lukasiewicz PD, Shields CR (1998), *A diversity of GABA receptors in the retina.* Semin Cell Dev Biol, **9**(3): p. 293–9.

56. MacDonald IM, Lines MA (2004), *Genetics and ARMD.* CMAJ, **170**(10): p. 1518–9.

57. Majumdar S, Heinze L, Haverkamp S, et al. (2007), *Glycine receptors of A-type ganglion cells of the mouse retina.* Vis Neurosci, **24**(4): p. 471–87.

58. Marc RE (1999), *Kainate activation of horizontal, bipolar, amacrine, and ganglion cells in the rabbit retina.* J Comp Neurol, **407**(1): p. 65–76.

59. Marc RE (1999), *Mapping glutamatergic drive in the vertebrate retina with a channel-permeant organic cation.* J Comp Neurol, **407**(1): p. 47–64.

60. Marc RE, Jones BW (2003), *Retinal remodeling in inherited photoreceptor degenerations.* Mol Neurobiol, **28**(2): p. 139–47.

61. Marc RE, Jones BW, Anderson JR, et al. (2007), *Neural reprogramming in retinal degeneration.* Invest Ophthalmol Vis Sci, **48**(7): p. 3364–71.
62. Marc RE, Jones BW, Watt CB, Strettoi E (2003), *Neural remodeling in retinal degeneration.* Prog Retin Eye Res, **22**(5): p. 607–55.
63. Masland RH, Ames A, 3 rd (1976), *Responses to acetylcholine of ganglion cells in an isolated mammalian retina.* J Neurophysiol, **39**(6): p. 1220–35.
64. Maubaret C, Hamel C (2005), *Genetics of retinitis pigmentosa: metabolic classification and phenotype/genotype correlations.* J Fr Ophtalmol, **28**(1): p. 71–92.
65. McMahon MJ, Packer OS, Dacey DM (2004), *The classical receptive field surround of primate parasol ganglion cells is mediated primarily by a non-GABAergic pathway.* J Neurosci, **24**(15): p. 3736–45.
66. Mehenti NZ, Tsien GS, Leng T, et al. (2006), *A model retinal interface based on directed neuronal growth for single cell stimulation.* Biomed Microdevices, **8**(2): p. 141–50.
67. Michaelides M, Hunt DM, Moore AT (2003), *The genetics of inherited macular dystrophies.* J Med Genet, **40**(9): p. 641–50.
68. Molnar A, Werblin F (2007), *Inhibitory feedback shapes bipolar cell responses in the rabbit retina.* J Neurophysiol, **98**(6): p. 3423–35.
69. Mullen RJ, LaVail MM (1976), *Inherited retinal dystrophy: primary defect in pigment epithelium determined with experimental rat chimeras.* Science, **192**(4241): p. 799–801.
70. Nakajima Y, Iwakabe H, Akazawa C, et al. (1993), *Molecular characterization of a novel retinal metabotropic glutamate receptor mGluR6 with a high agonist selectivity for l-2-amino-4-phosphonobutyrate.* J Biol Chem, **268**(16): p. 11868–73.
71. Namekata K, Okumura A, Harada C, et al. (2006), *Effect of photoreceptor degeneration on RNA splicing and expression of AMPA receptors.* Mol Vis, **12**: p. 1586–93.
72. Nawy S, Jahr CE (1990), *Suppression by glutamate of cGMP-activated conductance in retinal bipolar cells.* Nature, **346**(6281): p. 269–71.
73. Nehilla BJ, Popat KC, Vu TQ, et al. (2004), *Neurotransmitter analog tethered to a silicon platform for neuro-BioMEMS applications.* Biotechnol Bioeng, **87**(5): p. 669–74.
74. Nicholson C, Phillips JM (1981), *Ion diffusion modified by tortuosity and volume fraction in the extracellular microenvironment of the rat cerebellum.* J Physiol, **321**: p. 225–57.
75. Noell WK, Pewitt EB, Cotter JR (1989), *ERG of the pigmented rdy rat at advanced stages of hereditary retinal degeneration.* Prog Clin Biol Res, **314**: p. 357–75.
76. O'Hearn TM, Sadda SR, Weiland JD, et al. (2006), *Electrical stimulation in normal and retinal degeneration (rd1) isolated mouse retina.* Vision Res, **46**(19): p. 3198–204.
77. Okada M, Okuma Y, Osumi Y, et al. (2000), *Neurotransmitter contents in the retina of RCS rat.* Graefes Arch Clin Exp Ophthalmol, **238**(12): p. 998–1001.
78. Orr HT, Cohen AI, Carter JA (1976), *The levels of free taurine, glutamate, glycine and gamma-amino butyric acid during the postnatal development of the normal and dystrophic retina of the mouse.* Exp Eye Res, **23**(4): p. 377–84.
79. Perlman I (1978), *Dark-adaptation in abnormal (RCS) rats studied electroretinographically.* J Physiol, **278**: p. 161–75.
80. Finlayson PG, Iezzi R (2010), Glutamate stimulation of retinal ganglion cells in normal and S334ter rat retinas: a candidate for a neurotransmitter-based retinal prosthesis. Invest Ophthalmol Vis Sci, **51**: p. 3619–28.
81. Peterman MC, Bloom DM, Lee C, et al. (2003), *Localized neurotransmitter release for use in a prototype retinal interface.* Invest Ophthalmol Vis Sci, **44**(7): p. 3144–9.
82. Peterman MC, Mehenti NZ, Bilbao KV, et al. (2003), *The artificial synapse chip: a flexible retinal interface based on directed retinal cell growth and neurotransmitter stimulation.* Artif Organs, **27**(11): p. 975–85.
83. Peterman MC, Noolandi J, Blumenkranz MS, Fishman HA (2004), *Fluid flow past an aperture in a microfluidic channel.* Anal Chem, **76**(7): p. 1850–6.
84. Peterman MC, Noolandi J, Blumenkranz MS, Fishman HA (2004), *Localized chemical release from an artificial synapse chip.* Proc Natl Acad Sci USA, **101**(27): p. 9951–4.

85. Pinilla I, Lund RD, Lu B, Sauve Y (2005), *Measuring the cone contribution to the ERG b-wave to assess function and predict anatomical rescue in RCS rats.* Vision Res, **45**(5): p. 635–41.

86. Pu M, Xu L, Zhang H (2006), *Visual response properties of retinal ganglion cells in the Royal College of Surgeons dystrophic rat.* Invest Ophthalmol Vis Sci, **47**(8): p. 3579–85.

87. Rao-Mirotznik R, Buchsbaum G, Sterling P (1998), *Transmitter concentration at a three-dimensional synapse.* J Neurophysiol, **80**(6): p. 3163–72.

88. Roska B, Nemeth E, Orzo L, Werblin FS (2000), *Three levels of lateral inhibition: A space-time study of the retina of the tiger salamander.* J Neurosci, **20**(5): p. 1941–51.

89. Saifuddin U, Vu TQ, Rezac M, et al. (2003), *Assembly and characterization of biofunctional neurotransmitter-immobilized surfaces for interaction with postsynaptic membrane receptors.* J Biomed Mater Res A, **66**(1): p. 184–91.

90. Sauve Y, Lu B, Lund RD (2004), *The relationship between full field electroretinogram and perimetry-like visual thresholds in RCS rats during photoreceptor degeneration and rescue by cell transplants.* Vision Res, **44**(1): p. 9–18.

91. Schmidt M, Humphrey MF, Wassle H (1987), *Action and localization of acetylcholine in the cat retina.* J Neurophysiol, **58**(5): p. 997–1015.

92. Schmidt SY, Berson EL (1978), *Taurine uptake in isolated retinas of normal rats and rats with hereditary retinal degeneration.* Exp Eye Res, **27**(2): p. 191–8.

93. Sharpe LT, Stockman A (1999), *Rod pathways: the importance of seeing nothing.* Trends Neurosci, **22**(11): p. 497–504.

94. Shiells RA, Falk G (1990), *Glutamate receptors of rod bipolar cells are linked to a cyclic GMP cascade via a G-protein.* Proc Biol Sci, **242**(1304): p. 91–4.

95. Slaughter MM, Miller RF (1981), *2-Amino-4-phosphonobutyric acid: a new pharmacological tool for retina research.* Science, **211**(4478): p. 182–5.

96. Spitzer H, Almon M, Sherman I (1994), *A model for the early stages of motion processing based on spatial and temporal edge detection by X-cells.* Spat Vis, **8**(3): p. 341–68.

97. Stasheff SF (2008), *Emergence of sustained spontaneous hyperactivity and temporary preservation of OFF responses in ganglion cells of the retinal degeneration (rd1) mouse.* J Neurophysiol, **99**(3): p. 1408–21.

98. Stasi K, Naskar R, Thanos S, et al. (2003), *Benzodiazepine and kainate receptor binding sites in the RCS rat retina.* Graefes Arch Clin Exp Ophthalmol, **241**(2): p. 154–60.

99. Strettoi E, Mazzoni F, Damiani D, Novelli E (2008), *Structural consequences of inherited photoreceptor degeneration: a close look to the inner retina,* Invest Ophthalmol Visual Sci, 49, ARVO E-abstr #5181.

100. Strettoi E, Pignatelli V (2000), *Modifications of retinal neurons in a mouse model of retinitis pigmentosa.* Proc Natl Acad Sci USA, **97**(20): p. 11020–5.

101. Taylor WR (1999), *TTX attenuates surround inhibition in rabbit retinal ganglion cells.* Vis Neurosci, **16**(2): p. 285–90.

102. Tian N, Hwang TN, Copenhagen DR (1998), *Analysis of excitatory and inhibitory spontaneous synaptic activity in mouse retinal ganglion cells.* J Neurophysiol, **80**(3): p. 1327–40.

103. Ventriglia F, Di Maio V (2000), *A Brownian simulation model of glutamate synaptic diffusion in the femtosecond time scale.* Biol Cybern, **83**(2): p. 93–109.

104. Verweij J, Kamermans M, Spekreijse H (1996), *Horizontal cells feed back to cones by shifting the cone calcium-current activation range.* Vision Res, **36**(24): p. 3943–53.

105. Vu TQ, Chowdhury S, Muni NJ, et al. (2005), *Activation of membrane receptors by a neurotransmitter conjugate designed for surface attachment.* Biomaterials, **26**(14): p. 1895–903.

106. Wassle H, Koulen P, Brandstatter JH, et al. (1998), *Glycine and GABA receptors in the mammalian retina.* Vision Res, **38**(10): p. 1411–30.

107. Witkovsky P, Schutte M (1991), *The organization of dopaminergic neurons in vertebrate retinas.* Vis Neurosci, **7**(1–2): p. 113–24.

108. Yan C, Matsuda W, Pepperberg DR, et al. (2006), *Synthesis and characterization of an electroactive surface that releases gamma-aminobutyric acid (GABA).* J Colloid Interface Sci, **296**(1): p. 165–77.

109. Yang XL (2004), *Characterization of receptors for glutamate and GABA in retinal neurons.* Prog Neurobiol, **73**(2): p. 127–50.
110. Yang XL, Wu SM (1989), *Effects of prolonged light exposure, GABA, and glycine on horizontal cell responses in tiger salamander retina.* J Neurophysiol, **61**(5): p. 1025–35.
111. Zarbin MA, Wamsley JK, Palacios JM, Kuhar MJ (1986), *Autoradiographic localization of high affinity GABA, benzodiazepine, dopaminergic, adrenergic and muscarinic cholinergic receptors in the rat, monkey and human retina.* Brain Res, **374**(1): p. 75–92.
112. Zhang C, Bettler B, Duvoisin RM (1998), *Differential localization of GABA(B) receptors in the mouse retina.* Neuroreport, **9**(15): p. 3493–7.

Chapter 10
Synthetic Chromophores and Neural Stimulation of the Visual System

Elias Greenbaum and Barbara R. Evans

Abstract This chapter presents an overview of optical stimulation of neural cells by synthetic chromophores and their potential use in the field of artificial sight. The chromophores and techniques that are discussed include azo chromophores, photo release of caged neurotransmitters, pore blockers and photoisomerization, the channelrhodopsins, melanopsin, and the Photosystem I reaction center of green plants.

Abbreviations

ATR   All-trans retinal
ChR   Channel rhodopsin
Cy5   Red-emitting cyanine-based fluorescent dye
DIC   Differential interference contrast microscopy
FITC   Fluorescein isothiocyanate
PSI   Photosystem I reaction center
UV   Ultraviolet light

10.1 Introduction

Rods and cones contain the light-absorbing chromophores of the retina that trigger the primary events of vision. The light absorbing molecule in the discs of rod cells is rhodopsin, comprised of opsin, a protein, and 11-*cis*-retinal, a Vitamin A derivative. As illustrated in Fig. 10.1, absorption of a visible photon triggers the isomerization of

E. Greenbaum (✉)
Oak Ridge National Laboratory, Oak Ridge, TN 37831, USA
e-mail: greenbaum@ornl.gov

G. Dagnelie (ed.), *Visual Prosthetics: Physiology, Bioengineering, Rehabilitation*,
DOI 10.1007/978-1-4419-0754-7_10, © Springer Science+Business Media, LLC 2011

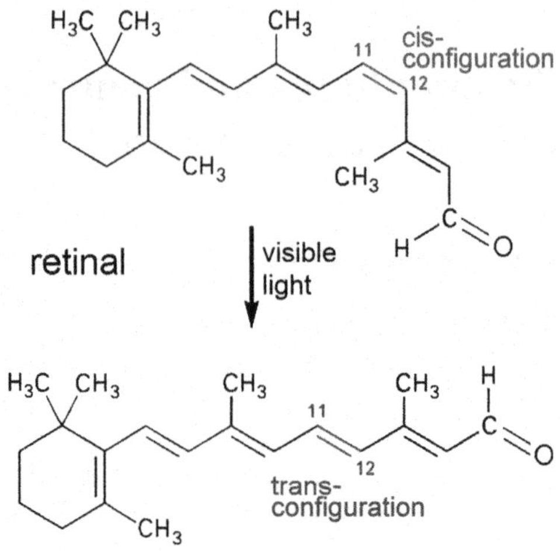

Fig. 10.1 Vision begins by photon absorption in the chromophore 11-*cis*-retinal with is converted to the all *trans* isomer

the 11-*cis* isomer to the all-*trans* isomer. This *cis-trans* isomerization activates a G protein cascade that sets in motion the molecular events in the rod outer segment that result in visual perception [31]. Visual diseases such as age-related macular degeneration or retinitis pigmentosa are characterized by loss of the first step in vision, the phototransduction cascade in the photoreceptor outer segment. Much of the remaining neural pathway from retina to brain remains intact. Stimulation of these surviving retinal neurons is the biomedical engineering basis of multielectrode retinal prosthetic devices [12]. However, considerations of geometry, stability and fabrication of electrodes plus power requirements and the physics of electric field propagation in conductive media place a practical upper limit on the number of electrodes. The use of synthetic chromophores for the optical stimulation of retinal neural cells presents an attractive, if challenging, alternative. Optical stimulation as a tool for studying neural systems is a well established idea. As noted by Zhang et al. "…it will be a physiologist's dream-come-true to simply sit back and let light beams stimulate and assay the operation of a well-defined excitable tissue, such as a neural circuit" [38]. Multiple approaches to optical stimulation of cells that are not normally light-sensitive are known. A logical extension of this work is the application of synthetic chromophores to in vivo stimulation of the visual system. This idea is the molecular analog of multielectrode prosthesis stimulation of neural cells and is the focus of this chapter.

Optical stimulation of neural cells can be viewed in at least two ways: as a useful experimental technique to expand our knowledge of neuroscience and mapping of neural pathways, or as a biomedical engineering approach to the development of molecular prosthetic structures that might be capable of replacing multielectrode visual prosthetic arrays. The contemplated advantages of using molecular chromophores for optical stimulation of the retinal cells are their nanometer size, direct interaction with the neural membrane, and ability for spectral tuning. External power sources, in principle, are not necessarily needed. The application of synthetic chromophores to artificial sight is the molecular analog and possible successor to multielectrode arrays for the stimulation of neural cells. Three broad approaches have been proposed: (1) chemical modification of ion channels, usually with derivatives of the photoisomerizable chromophore azobenzene [2, 32]; (2) photochemical release of signaling molecules [4, 20, 37]; and (3) application of light-sensitive proteins such as channelrhodopsin [3, 13, 19, 21] or Photosystem I reaction centers [10, 14].

10.2   Pioneering Experiments

10.2.1   Stimulation with No Chromophores

We begin by noting one optical stimulation technique that dispenses with synthetic chromophores entirely. Fork reported observation of light-induced neural activity when laser irradiation was applied to the abdominal ganglion of the marine mollusk *Aplysia californica* [9]. Neural cells were impaled with conventional microelectrodes and illuminated with a laser beam with a minimum spot size of 10 μm. Laser stimulation of the cells with blue (488 nm) or green (515 nm) light produced firing with the light pulses "on" in some cases and "off" in others. In other experiments, especially those with the addition of ouabain, firing occurred during the light pulses whereas in others, firing occurred when the laser beam was switched off. In this work, none of the cells was selected for photoreceptor activity. Fork concluded that the laser-induced signals were caused by a mechanism other than damage. However, detailed work by Hirase et al. [11] indicated that relatively low power laser irradiation can produce reactive oxygen species and higher powers can result in membrane damage. Nonetheless, a significant result of Fork's work is that intense local electromagnetic disturbances can induce neural activity in cells.

The intensity of the laser beams used for this prior work was high. For example, a typical response of a silent cell in normal seawater used a 12.5 mW beam at 488 nm. This corresponds to an irradiance of $1.6 \times 10^8 \, \text{W/m}^2$. For the experiments using ouabain, a beam power of 4.5 mW was used, corresponding to an irradiance of $5.7 \times 10^7 \, \text{W/m}^2$. The corresponding solar irradiance at noon on a clear day is of the order $1 \times 10^3 \, \text{W/m}^2$. This early work pointed in the correct direction for optical stimulation of neural cells. The use of synthetic chromophores with tailored light absorbing properties can be expected to greatly reduce the intensity of light that is required to achieve a specific optoneural effect.

10.2.2   Azo Chromophores

One workhorse for optical modulation of neural cells is the azo class of synthetic chromophores that are characterized by the light-induced *trans-cis* isomerization

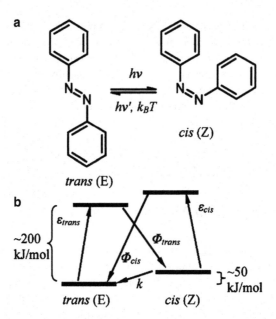

Fig. 10.2 (**a**) Azobenzene can be converted from the *trans* to the *cis* state photochemically, and will revert back to the stable *trans* state thermally. Alternately, the *cis* to *trans* conversion can be effectuated with a distinct wavelength of light. (**b**) Simplified state model for azobenzene chromophores. The extinction coefficients are denoted ε, whereas the quantum yields for the photoisomerizations are labeled Φ. The rate of thermal relaxation is denoted by k. Competition between these pathways determines the composition of the photo-stationary state [from ref. [36], Elsevier © 2006, used with permission]

(and its reversal) about the double nitrogen bond: the azo linkage, –N=N–. The parent molecule for the early studies of this chromophore is azobenzene, illustrated in Fig. 10.2, in which the azo linkage bridges two phenyl rings. Azobenzene and its myriad substituent derivatives are often referred to as "photoswitches." However, the term "switch" is not quite right as the word is commonly understood: a device with two stable states. Azobenzene does not have binary stability. Only the *trans* state is stable. The *cis* state is more energetic by $49 \, \text{kJ} \, \text{mol}^{-1}$ (in heptane) [7]. The rate of decay for substituted derivatives depends on the specific molecule. The lifetime of azobenzenes is on the order of hours, but is considerably less for aminoazobenzenes and pseudo-stilbenes [36]. Continuous irradiation of azobenzene produces mixtures of photostationary states of the *trans* and *cis* isomers whose relative concentration is wavelength dependent. 350 nm light is preferentially absorbed by the *trans* isomer and populates the *cis* state, whereas 450 nm light accelerates conversion of the *cis* form back to the *trans* state [30].

Lester and Nerbonne provided a succinct summary of the way in which physiological systems can be manipulated by light: a physiological parameter is monitored while photochemistry is used to alter the physiology of the system being monitored [18]. Deal, Erlanger and Nachmansohn showed how carbamylcholine-produced depolarization of the excitable membrane of the monocellular electroplaques preparation of *Electrophorus* can be regulated by light [6]. They worked with two photoisomerizable compounds: (1) *N-p*-phenylazophenyl-*N*-phenylcarbamylcholine chloride and (2) *p*-phenylazophenyltrimethylammonium chloride. The *trans* photostationary state of each predominates under 420 nm light, whereas the *cis* version is the majority species under 320 nm irradiation. Both isomers inhibit depolarization of the membrane. However, the *trans* isomer is a stronger inhibitor than the *cis* isomer. The result of Deal et al. is an early example of photoregulation of the potential difference across an excitable membrane by exposing electroplaques to light of appropriate wavelengths in the presence of a solution of carbamylcholine and either of the two compounds. The work illustrated coupling a *cis-trans* isomerization, the first step in the initiation of a visual impulse, with substantial changes (20–30 mV) in the potential difference across an excitable membrane.

Lester et al. [17] prepared a covalently bound photoisomerizable agonist and compared it with reversibly bound agonists at *Electrophorus* electroplaques. Light-flash experiments with tethered *3*-(α-bromomethyl)-3′-[α-(trimethylammonium) methyl]azobenzene (QBr) resemble those with the reversible photoisomerizable agonist, 3,3′,*bis*-[α-(trimethylammonium)methyl]-azobenzene (Bis-Q): the conductance is increased by *cis → trans* photoisomerizations and decreased by *trans → cis* photoisomerizations. As with Bis-Q, light-flash relaxations had the same rate constant as voltage-jump relaxations. Receptors with tethered *cis*-QBr have a channel duration severalfold briefer than with the tethered *trans* isomer. By comparing the agonist-induced conductance with the *cis/trans* ratio, Lester et al. concluded that each channel's activation is determined by the configuration of a single tethered QBr molecule.

Balasubramanian et al. embedded azobenzene and azobenzene-*p*-carboxylic acid methyl ester in a model membrane system that was the microemulsion obtained by the dispersion of water and hexadecane using amphipathic potassium oleate as the emulsifier and hexanol as the cosurfactant [1]. This work demonstrated light-induced alteration of the electrical conductivity in the birefringent lamellar multibilayer system that was attributed to the optical activity of the azo chromophores. This work also demonstrated light-induced alteration of the ester hydrolytic activity of α-chymotrypsin dissolved in the membrane containing azobenzene and azobenzene ester separately. Other applications of azo chromophores to biological systems have been studied [16, 28, 29, 35].

10.3   Current Research

10.3.1   Caged Neurotransmitters

As illustrated in Fig. 10.3, when illuminated with UV light or by multiphoton excitation [33], caged amino acid neurotransmitters are converted into biologically active amino acids that can rapidly initiate neurotransmitter action. These caged probes

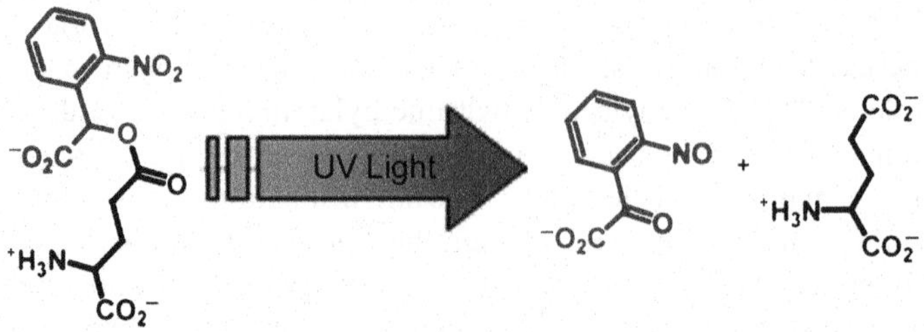

Fig. 10.3 UV light or multiphoton excitation of caged amino acid neurotransmitters can be converted into biologically active amino acids

provide a means of controlling the release – both spatially and temporally – of agonists for kinetic studies of receptor binding or channel opening. The technique of rapid light-induced release of signaling molecules [34] descends from the flash photolysis experiments of Norrish and Porter [24]. Calloway and Katz pioneered a photochemical approach for high-spatial-resolution mapping of functional circuitry in living mammalian brain slices [4]. Photostimulation was achieved by bathing brain slices in a molecularly caged form of the neurotransmitter glutamate [L-glutamic acid alpha-(4,5-dimethoxy-2-nitrobenzyl) ester], which was then converted to the active form by brief pulses (<1 ms) of ultraviolet irradiation. Using this technique, the locations of neurons making functional synaptic connections to a single neuron were revealed by photostimulation of highly restricted areas of the slice (50–100 μm in diameter) while maintaining a whole-cell recording of the neuron of interest.

10.3.2   Pore Blocker and Photoisomerization

Building on the work of Lester et al. [17], Banghart et al. [2] used structure-based design to develop a chemical gate that confers light sensitivity to an ion channel for remote control of neuronal firing using a pore blocker and photoisomerizable azobenzene structure. Figure 10.4 illustrates the basic idea. Bistable positioning of the pore blocker was achieved with light of two different wavelengths. Absorption of a 500 nm photon triggered a *cis-trans* isomerization. The ~1.7 nm length of the *trans* isomer moved the blocker to the pore of the ion channel. Conversely, absorption of a near UV 380 nm photon triggered a *trans-cis* isomerization, The ~1.0 nm length of the *cis* isomer caused retraction of the pore blocker. The light-activated gate was covalently linked to the ion channel and the ion channel was integral to the neuronal cell membrane. The control over individual neurons was spatially accurate and did not rely on diffusible ligands. Also, the gate could be reversibly photo switched, allowing recurrent control of neural activity. Inside-out patches from an oocyte were treated with 100 μM of the triad maleimide + azo linkage chromophores – triethanolamine for 30 min. The patch showed a large Shaker current in 380 nm light (*cis*) and almost complete block in 500 nm light (*trans*). Current block in the dark followed a biexponential time course with $\tau_1 = 0.49$ min and $\tau_2 = 4.79$ min.

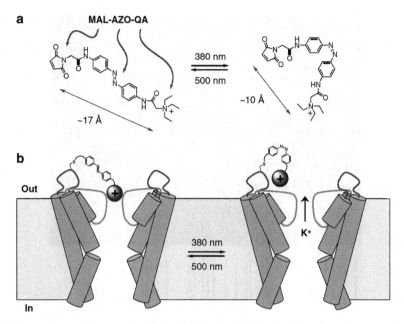

Fig. 10.4 Application of azobenzene chromophores to the gating of ionic currents through modified Shaker channels. MAL is maleimide for cystine tethering. QA is a quaternary ammonium group to block the channel [from ref. [2], Nature Publishing Co., © 2004, used with permission]

Volgraf et al. [32] applied the azobenzene technique to a ligand-gated ion channel, the ionotropic glutamate receptor (iGluR). Using structure-based design, they modified the ligand-binding domain to develop a light-activated channel. An agonist was covalently tethered to the protein through an azobenzene moiety, which functioned as the optical switch. The agonist was reversibly presented to the binding site upon photoisomerization, initiating domain closure and channel gating. Photoswitching occurred on a millisecond timescale, with channel conductances that reflect the photostationary state of the azobenzene at a given wavelength.

10.3.3   The Channelrhodopsins

Nagel et al. have shown that Channelrhodopsins 1 and 2 (ChR1 and ChR2) are involved in generation of photocurrents of the green alga *Chlamydomonas reinhardtii*. ChR1 is a light-gated proton channel [22]. ChR2, on the other hand, is a directly light-switched cation-selective ion channel [23]. It opens rapidly after absorption of a photon to generate a large permeability for monovalent and divalent cations. Nagle et al. have demonstrated that ChR2 may be used to depolarize cells by illumination [23]. Boyden et al. [3] and Li et al. [19] achieved temporally precise, noninvasive control in well-defined neuronal populations by adapting the naturally occurring algal protein

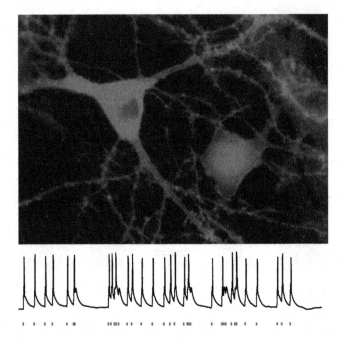

Fig. 10.5 Neurons expressing yellow fluorescent protein-tagged Channelrhodopsin-2 and a voltage trace showing photo-stimulation-elicited spikes. From D. Evanko (2005), Nat Methods, 2: p. 726–7. Nature Pub. Co. © 2005. Used with permission

Channelrhosopsin-2 (ChR2), a rapidly gated light-sensitive cation channel. The technique used lentiviral gene delivery in combination with high-speed optical switching to photostimulate mammalian neurons. The work demonstrated reliable, millisecond-timescale control of neuronal spiking, as well as control of excitatory and inhibitory synaptic transmission. Figure 10.5 illustrates neurons expressing yellow fluorescent protein-tagged Channelrhodopsin-2 and a voltage trace showing photo-stimulation-elicited spikes. The first 315 amino-acid residues of *C. reinhardtii* Channelrhodopsin-2 coupled to retinal can be used to impart fast photosensitivity [3, 13, 19, 23]. ChR2 is a seven-transmembrane protein with a molecule of all-trans retinal (ATR) bound at the core as a photosensor [23]. Upon illumination with ~470 nm blue light, ATR triggers a conformational change to open the channel pore. Since ChR2 is a light-sensitive ion channel, the expected fast response was indeed observed, within 50 μs of illumination [3]. Combining ChR2 with fast light switching made it possible to activate neurons with the temporal precision of single action potentials [3].

10.3.4 Melanopsin

Another route that has been examined is the increase and relocation of intrinsic mammalian visual receptors through transgenic ecotopic expression to restore photosensitivity to the retina. Melanopsin is a retinal-containing, photosensitive protein which is expressed at low levels in neuronal cells, including a small subset of the retinal ganglia. It does not directly act to generate a membrane potential, but transfers the visual stimulus through a signaling pathway. Utilization of melanopsin would have the advantages of using an intrinsic mammalian retinal protein with a visual pigment similar to that of the natural photoreceptors, but has the disadvantage of a slower light response time. Through transfection with a viral vector construct in adeno-associated virus, high levels of recombinant melanopsin were introduced into the retinal cells of mice homozygous for the *rd* mutation, which results in complete loss of rod photoreceptors. The transduction of the *rd* mice with the ectopic melanopsin restored light response as determined by behavioral tests of live mice and light-stimulated action potentials of isolated retinal cells [39].

10.3.5 Nanoscale Photovoltaics: The Photosystem I Reaction Center

As illustrated in Fig. 10.6, the photosynthetic membranes of green plants contain two molecular photovoltaic structures, Photosystems I and II (PSI and PSII) that are serially connected in an electron transport chain that drive the endergonic reactions of photosynthesis. The bioenergetic properties of PSI have been reviewed by Chitnis [5]. Photon absorption in PSI triggers a charge separation that generates a voltage across the photosynthetic membrane. This voltage is the source of Gibbs energy that drives the energetically uphill reactions of photosynthesis. It is possible to isolate PSI reaction centers and preserve their full photovoltaic properties [8, 15]. It has been proposed to fuse PSI reaction centers in membranes in close proximity to voltage-gated ions channels and to use the photovoltaic properties of PSI to gate these channels [10]. One example of the idea is illustrated in Fig. 10.7. PSI located

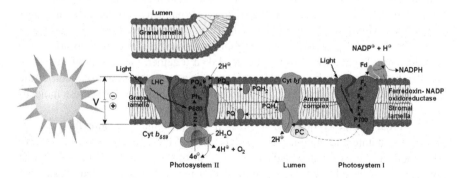

Fig. 10.6 Schematic illustration of the photosynthetic membrane. Photosystems I and II are integral membrane nanoscale molecular photovoltaic structures. PSI can be used to impart photoactivity to mammalian cells

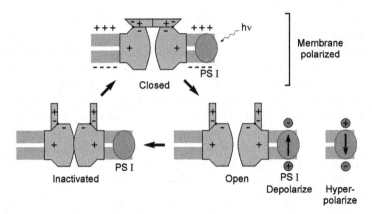

Fig. 10.7 Schematic illustration of PSI adjacent to a voltage-gated ion channel. One or more PSI reaction centers may be able to trigger the ion channels. Depending on orientation, PSI can depolarize or hyperpolarize the membrane

in close proximity to voltage-gated ion channels, either in the membrane or externally at the lipid–water interface, may be capable of generating a local electrical disturbance of sufficient magnitude to create an excitatory postsynaptic potential and generate an action potential. Kuritz et al. [14] have shown that PSI-proteoliposomes incubated with retinoblastoma cells imparted optical activity to the cells as measured by the light-induced slow movement of calcium ions into the cells. Pennisi et al. have performed experimental and theoretical studies on the incorporation of PSI reaction centers in human cells and lipid vesicles [25–27]. In particular, new methods of delivery and detection of PSI in the membrane of human cells have been developed (Fig. 10.8) [27]. Purified fractions of PSI were reconstituted in proteoliposomes that were used as vehicles for the membrane incorporation. A fluorescent impermeable dye was entrapped in the vesicles to qualitatively analyze the nature of the vesicle–cell interaction. After incorporation, the localization and orientation of the complexes in the membrane was studied using immunofluorescence microscopy. The results showed complexes oriented as in native membranes, which were randomly distributed in clusters over the entire surface of the cell. Additionally, analysis of cell viability showed that the incorporation process does not damage the cell membrane. Taken together, the results of this work suggest that the mammalian cellular membrane is a reasonable environment for the incorporation of PSI complexes, which opens the possibility of using these molecular photovoltaic structures for optical control of cell activity.

The direct use of chlorophyll itself as a visual pigment in animals has been observed in nature and attempted in the laboratory. Chlorophyll-like pigments have been found to be associated with the rhodopsin of deep sea fish [40, 41]. These pigments are believed to be used for dim light vision, as this would enable using the red part of the incident light spectrum, which has tenfold less loss due to scattering. Based on these observations, direct utilization of chlorophyll-derived pigments for

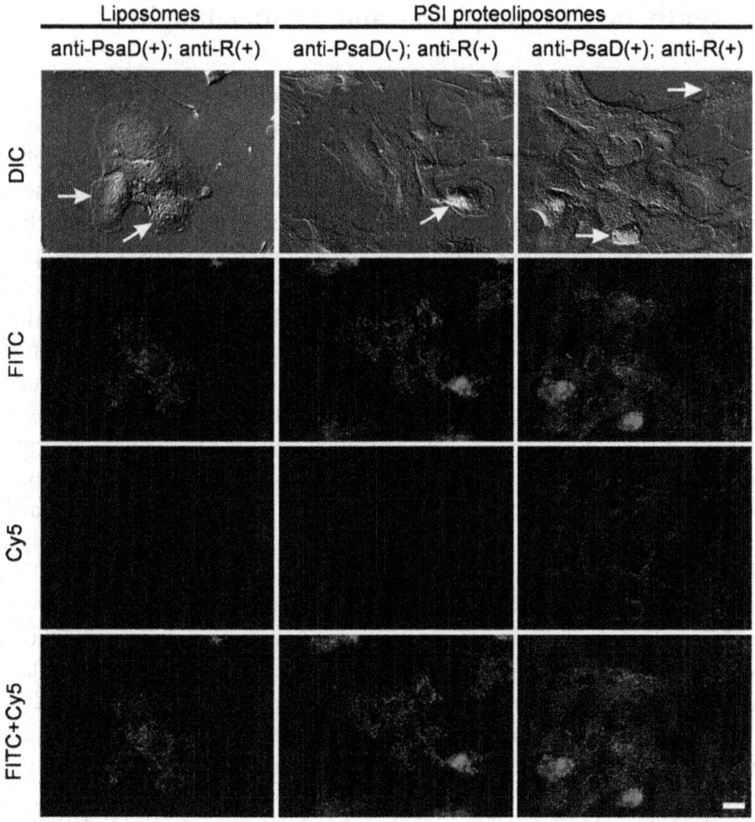

Fig. 10.8 Simultaneous evaluation of pyranine uptake and immunodetection of PSI in adipose tissue-derived stem cells. The images were obtained using differential interference contrasting (DIC; *first row*), and in the fluorescein isothiocyanate (FITC) and Cy5 fluorescent dye channels (*second* and *third row*, respectively). In this experiment, pyranine fluorescence is detected with the FITC channel and the secondary antibody fluorescence with the Cy5 channel. The images corresponding to the Cy5 channel clearly show that PSI complexes are associated with the membrane of the cells from the experimental sample in contrast with the controls. The *fourth row* has the merged images from the fluorescence channels, where it is possible to see that there is no overlap between the FITC channel (indicative of cytoplasmic localization) and the Cy5 channel (indicative of membrane localization). Individual cells that are separated from the rest and can be more clearly visualized are indicated with *arrows* in the DIC images. Scale bar = 30 μm. From [27], with permission. Biomedical Engineering Society © 2008

enhancement of vision in mammals has been attempted [42]. Dark-adapted mice were injected with the water-soluble chlorophyll derivative chlorin-e_6, resulting in its accumulation in the outer segment of the retina. Comparison of the response of chlorin-e_6-injected and control live mice to red light indicated that they responded to red light. The electrophysiological response to red light of the melanopsin-expressing retinal ganglion cells was doubled in intensity compared to controls.

10.4   Synthetic Chromophores and Artificial Sight

We presented a brief review of the three main techniques for imparting optical activity to mammalian cells: (1) light-mediated untethering (or "uncaging") of chemically modified signaling molecules; (2) chemical modification of ion channels and receptors to render them light-responsive and (3) introduction of light-sensitive proteins into nonphotoactive cells. These are post-electrode prostheses techniques and ideas in laboratory methodology for the study of excitable cells. In principle, they offer the ability to target multiple cells of a specific class simultaneously. External electrodes are limited in their spatial resolution for heterogeneous tissue. Although intracellular electrodes can target specific neurons, they don't lend themselves to simultaneous targeting of multiple cells of a specific subclass. Moreover, mechanical electrodes are intrusive structures in the context of excitable tissue. Intelligently designed molecular scale activators powered by photon absorption in synthetic chromophores can, in principle, blend into the membranes of excitable tissue with linear dimensions that are compatible with the scale-length and fine structure of the tissue.

The field of synthetic chromophores and its application to artificial sight is motivated by advances that have been made with multielectrode retinal prosthesis arrays. Numerous studies have reported that stimulation of neurons in the visual pathway evokes the perception of light. It is assumed that analogous stimulation at the molecular level will mimic the action of electrodes, with the added advantage of nanoscale resolution and auto-power by the photons that trigger the neural activity. There is, at present, no clinical data to support this assumption. Moreover, in order for synthetic chromophores to be relevant to real-world applications they need to be stable or easily rejuvenated and operate at ambient wavelengths and light intensities, either on their own or in conjunction with optoelectronic signal conditioning devices. These are challenging areas of research and biomedical engineering that are currently in early stages of development.

Acknowledgments   The author thanks M. S. Humayun, J. D. Weiland, T. Kuritz, I. Lee, C. P. Pennisi, C. A. Sanders, B. R. Evans, and H. M. O'Neill for advice, support and discussions. This work was supported by the Office of Biological and Environmental Research, U.S. Department of Energy. Oak Ridge National Laboratory is managed by UT-Battelle, LLC for the U. S. Department of Energy under Contract No. DE-AC05-00OR22725.

References

1. Balasubramanian D, Subramani S, Kumar C (1975), *Modification of a model membrane structure by embedded photochrome*. Nature, **254**(5497): p. 252–4.
2. Banghart M, Borges K, Isacoff E, et al. (2004), *Light-activated ion channels for remote control of neuronal firing*. Nat Neurosci, **7**(12): p. 1381–6.
3. Boyden ES, Zhang F, Bamberg E, et al. (2005), *Millisecond-timescale, genetically targeted optical control of neural activity*. Nat Neurosci, **8**(9): p. 1263–8.

4. Callaway EM, Katz LC (1993), *Photostimulation using caged glutamate reveals functional circuitry in living brain slices*. Proc Natl Acad Sci USA, **90**(16): p. 7661–5.

5. Chitnis PR (2001), *Photosystem I: function and physiology*. Annu Rev Plant Physiol Plant Mol Biol, **52**: p. 593–626.

6. Deal WJ, Erlanger BF, Nachmansohn D (1969), *Photoregulation of biological activity by photochromic reagents. 3. Photoregulation of bioelectricity by acetylcholine receptor inhibitors*. Proc Natl Acad Sci USA, **64**(4): p. 1230–4.

7. Dias AR, Dapiedade MEM, Simoes JAM, et al. (1992), *Enthalpies of formation of cis-azobenzene and trans-azobenzene*. J Chem Thermodyn, **24**(4): p. 439–47.

8. Evans BR, O'Neill HM, Hutchens SA, et al. (2004), *Enhanced photocatalytic hydrogen evolution by covalent attachment of plastocyanin to photosystem I*. Nano Lett, **4**(10): p. 1815–9.

9. Fork RL (1971), *Laser stimulation of nerve cells in aplysia*. Science, **171**(3974): p. 907–8.

10. Greenbaum E, Humayun MS, Kuritz T, et al. (2001), *Application of photosynthesis to artificial sight*. In: *IEEE Engineering in Medicine and Biology*. Istanbul, Turkey: IEEE.

11. Hirase H, Nikolenko V, Goldberg JH, Yuste R (2002), *Multiphoton stimulation of neurons*. J Neurobiol, **51**(3): p. 237–47.

12. Humayun MS, Weiland JD, Chader G, Greenbaum E (2007), *Artificial sight: basic research, biomedical engineering, and clinical advances*. New York: Springer.

13. Ishizuka T, Kakuda M, Araki R, Yawo H (2006), *Kinetic evaluation of photosensitivity in genetically engineered neurons expressing green algae light-gated channels*. Neurosci Res, **54**(2): p. 85–94.

14. Kuritz T, Lee I, Owens ET, et al. (2005), *Molecular photovoltaics and the photoactivation of mammalian cells*. IEEE Trans Nanobioscience, **4**(2): p. 196–200.

15. Lee I, Lee JW, Stubna A, Greenbaum E (2000), *Measurement of electrostatic potentials above oriented single photosynthetic reaction centers*. J Phys Chem B, **104**(11): p. 2439–43.

16. Lee WS, Ueno A (2001), *Photocontrol of the catalytic activity of a beta-cyclodextrin bearing azobenzene and histidine moieties as a pendant group*. Macromol Rapid Commun, **22**(6): p. 448–50.

17. Lester HA, Krouse ME, Nass MM, et al. (1980), *Covalently bound photoisomerizable agonist – comparison with reversibly bound agonists at electrophorus electroplaques*. J Gen Physiol, **75**(2): p. 207–32.

18. Lester HA, Nerbonne JM (1982), *Physiological and pharmacological manipulations with light-flashes*. Annu Rev Biophys Bioeng, **11**: p. 151–75.

19. Li X, Gutierrez DV, Hanson MG, et al. (2005), *Fast noninvasive activation and inhibition of neural and network activity by vertebrate rhodopin and green algae channelrhodopsin*. Proc Natl Acad Sci USA, **102**(49): p. 17816–21.

20. Melyan Z, Tarttelin EE, Bellingham J, et al. (2005), *Addition of human melanopsin renders mammalian cells photoresponsive*. Nature, **433**(7027): p. 741–5.

21. Nagel G, Brauner M, Liewald JF, et al. (2005), *Light activation of channelrhodopsin-2 in excitable cells of Caenorhabditis elegans triggers rapid behavioral responses*. Curr Biol, **15**(24): p. 2279–84.

22. Nagel G, Ollig D, Fuhrmann M, et al. (2002), *Channelrhodopsin-1: a light-gated proton channel in green algae*. Science, **296**(5577): p. 2395–8.

23. Nagel G, Szellas T, Huhn W, et al. (2003), *Channelrhodopsin-2, a directly light-gated cation-selective membrane channel*. Proc Natl Acad Sci USA, **100**(24): p. 13940–5.

24. Norrish RGW, Porter G (1949) *Chemical Reactions produced by very high light intensities*. Nature, **164**: p. 658

25. Pennisi CP, Greenbaum E, Yoshida K (2008), *Spatial distribution of the electric potential from Photosystem I reaction centers in lipid vesicles*. IEEE Trans Nanobiosci, **7**(2): p. 164–71.

26. Pennisi CP, Greenbaum E, Yoshida K (2010), *Analysis of light-induced transmembrane ion gradients and membrane potential in Photosystem I proteoliposomes*. Biophys Chem, **146**: p. 13–24.

27. Pennisi CP, Jensen PE, Zachar V, et al. (2009), *Incorporation of photosynthetic reaction centers in the membrane of human cells: toward a new tool for optical control of cell activity*. Cell Mol Bioeng, **2**(1): p. 156–65.

28. Pieroni O, Houben JL, Fissi A, Costantino P (1980), *Reversible conformational-changes induced by light in poly(l-glutamic acid) with photochromic side-chains*. J Am Chem Soc, **102**(18): p. 5913–5.

29. Sisido M, Ishikawa Y, Itoh K, Tazuke S (1991), *Helically arranged azobenzene chromophores along a polypeptide-chain. 1. synthesis and circular-dichroism*. Macromolecules, **24**(14): p. 3993–8.

30. Standaert RF, Park SB (2006), *Abc amino acids: design, synthesis, and properties of new photoelastic amino acids*. J Org Chem, **71**(21): p. 7952–66.

31. Stryer L (1995), *Biochemistry*, 4 ed. New York: W. H. Freeman and Company. p. 1064.

32. Volgraf M, Gorostiza P, Numano R, et al. (2006), *Allosteric control of an ionotropic glutamate receptor with an optical switch*. Nat Chem Biol, **2**(1): p. 47–52.

33. Wang SS, Khiroug L, Augustine GJ (2000), *Quantification of spread of cerebellar long-term depression with chemical two-photon uncaging of glutamate*. Proc Natl Acad Sci USA, **97**(15): p. 8635–40.

34. Wilcox M, Viola RW, Johnson KW, et al. (1990), *Synthesis of photolabile precursors of amino-acid neurotransmitters*. J Org Chem, **55**(5): p. 1585–9.

35. Willner I, Rubin S (1996), *Control of the structure and functions of biomaterials by light*. Angew Chem Int Ed Engl, **35**(4): p. 367–85.

36. Yager KG, Barrett CJ (2006), *Novel photo-switching using azobenzene functional materials*. J Photochem Photobiol A Chem, **182**(3): p. 250–61.

37. Zemelman BV, Nesnas N, Lee GA, Miesenbock G (2003), *Photochemical gating of heterologous ion channels: remote control over genetically designated populations of neurons*. Proc Natl Acad Sci USA, **100**(3): p. 1352–7.

38. Zhang F, Wang LP, Boyden ES, Deisseroth K (2006), *Channelrhodopsin-2 and optical control of excitable cells*. Nat Methods, **3**(10): p. 785–92.

39. Lin B, Koizumi A, Tanaka N, Panda S, Masland RH (2008), *Restoration of visual function in retinal degeneration mice by ectopic expression of melanopsin*. Proc Natl Acad Sci USA **105**(41): 16009–14.

40. Partridge JC, Douglas RH (1995) *Far-red sensitivity of dragon fish*. Nature **375**: p. 21–2.

41. Douglas RH, Partridge JC, Dulai K, Hunt D, Mullineaux CW, Tauber AY, Hynninen PH (1998), *Dragon fish see using chlorophyll*. Nature, **393**: p. 423–4.

42. Washington I, Zhou J, Jockusch S, Turro NJ, Nakanishi K, Sparrow JR (2007), *Chlorophyll derivatives as visual pigments for super vision in the red*. Photochem Photobiol Sci, **7**: p. 775–9.

Chapter 11
Biophysics/Engineering of Cortical Electrodes

Philip R. Troyk

Abstract This chapter provides a description of how microelectrodes are used to form an artificial interface to the cortex. Microelectrodes inserted into the cortex are called "intracortical electrodes" and are anticipated for use in cortical visual prostheses. Owing to the nature of the cortical environment, the design and use of these electrodes pose challenges for the clinical deployment of cortical prostheses. The combined effects of electrode charge injection and effects of the *in vivo* environment are discussed.

Abbreviations

Ag|AgCl	Silver–silver chloride
AIROF	Activated iridium oxide film
C	Capacitance
CSC	Charge storage capacity
CSC_A	Anodic charge storage capacity
CSC_C	Cathodic charge storage capacity
CV	Cyclic voltammetry
I	Current
Ir	Iridium
IR	Infrared
PEDOT	Polyethylenedioxythiophene
Pt	Platinum
R	Resistance
Redox	Reduction-oxidation chemical reaction
SIROF	Sputtered iridium oxide film
V	Voltage

P.R. Troyk (✉)
Department of Biomedical Engineering,
Pritzker Institute of Biomedical Science and Engineering, Illinois Institute of Technology,
3255 S. Dearborn, WH 314, Chicago, IL 60616, USA
e-mail: troyk@iit.edu

G. Dagnelie (ed.), *Visual Prosthetics: Physiology, Bioengineering, Rehabilitation*,
DOI 10.1007/978-1-4419-0754-7_11, © Springer Science+Business Media, LLC 2011

11.1   Background

Electrical stimulation of the visual cortex has been used for investigative studies of the visual system and visual prostheses research since the first half of the twentieth century. These have all been based upon the biophysical phenomena of passing electrical charge within neural tissue for the purpose of activating the neural networks, and producing visual sensations commonly called phosphenes. For a visual prosthesis, the underlying assumption is that the electrical stimulation can be organized into spatiotemporal patterns that can manipulate the stimulated neural substrate in a manner that exploits the natural tuning properties of the visual system. By strategic patterning of the stimulation, it is assumed that an image, captured by electronic means, can be translated into a visual sensation that mimics biological vision. To date, this assumption remains unproven.

To a large extent, the primary limitation in designing, and deploying, all visual prostheses is the inability to implant an artificial neural interface that accommodates the density and scale of the visual system at the retina, optic nerve, or primary visual cortex. Despite the huge advances seen within the electronics industry, over the past 75 years, over this same time period the state-of-the-art for interfacing to neural tissue has not significantly changed. Electrical currents are passed into neural tissue through metal electrodes placed near the target neural tissue.

For stimulation of the visual cortex, sub-dural electrodes have been previously used on the surface of brain, with limited success [7, 18–20]. Relative to surface electrodes, intracortical electrodes penetrate the visual cortex and use smaller electrical currents in closer physical proximity to the cortical neurons, and it is generally accepted that their use, with exposed tip sizes on the order of the target neurons, has a significantly higher likelihood of success for the design of a cortical visual prosthesis. Compared to epi- and subretinal electrodes, implanted intracortical electrodes are surrounded by a very different medium, and this has contributed to notable differences in their *in vivo* behavior. Owing to their small size, and the need to stabilize their position within the cortical tissue, the functional understanding and mechanical design of intracortical electrodes have been particularly challenging.

11.2   Physical Structure of Intracortical Electrodes

Intracortical metal electrodes are typically fabricated from rigid shafts designed to penetrate the visual cortex. The shaft of the electrode is often insulated with a biocompatible polymer, e.g. Parylene-C. At the tip of the electrode is an exposed noble metal surface, commonly platinum or iridium. The surface area of the exposed tip is carefully controlled, during manufacture, so as to target a pool of neurons within a predefined semi-spherical volume surrounding the tip, while allowing for the safe transfer of charge.

The material used for the shaft can vary. Silicon, polymers, and bare metal wires have been used. In its simplest form, the electrode is comprised of a metal wire whose tip has been etched to a controlled-geometry point. More complicated structures use metal-tipped silicon shafts, and thin-film-fabricated multi-dimensional silicon shanks that contain multiple surface-deposited metal electrode sites [3, 44, 47]. The length of the electrodes is typically between 1.5 and 2 mm in order to target cortical neuronal layer IV. However, in practice it is difficult to assure the depth penetration, or maintenance of the tip position within the cortex.

Singular discrete-wire intracortical electrodes have historically been used for both recording and stimulation of cortical neurons. Commonly, such electrodes are fabricated by cutting off the end of an insulated small diameter (25–50 μm) noble-metal (typically platinum-iridium) wire. More sophisticated designs use controlled etching of the bare metal wire in order to obtain a precise tip shape, polymer insulation of the electrode's shaft, and final laser ablation to precisely expose the metal tip. Positional stability of the singular intracortical electrode tip in the brain is crucial if it is desired to consistently record from, or stimulate, a particular neuronal pool. Gualtierotti and Bailey [24] are credited with being the first to describe a "neutral buoyancy" microelectrode. In their concept, the intracortical electrode needed to "float" on the surface of the cortex so that the normal movement of the brain would not disrupt the position of the electrode relative to the target neurons. Their design was not practical for mass production, however subsequent design have strived to retain this principle of minimal mass and mechanical floating.

The "hat pin" intracortical electrode design was developed by Salcman and Bak [39], and was used in several human visual prosthesis investigative experiments [4, 40]. This design, a derivative of the Gualtierotti and Bailey electrode, is comprised of a 37-μm diameter iridium wire microwelded to a 25-μm diameter gold wire lead, with the electrode tip etched to produce a radius of 1–5 μm. The electrode shaft is coated with a 3–4 μm thick layer of Parylene-C insulation. A dual-beam excimer laser is used to control the exposure of the metal tip. The junction between the electrode shaft and the connecting gold lead wire is reinforced with an epoxy ball, with the resulting structure resembling a hat-pin, as shown in Fig. 11.1. In some versions of this design, two electrodes are incorporated within a single epoxy ball to produce an electrode doublet. Insertion of the electrode structure into the cortex can be accomplished by hand, using surgical forceps.

Other variations of this basic wire-type electrode design have evolved over the past 25 years that use blunter tips with controlled-cone shapes. In each case, the designers strived to produce electrodes that were consistent in shape, length, and tip exposure, with the goal of minimizing tissue insertion damage and preserving the underlying neuronal substrate.

It has generally been accepted that a viable cortical visual prosthesis will require hundreds, possibly thousands, of intracortical electrodes, and while earlier experiments were successfully performed using singular intracortical electrodes, the surgical difficulty associated with implantation of individual electrodes motivated the design of electrode arrays. Cortical electrode arrays are comprised of a group of electrodes whose relative position and cortical penetration are maintained by a superstructure. The number of electrodes contained within an array can vary from 16 to over 100, depending upon the application and manufacturing method.

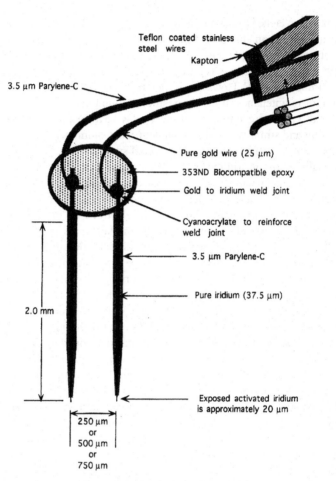

Fig. 11.1 Dual hat pin electrode as described by Salcman and Bak. From Schmidt et al. [40]

Interconnection between the array's electrodes and electronic circuitry used for generating the stimulation currents becomes more challenging as the number of the electrodes within the array increases.

Cortical insertion of the intracortical electrode arrays is most often performed using an array-specific high-speed insertion tool in order to minimize the "bed-of-nails" effect. Slow insertion of an array, through the pia, can cause unacceptable cortical deformation and significant micro-bleeding, thus damaging the target neuronal pool. These conditions are avoided by using rapid insertion, and depending upon the number electrodes within the array, and the electrode tip shapes, speeds from 1 to 10 m/s are used. Using rapid insertion, the array can directly penetrate blood vessels with little to no resultant bleeding.

A smaller electrode physical tip size offers the promise of selective stimulation of a small pool of neurons, and the design of the intracortical stimulation electrode is most often faced with a compromise between the desire for a small-geometry tip, and a limitation in the charge per unit area (charge density) that can be safely injected into the tissue without causing damage to the electrode or the surrounding neuronal tissue. As the tip area is made smaller, the safe charge, and charge density, limit must be correspondingly reduced, albeit most often in a non-linear manner. Typically, intracortical electrodes are made with tip areas of less than $2,000\,\mu m^2$, and for some studies electrodes as small as $200\,\mu m^2$ have been used [40].

Despite the variations in electrode shape, length, or array composition, the basic interface between the electrode and the neuronal tissues remains that of a metal surface through which electrical charge is passed, and it is the nature of this interface that occupies the efforts of numerous research laboratories.

11.3 Charge Injection Using Intracortical Electrodes

11.3.1 The Intracortical Electrode as a Transducer

Electrical stimulation of neurons is accomplished by depolarizing the neuronal membrane through the flow of ionic current between two electrodes, and typically this is accomplished by injecting pulses of ionic current through the neuronal tissue that surrounds the electrode. Use of pulses to initiate the neuronal activation derives directly from the capacitive nature of the neuronal membrane. The fundamental function of the intracortical electrode is to act as a transducer for converting electronic current, flowing from the stimulator circuitry to the electrode, into ionic current that flows within the biological tissues. In order to elicit a neuronal response, a threshold membrane polarization must be reached, and for a specific electrode geometry this defines a threshold stimulation charge that the electrode must support for each stimulation pulse. For a metal electrode to support the threshold charge injection, suitable processes occurring at the electrode-tissue interface are required to cause the necessary charge-carrier conversion. These processes can be capacitive, or faradaic. In the former, a capacitive interface, formed by either the electrode-electrolyte double-layer, or a dielectric layer, is charged and discharged. For faradaic reactions, electrochemical reactions involve metal-specific charge species that are oxidized and reduced. In order to protect the electrode and the surrounding tissue from deterioration, these reactions must be reversible and limit the injection of reaction by-products into the tissue.

Capacitive-type electrodes [25], inject charge exclusively through capacitive charging and discharging, and therefore they are conceptually attractive, for use as intracortical electrodes, because they avoid the use of faradaic reactions. However, presently-known capacitive electrodes do not have sufficiently high charge-storage capacities to make them useful for intracortical stimulating electrodes. The use of surface-roughening, porous-material electrode coatings, and high-dielectric constant films, such as Ta_2O_5 or TiO_2, have been investigated in an attempt to increase capacitive electrode charge capacities to the required stimulation threshold levels.

However, even with these techniques, the charge injection capacity of current capacitive-type electrodes is not adequate for small area ($<2,000$ μm^2) intracortical electrodes [37, 38].

Faradaic-type electrodes, accomplish the charge-carrier conversion through reduction and oxidation (redox) of surface species. One metal widely used for numerous stimulation electrodes, including intracortical microelectrodes, is Platinum (Pt) or a Pt-based alloy. While the precise nature of the reactions utilized by a Pt electrode is unclear, it is generally accepted that charge transfer occurs substantially by H-atom plating and stripping, with the double-layer capacitance contributing less than 15% to the total charge injection – Fig. 11.2. To improve the ability of an electrode to act as a transducer and inject higher levels of ionic current into the tissue, coatings such as activated-iridium-oxide-film (AIROF) have been used.

In order to initiate the faradaic reaction, a reaction-activation voltage drop across the metal-tissue (metal-electrolyte) interface must be achieved, and this voltage drop is commonly called the "electrode polarization." The voltage drop is caused by the flow of current, electronic and ionic, at the metal-electrolyte interface. Initially, the current flow, and charge transfer, is supported by the charging, or discharging, of the double-layer capacitance. As the charge in this capacitance is quickly exhausted, the resulting increase in electrode polarization initiates the first available redox reaction capable of supporting the charge injection. As this initial reaction is exhausted, either due to unavailability of counter ions or reaction rate limitations, a new reaction must be recruited. The surface conditions of the metal electrode might cause the initiation and exhaustion of several reactions as the polarization increases and the charge injection is continuously supported.

In theory, the charge injection process is reversible under the assumption that the redox reactions which are utilized are reversible. Therefore, it is common to use two phases for the stimulation waveform. The first phase is designed to stimulate the

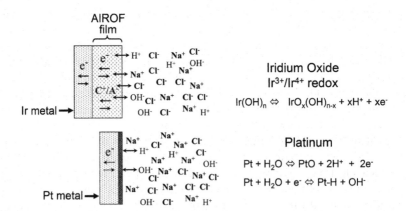

Fig. 11.2 Depiction of charge injection reactions taking place at the surface of Iridium Oxide (*top*) and Platinum (*bottom*) intracortical electrodes. The AIROF film provides a buffer zone in which reversible redox reactions can take place and permit enhanced charge injection

target neuronal pool. The second phase is designed to reverse the faradaic reactions that were utilized in the first phase by reversing the current in the electrode-electrolyte interface. Most often, but not exclusively, stimulation waveforms are generated by electronic current sources with the magnitude and duration of the first phase being replicated with an opposite magnitude and identically-timed pulse in the second phase. Owing to the simplicity in producing rectangular current pulses, the biphasic-balanced-constant-current rectangular stimulation waveform has become an historical standard for most neural stimulation systems [21, 22]. In order for an intracortical electrode to act as a stimulating transducer for chronic stimulation of neural tissue, the reversibility of the reactions must be assured, or else redox reaction by-products and unacceptable local pH shifts may damage the tissue and render the surrounding neuronal pool unusable [2, 9, 14, 36]. While some biological damage is unavoidable, due to mechanical damage resulting from the electrode insertion and electrical damage resulting from continuous charge injection, the damage must be self-limiting in order for the intracortical electrode to be a critical component for a cortical stimulation system.

11.3.2 Charge Injection Limits

Unfortunately, the total charge capacity of the reversible reactions that are available for a bare Pt intracortical electrode is often insufficient to initiate the necessary neuronal response. While Pt has been used quite successfully for heart pacers, cochlear implants, and other implanted neural stimulators, the extremely small size of intracortical electrodes limits the useable electrode charge capacity to well below the charge-per-pulse required for stimulation of cortical neurons. Despite this, it is quite easy to continue driving a Pt electrode with electronic current past the point at which the reversible reactions are exhausted, thus necessitating the recruitment of irreversible reactions. The most frequently recruited irreversible reactions are oxidation and reduction of water. From an electrochemical standpoint, water decomposition can provide an almost inexhaustible supply of charge carriers within the tissue, albeit with corrosion of metal. From a biological standpoint, decomposition of water as a means of neural stimulation is accompanied by huge local pH shifts, migration of metal ions into the tissue, and evolution of hydrogen and oxygen gasses. It is considered unacceptable to initiate water decomposition during cortical neural stimulation. Therefore, considerable research has been dedicated to understanding how to avoid the use of irreversible reactions during neural stimulation.

It is well-known that for a given electrode design, consisting of a particular metal type and geometric shape, there exists a window of polarization for which the electrode will not experience water redox reactions. This window is commonly called the "water window," and, for many metals, is roughly within the range of +0.8 to −0.6 V with respect to a Ag|AgCl reference electrode. As long as the electrode polarization is held to a value within the water window there is a reasonable expectation that water decomposition will not occur. This should not be interpreted to mean that all reactions occurring within the water window are safe: There is some suspicion that for many platinum electrode designs, some reactions taking place within the water window are not entirely reversible.

Understanding the conditions that drive an electrode's polarization outside of the water window is often difficult, and for many years researchers attempted to define, a priori, the maximum charge density per stimulation pulse that a particular metal, or electrode coating, could support. This quantity became known as the maximum injectable charge density and is expressed in units of charge/area, frequently using mC/cm^2. Establishment of the maximum injectable charge density for a particular electrode design is difficult due to the uncertainties about the relationship between charge injection and electrode polarization. Often, limits were defined based upon empirical *in vitro* studies, in which the physical condition of the electrode was examined following an extended pulsing regime [6]. In companion studies, electrodes were pulsed *in vivo* using predefined charge densities with post implantation histology being performed upon the surrounding tissue to examine adverse effects upon the local neurons or migration of metal ions [29]. Unfortunately, studies for particular electrode designs, using particular electrode metals or coatings, were often extrapolated to define the *general* charge injection limits for that same metal, or coating, *type* without regard to the electrode geometry. This created much confusion about how electrodes could be designed and used for chronic cortical stimulation so as to avoid both deterioration of the electrode and damage to the surrounding biological tissue.

In addition to damage that might be imparted to the surrounding tissue from the use of irreversible reactions, neurons can also be adversely affected, or even damaged, by over-excitation, even when operating an electrode within the water window. These changes in neuronal sensitivity and viability were observed through studies at the Huntington Medical Research Institutes [29]. From a functionality standpoint, over excitation of cortical neurons can produce a depression in neuronal activity and sensitivity. Once they are excessively stimulated, the neurons become less sensitive to subsequent stimulation pulses, thus shifting the firing threshold of the target neuronal pool. Depression can occur from either individual electrodes, near a small pool of neurons, being driven above the charge-induced depression level, or by ensembles of electrodes whose individual sub-depression charge injection levels summate causing distributed depression within a larger neuronal pool. In general, histological studies of depressed neurons do not show physical damage [29]. Yet, recovery of the depressed neurons can require hours, or even days, after cessation of the stimulation. Beyond the depression-induced threshold, physical damage to cortical neurons has been observed as a result of stimulation pulsing even when there is no evidence of driving the electrode beyond the water window [30]. This condition results in a much more serious impact upon the target neuronal pool since it is irreversible.

The combined effects of reaction-induced electrode or tissue damage and the stimulation-induced tissue depression or damage are hard to quantify. To date, no studies have established a priori charge injection limits for intracortical electrodes that consider both electrochemical and stimulation-induced effects. Furthermore, the temptation to apply results from studies of one electrode design to other electrode designs that use the same metal or coating often results in the discovery of electrode failure, or damage to the surrounding neurons, only after the electrode is implanted and chronically stimulated. Using the stability of the functional response of the stimulated neurons as a measure of safe charge injection is risky, since shifts in neuronal thresholds may occur only after damage to either the electrode or the tissue has already occurred.

11.4 Intracortical Electrode Coatings

Recognizing the charge injection limitations of intracortical metal electrodes, the National Institutes of Health, under the administration of the Neural Prosthesis Program, funded research to identify coatings that could be placed over metal electrodes towards the goal of limiting the charge injection reactions to within the water window [30]. This resulted in the identification and development of Activated IRidium Oxide Film (AIROF) at EIC Laboratories (Norwood, MA). AIROF is a faradaic coating that is based upon the electrochemical growth of a three-dimensional film of hydrated iridium oxide [1, 30]. Presently, AIROF, and Sputtered IRidium Oxide Film (SIROF) [15, 27, 41] have emerged as the preferred coating materials for many neural stimulating electrodes.

AIROF is formed upon pure iridium metal electrodes using an electrochemical activation process. The attractiveness of AIROF as a stimulation electrode coating was recognized by Brummer and first reported in 1983 [8, 35]. Charge injection limits for AIROF electrodes in physiological buffer were reported in 1988 by Beebe and Rose [6].

The electrochemistry of the activation process has been studied extensively and models to explain the observed accumulation of oxide and the charge propagation mechanism have been suggested [10, 11, 30, 32]. It is known that thick anodic oxide films can be formed on the surface of an iridium electrode by continuously cycling the electrode potential with a triangular or square waveform in an aqueous electrolyte. The potential limits are typically between values slightly positive of hydrogen evolution and just negative of the onset of oxygen evolution. It has been shown that AIROF formation is influenced by the chemical composition of the electrolyte; the geometry and morphology of the iridium metal substrate; and, the duration and form of the voltage/current activation waveform [23, 45]. The electrolyte composition influences the rate of formation as well as oxide morphology through the pH, the ionic strength, the conductivity and structure of the double layer at the Ir/electrolyte interface [17].

The benefit of using AIROF on intracortical electrodes stems from the premise that the redox reactions needed for charge transfer from the electrode to the tissue can take place exclusively within the film. Thus the film serves as a buffer zone for charge transfer between the metal surface and the biological electrolyte. AIROF is known for demonstrating significantly higher maximum charge injection limits, and this high charge capacity is obtained from a reversible Ir^{3+}/Ir^{4+} valence transition that takes place within the film [8, 35] as depicted in Fig. 11.2. By restricting the redox reactions within the film, and utilizing a known high-charge capacity reaction, an increase in the injectable charge capacity and significantly improved safety and consistency of neural stimulation can be obtained. For cathodal-first stimulation pulses, the ability of AIROF to inject charge can be further be increased by applying a positive bias of 0.4–0.8 V (vs. Ag|AgCl) prior to the stimulation pulse [6]. The bias acts to convert the AIROF from a mixed Ir^{3+}/Ir^{4+} valence state to the Ir^{4+} valence state, not only making the film significantly more electronically conductive, but also richer in the Ir^{4+} needed for reduction during the cathodal phase. In some cases, the use of bias allows for as much as a factor of three increase in charge capacity [16].

It is not surprising that iridium oxide films have emerged as the preferred coatings for intracortical, and other neural prosthesis, electrodes. AIROF has been shown, *in vitro*, to allow for about 10–20 times the maximum injectable charge, when compared to bare Pt, achieving a charge density limit of up to 3.5 mC/cm^2 for anodally-biased cathodal-first pulses [6]. However, the use of AIROF, rather than bare metal, is fraught with some additional peril. AIROF is susceptible to damage if the electrode polarization moves outside of the water window. Initiating water decomposition reactions can cause the AIROF to delaminate from the underlying metal surface, thus rendering the electrode non-usable for continued charge injection. While it is generally regarded that it is not viable, for any electrode, to inject charge outside of the water window, there is often uncertainty about the voltage and current conditions for which the electrode polarization exceeds the water window limits. If using a Pt electrode, a momentary transgression of the water window limits may cause highly undesirable reactions and residual by-products that enter the tissue. Yet the surface of the electrode may remain relatively unharmed. For the AIROF electrode, the films acts as a buffer zone that protects the tissue, and therefore reactions outside of the water window can potentially damage the AIROF in an irreversible manner.

11.5 Characterization of Intracortical Electrodes

11.5.1 Cyclic Voltammetry

Since the faradaic reactions used for electrode/tissue charge transfer are initiated by polarization of the electrode-electrolyte interface, it is useful to use an analytical method for examining how this interface behaves within, and outside of, the water window. Cyclic voltammetry (CV) is a commonly-utilized method for accessing the nature and behavior of stimulating electrodes. As derived from standard electrochemical methods, CV uses three-electrodes within an electrolyte. The potential of the intracortical electrode, with respect to a reference electrode, is periodically swept between two predetermined potential limits, usually at the water window boundaries, while measuring the current that flows between the intracortical electrode and a larger counter electrode. The potential sweep shifts the electrode-electroyte interface through the full range of reversible redox reactions while the measured current provides an indication of the capacity and rate of these reactions. Integration of the CV current waveform is often used to calculate a total charge storage capacity (CSC), for both anodic-(CSC_A) or cathodic-(CSC_C) first stimulation. Typically, CV electrode measurements are made at sweep rates that are much slower than the voltage changes which the electrode-electrolyte interface experiences during a typical stimulation pulse, with CV sweeps typically on the order of 50 mV/s. Since the charge injection redox reactions are rate dependent, it is important to understand that CSC values are always larger than maximum charge injection values for any given electrode. Typically, less than 20% of the CSC can be utilized during a stimulus pulse. In this regard, review of the literature can often become confusing when comparing reported values of CSC to reported values of charge injected *in vitro* and *in vivo*. In other words, only a fraction of the CSC can be accessed during a short duration stimulus pulse. The CV measurement is highly sensitive to the condition of the electrode-electrolyte interface, the morphology the electrode coating, the electrode surface roughness, the geometric shape of the electrode tip, and the nature of the electrolyte. For any electrode metal, or coating, the shape of the CV can vary dramatically, depending upon how the electrode is fabricated and in what electrolyte the measurement is performed, even though the nature of the redox reactions themselves remains the same.

11.5.2   Electrode Stimulation Voltage Waveforms

Stimulation of cortical neural tissue is most commonly accomplished by driving the electrode with a two-phase waveform that consists of a first neural-stimulation phase and a second charge-recovery phase. Typically, each of these phases are generated by constant-current electronic circuits producing rectangular pulses. Often the first phase consists of a cathodal (negative) constant current pulse, followed by a second phase anodal (positive) constant current pulse as depicted in Fig. 11.3. In Fig. 11.3, a highly simplified model for an intracortical electrode is presented consisting of a series resistive-capacitive network. While simplistic, this model does allow for a first-order understanding of the relationship between the electrode-electrolyte interface and the voltage/current waveforms. The resistive component is commonly called: the access resistance, and the capacitive component is commonly called: the electrode pseudocapacitance. These are, of course, merely lumped-model approximations for the electrical and electrochemical processes that take place during a stimulation pulse.

Referring to Fig. 11.3, during the first cathodal phase, constant current is forced through the electrode for the purpose of activating near-by cortical neurons. At the leading edge of the current pulse, an immediate voltage drop across

Fig. 11.3 Depiction of a model for a stimulator passing balanced biphasic current from a microelectrode to a counter electrode. The components of the electrode voltage excursion waveform are identified and related to the electrode model R and C

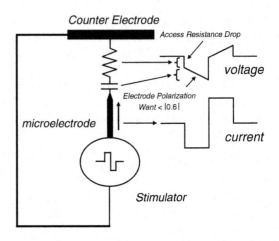

the electrode-electrolyte interface is observed. For this simplified model, this leading-edge drop is caused by the *IR* drop on the access resistance. In accordance with circuit theory, the voltage on the capacitor remains unchanged at the current pulse leading edge. As current is forced through the electrode, the capacitance, *C*, charges in a time-linear manner, deriving from $I = C(dv/dt)$. In the model of Fig. 11.3, the charging of this capacitor (*dv*) represents the electrode polarization, and the redox reactions should remain within the water window provided that $dv < 0.6$ V.

At the end of the first phase, the current changes from cathodal to anodal as the second charge-recovery phase is initiated. For the lumped model, the magnitude of the first-phase trailing edge step of the voltage waveform is twice that of the first-phase leading edge because the summation of the turning off of the cathodal current and the turning on of the anodal current produce a current step of twice that of the leading edge. This voltage step is the drop across the access resistance, *R*. During the second phase, anodal current is forced through the electrode in an attempt to restore the electrode to the pre-stimulus condition. In the simple model of Fig. 11.3, use of equal (but opposite) first and second phase currents, with equal pulse durations, produces equal first and second phase charges, thus exactly returning the electrode to the pre-stimulus voltage level in anticipation of the next stimulus pulse. During the interval between biphasic stimulation pulses, some method of electrode voltage control is typically employed to assure that the electrode potential remains stable, at a pre-determined level, so that for repeated stimulation pulses the electrode can stimulate neurons in a consistent manner.

In practice, the simplistic model of Fig. 11.3 fails to account for important aspects of the electrode's charge injection process. These include: (1) Multiple contributions to the access resistance drop that are inconsistent with an ideal resistor model, (2) Non-linear behavior of the electrode polarization that is inconsistent with an ideal capacitor model, and (3) Imbalances in the stimulator phase charges.

11.5.3 Non-ideal Access Resistance Behavior

Historically, the leading edge voltage drop was attributed to the electrolyte resistance caused by limitations in ionic conductivity of the electrolyte. Thus it was common practice to subtract the entire leading edge drop from the total electrode voltage excursion, during the first phase, as a means of determining the electrode polarization. However, the leading edge drop can include other effects besides simple electrolyte resistance, specifically, concentration polarization near the electrode-electrolyte interface. Concentration polarization is essentially caused by a depletion in electrolyte charge carriers (counter ions) at the onset of the current pulse. For coated metal electrodes, such as AIROF, near instantaneous changes in film conductivity at the leading edge of the current can be a secondary contribution to the access voltage drop.

11.5.4 Non-linear Electrode Polarization

Based upon the earlier discussion, it is obvious that the dynamics of charge injection via redox reactions cannot be directly compared to the charging and discharging of an ideal capacitor. Owing to the complex geometric shape of the electrode tip, and the highly non-uniform current densities, as well as the range of possible of redox reactions that might be experienced, the behavior of the electrode-electrolyte interface might be better explained by a set of distributed RC networks, however even this remains an oversimplification. Rather, the behavior of the electrode during what is often called the electrode polarization phase, or the capacitive charging phase, is driven by the rates of one or more reactions, the changes in interfacial and film conductivity, and the closeness of the electrode voltage to the edge of the water window. Strictly speaking, the electrode polarization is comprised of the reaction activation overpotential and a shift in the electrode equilibrium potential. However, these components cannot be easily derived from the stimulus voltage excursion waveform.

11.5.5 Determining Electrode Safety

The uncertainties in determining the components, and magnitude, of the leading-edge access voltage drop make the estimation and prediction of electrode polarization, during any given stimulus pulse, difficult. Often the leading edge drop is by far the largest component of the total voltage excursion experienced by an electrode during a stimulation pulse. Simply subtracting the measured access voltage from the total voltage excursion is most often inadequate for estimating whether the electrode polarization is within the water window. It is unclear how much of the leading edge access voltage drop is truly caused by a benign resistive drop, and how much is caused by an interfacial process that might contribute to undesirable redox reactions.

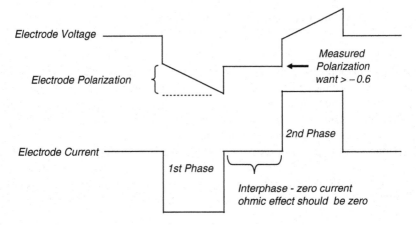

Fig. 11.4 Depiction of an alternate current waveform for delivering stimulation pulses to microelectrodes. During the zero-current interphase portion of the waveform the effect of access resistance upon the voltage waveform is eliminated, and the residual voltage, during the interphase, is a reasonable measure of the electrode polarization. To remain within the water window the measured polarization should be more positive than −0.6 V with respect to Ag|AgCl

Too often, the injectable charge capacity of a particular electrode is estimated from the use of a priori published material type-based charge densities, and this approach does take into account the actual dynamic behavior of the electrode since the wrong parameter, i.e. charge density rather than electrode polarization, is being considered. An alternate method of estimating the electrode polarization is depicted in Fig. 11.4 and involves adding a third interphase region to the stimulation waveform. If a short period of zero-current is imposed between the first and second phases, then a measurement of the electrode voltage during this time of zero-current should be free from true *IR* drops, and should be a better estimate of the polarization caused by the delivery of charge during the first phase of the biphasic waveform [13]. The disadvantage to this approach is that the measurement is made after the polarization has occurred. There exists some debate about whether a typical AIROF intracortical electrode can tolerate single-stimulus conditions that transgress the water window without damage, and use of the interphase voltage measurement as a continuous measure of electrode safety may be inadequate to protect an AIROF electrode. However, there presently exists no implantable stimulator that uses leading-edge voltage measurements in a predictive manner to protect either AIROF, or bare metal, intracortical electrodes from damage.

11.6 Contrasts of *In Vitro* and *In Vivo* Behavior

Most available data for intracortical electrodes come from *in vitro* studies that were carried out in model physiological fluid. Based upon those studies, the maximum injectable charge capacity for AIROF intracortical electrodes whose tip areas are under 2,000 μm² is well within the anticipated stimulation charge thresholds for visual cortex neurons. For example, Schmidt et al. [40] observed that stable phosphenes could be obtained in a human volunteer when using 0.4–4.6 nC/phase of stimulation, whereas *in vitro* measurements of 2,000 μm² AIROF electrodes in phosphate-buffered saline typically show up to ten times this required charge capacity, while maintaining operation within the water window. This led to the historical conclusion that AIROF intracortical electrodes were more than adequate for long-term stimulation of the visual cortex in visual protheses.

More recently, this view has been challenged, as *in vivo* studies of AIROF and SIROF electrodes have been performed. Cogan et al., compared the *in vitro* and *in vivo* charge injection behavior of large area (~125,000 μm²) AIROF electrodes intended for a retinal visual prosthesis [12], and found that the charge capacity of the electrodes, once implanted subretinally in rabbits, required three times the total electrode voltage excursion as had been observed *in vitro*, for delivery of the same charge. Hu et al. [26] compared the performance of 2,000 μm² intracortical electrodes implanted within the cortex of a zebra finch with their performance in dilute phosphate buffered saline, and found a factor of four decrease in their charge injection capacity *in vivo* for equal *in vitro* and *in vivo* voltage excursions. Even more disturbing, is the observation that electrode polarization, *in vivo*, appears to increase by a factor of two, over that seen *in vitro*, for equal charge injection [12].

Figure 11.5 shows a dramatic demonstration of the loss of charge capacity, relative to *in vitro* behavior, for intracortical electrodes placed within the *in vivo* cortical environment. Two electrodes were measured *in vitro*, immediately placed *in vivo*, then immediately replaced into the *in vitro* environment. The stimulator circuitry was specially designed to limit the total electrode voltage to less than ±0.6 V (water window) in order to prevent electrode damage. On the left of Fig. 11.5 are shown the pulse voltage excursions for the two AIROF intracortical

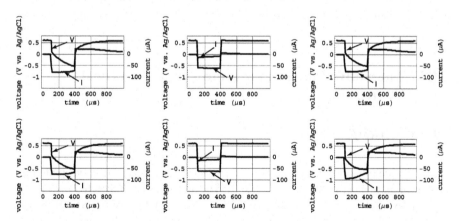

Fig. 11.5 AIROF intracortical electrodes tested in vitro and in vivo. Electrodes were transferred between a beaker of PBS and the cortex of a Zebra Finch during the same experiment. In vitro current and voltage excursions for two electrodes are shown on the right and left set of plots. In vivo waveforms are shown in the center plots. Note the dramatic decrease in the in vivo injectable charge capacity, relative to the in vitro behavior, as seen in the center plots by the larger voltage excursions for the smaller stimulation currents

electrodes, of area $2,000 \, \mu m^2$, while placed within phosphate-buffered saline. The voltage and current are indicated on each of the plots as (I) and (V). On the left, it can be seen that for approximately −0.6 V of *total* voltage excursion, the electrodes are able to support ~75 μA for 300 μs (22.5 nC). In the center plots, the same two electrodes are shown placed within the cortex of a Zebra Finch. Note that for the same voltage excursion as was seen, *in vitro*, *in vivo* less than 10 μA is supported for 300 μs (3 nC), and the leading edge access voltage has significantly increased. On the right one can see the two final *in vitro* plots, with the charge capacity restored to the original *in vitro* values.

These studies, and others, lead to the conclusion that there are notable differences between the *in vitro* and *in vivo* environment that significantly impact the ability of an intracortical electrode to act as a charge transfer transducer for a cortical visual prosthesis. While the behavioral differences with respect to electrode voltage excursion seem self-evident, the causes of them remain unclear. Yet, there are identifiable and unique characteristics of the *in vivo* environment: (1) The presence of proteins and other organic species, (2) The presence of cells that could impede the mobility and transport of counter ion charge carriers, (3) Encapsulation of the electrode surface by bio-molecules. For the Zebra Finch experiment of Fig. 11.5, described above, there was insufficient time for cell growth or significant encapsulation of the electrodes. Indeed, only minor washing of the electrode was performed when transferring from the *in vivo* to the *in vitro* environment. One hypothesis is that of the three factors listed above, reduced mobility of counter ions seems the most likely. More studies are needed in order to explain, and understand how to minimize the adverse effects of the *in vivo* environment upon electrode charge injection. Whatever the causes, the capabilities of present-day intracortical electrode technology seem marginal, but probably adequate, for the demands of current cortical visual prostheses designs. However, future visual prosthesis designs may very well exceed currently-achievable injectable charge capabilities.

11.7 Alternative Coatings for Improving Intracortical Electrodes

11.7.1 SIROF

Iridium oxide films can also be applied to metal surfaces using reactive ion sputtering within an oxidizing plasma [15, 27, 41, 43]. The performance of SIROF *in vitro* and *in vivo* favorably compares to that of AIROF. One advantage of the SIROF is the physical robustness of the deposited films, and this might make them more resistant than AIROF to short-term operation outside of the water window. Additionally, the ability to sputter the films on a variety of base metals while using masking, or other patterning techniques might make SIROF more adaptable to a wider range of physical intracortical electrode designs.

11.7.2 PEDOT

Polyethylenedioxythiophene is an emerging electrode coating material that is based upon an electrically conducting polymer. Earlier work on other electrically conducting polymers was less than promising, however more recently there has been the suggestion that such polymers might be modified with biomolecules or nerve growth factors in order to promote the functionality of the interface between the electrode and surrounding neurons. In theory, if neurons could be attracted nearer to the electrodes than is ordinarily seen, lower stimulation thresholds might be achieved. PEDOT-coated electrodes, that are characterized by reduced impedances have been used for chronic recording studies [28]. Preliminary work towards the possibility of performing *in vivo* polymerization of PEDOT has been reported [33, 34, 46]. Charge injection capabilities of PEDOT, *in vitro*, are on the order of AIROF [31], although *in vivo* studies are currently lacking. The nature of charge injection by PEDOT with respect to capacitive or faradaic means has not been fully explained. There is some suggestion that for potentials more positive than $-0.6\,V$, the primary process may be capacitive although present studies are not conclusive.

11.7.3 Carbon Nanotube Coatings

Carbon Nanotubes have been suggested as an intracortical electrode coating due to the extremely high increase in surface made possible by the presence of the tubes [42]. This work is in the early stages of research. The basic principle is that by dramatically increasing the surface area of the electrode, via the three dimensional structure, a corresponding increase in the double layer capacitance would result. Earlier attempts to use more conventional means of surface roughening and creation of porous structures for metal electrodes showed similar increases in surface area and *in vitro* charge injection. However, once placed into the biological environment adsorption of biomolecules seemed to clog the pores and defeat the strategy. Whether carbon nanotubes would be susceptible to similar effects is presently unknown. It may be possible to chemically alter the nanotubes so as to improve their biocompatibility [5].

11.8 Conclusion

Present-day technology for fabricating intracortical microelectrodes still relies upon mechanically stabilizing a metal, or a coated metal surface, near a pool of target neurons. In this regard, the basic technology for transferring charge to cortical tissue has not changed within the past century. A better understanding of how a metal electrode functions as a charge-transfer transducer has facilitated the establishment of relatively safe driving strategies for chronic stimulation of cortical neurons. Whether this technology is sufficient for the deployment of a cortical visual prosthesis remains untested. The sensory functionality demonstrated by cortical prostheses that used larger electrodes on the surface of the brain was disappointing from the standpoint of the users' abilities to integrate stimuli into coherent visual perceptions [18]. Despite some of the limitations in current electrode technology, it is expected that modern versions of these earlier surface electrode visual prostheses, that utilize intracortical electrodes, may be used in human trials within the next 5 years.

References

1. Agnew WF, McCreery DB (1990), *Considerations for safety with chronically implanted nerve electrodes.* Epilepsia, **31**(2): p. S27–32.
2. Agnew WF, Yuen TG, McCreery DB, Bullara LA (1986), *Histopathologic evaluation of prolonged intracortical electrical stimulation.* Exp Neurol, **92**(1): p. 162–85.
3. Anderson DJ, Najafi K, Tanghe SJ, et al. (1989), *Batch-fabricated thin-film electrodes for stimulation of the central auditory system.* IEEE Trans Biomed Eng, **36**(7): p. 693–704.
4. Bak M, Girvin JP, Hambrecht FT, et al. (1990), *Visual sensations produced by intracortical microstimulation of the human occipital cortex.* Med Biol Eng Comput, **28**(3): p. 257–9.
5. Banerjee S, Kahn MG, Wong SS (2003), *Rational chemical strategies for carbon nanotube functionalization.* Chemistry, **9**(9): p. 1898–908.
6. Beebe X, Rose TL (1988), *Charge injection limits of activated iridium oxide electrodes with 0.2 ms pulses in bicarbonate buffered saline.* IEEE Trans Biomed Eng, **35**(6): p. 494–5.
7. Brindley G, Lewin W (1968), *The sensations produced by electrical stimulation of the visual cortex.* J Physiol, **196**: p. 479–93.
8. Brummer SB, Robblee LS, Hambrecht FT (1983), *Criteria for selecting electrodes for electrical stimulation: Theoretical and practical considerations.* Ann NY Acad Sci, **405**: p. 159–71.
9. Brummer SB, Turner MJ (1977), *Electrochemical considerations for safe electrical stimulation of the nervous system with platinum electrodes.* IEEE Trans Biomed Eng, **24**(1): p. 59–63.
10. Buckely DN, Burke LD (1975), *The oxygen electrode part 5 – Enhancement of charge capacity of an iridium surface in the anodic region.* J Chem Soc Faraday Trans, **71**: p. 1447–459.
11. Burke LD, Scannell RA (1984), *An investigation of hydrous oxide growth on iridium in base.* J Electroanal Chem, **175**: p. 119–41.
12. Cogan SF (2006), *In vivo and in vitro differences in the charge-injection and electrochemical properties of iridium oxide electrodes.* Conf Proc IEEE Eng Med Biol Soc, **1**: p. 882–5.
13. Cogan SF (2008), *Neural stimulation and recording electrodes.* Annu Rev Biomed Eng, **10**: p. 275–309.
14. Cogan SF, Guzelian AA, Agnew WF, et al. (2004), *Over-pulsing degrades activated iridium oxide films used for intracortical neural stimulation.* J Neurosci Methods, **137**(2): p. 141–50.
15. Cogan SF, Plante TD, Ehrlich J (2004), *Sputtered iridium oxide films (SIROFs) for low-impedance neural stimulation and recording electrodes.* Conf Proc IEEE Eng Med Biol Soc, **6**: p. 4153–6.
16. Cogan SF, Troyk PR, Ehrlich J, et al. (2006), *Potential-biased, asymmetric waveforms for charge-injection with activated iridium oxide (AIROF) neural stimulation electrodes.* IEEE Trans Biomed Eng, **53**(2): p. 327–32.
17. Conway BE (1991), *Transition from 'supercapacitor' to 'battery' behavior in electrochemical energy storage.* J Electrochem Soc, **138**: p. 1539–48.
18. Dobelle WH (2000), *Artificial vision for the blind by connecting a television camera to the visual cortex.* ASAIO J, **46**(1): p. 3–9.

19. Dobelle WH, Mladejovsky MG (1974), *Phosphenes produced by electrical stimulation of human occipital cortex, and their application to the development of a prosthesis for the blind.* J Physiol, **243**(2): p. 553–76.
20. Dobelle WH, Mladejovsky MG, Evans JR, et al. (1976), *"Braille" reading by a blind volunteer by visual cortex stimulation.* Nature, **259**(5539): p. 111–2.
21. Donaldson ND, Donaldson PE (1986), *When are actively balanced biphasic ('Lilly') stimulating pulses necessary in a neurological prosthesis? I. Historical background; Pt resting potential; Q studies.* Med Biol Eng Comput, **24**(1): p. 41–9.
22. Donaldson ND, Donaldson PE (1986), *When are actively balanced biphasic ('Lilly') stimulating pulses necessary in a neurological prosthesis? II. pH changes; noxious products; electrode corrosion; discussion.* Med Biol Eng Comput, **24**(1): p. 50–6.
23. Gottesfeld S, McIntyre JDE (1979), *Electrochromism in anodic iridium oxide films II. pH effects on corrosion stability and the mechanisms of coloration and bleaching.* J Electrochem Soc, **126**: p. 742–50.
24. Gualtierotti T, Bailey P (1968), *A neutral buoyancy micro-electrode for prolonged recording from single nerve units.* Electroencephalogr Clin Neurophysiol, **25**(1): p. 77–81.
25. Guyton DL, Hambrecht FT (1974), *Theory and design of capacitor electrodes for chronic stimulation.* Med Biol Eng, **12**(5): p. 613–20.
26. Hu Z, Troyk PR, Brawn TP, et al. (2006), *In vitro and in vivo charge capacity of AIROF microelectrodes.* Conf Proc IEEE Eng Med Biol Soc, **1**: p. 886–9.
27. Klein JD, Clauson SL, Cogan SF (1989), *Morphology and charge capacity of sputtered iridium oxide films.* J Vac Sci Technol, **A7**: p. 3043–7.
28. Ludwig KA, Uram JD, Yang J, et al. (2006), *Chronic neural recordings using silicon microelectrode arrays electrochemically deposited with a poly(3,4-ethylenedioxythiophene) (PEDOT) film.* J Neural Eng, **3**(1): p. 59–70.
29. McCreery DB, Agnew WF, Bullara LA (2002), *The effects of prolonged intracortical microstimulation on the excitability of pyramidal tract neurons in the cat.* Ann Biomed Eng, **30**(1): p. 107–19.
30. McCreery D, Pikov V, Troyk PR (2010), *Neuronal loss due to prolonged controlled-current stimulation with chronically implanted microelectrodes in the cat cerebral cortex.* J Neural Eng, **7**(3): p. 036005.
31. Nyberg T, Shimada A, Torimitsu K (2007), *Ion conducting polymer microelectrodes for interfacing with neural networks.* J Neurosci Methods, **160**(1): p. 16–25.
32. Pickup PG, Birss VI (1987), *A model for anodic hydrous oxide growth at iridium.* J Electroanal Chem, **220**: p. 83–100.
33. Richardson-Burns SM, Hendricks JL, Foster B, et al. (2007), *Polymerization of the conducting polymer poly(3,4-ethylenedioxythiophene) (PEDOT) around living neural cells.* Biomaterials, **28**(8): p. 1539–52.
34. Richardson-Burns SM, Hendricks JL, Martin DC (2007), *Electrochemical polymerization of conducting polymers in living neural tissue.* J Neural Eng, **4**(2): p. L6–13.
35. Robblee LS, Lefko JL, Brummer SB (1983), *Activated Ir: An electrode suitable for reversible charge injection in saline.* J Electrochem Soc, **130**: p. 731.
36. Robblee LS, McHardy J, Agnew WF, Bullara LA (1983), *Electrical stimulation with Pt electrodes. VII. Dissolution of Pt electrodes during electrical stimulation of the cat cerebral cortex.* J Neurosci Methods, **9**(4): p. 301–8.
37. Robblee LS, Rose TL (1990), *Electrochemical guidelines for selection of protocols and electrode materials for neural stimulation,* in *Neural Prostheses: Fundamental Studies,* Agnew WF, McCreery DB, Editors. Prentice Hall: Englewood Cliffs, NJ. p. 25–66.
38. Rose TL, Kelliher EM, Robblee LS (1985), *Assessment of capacitor electrodes for intracortical neural stimulation.* J Neurosci Methods, **12**(3): p. 181–93.
39. Salcman M, Bak MJ (1976), *A new chronic recording intracortical microelectrode.* Med Biol Eng, **14**(1): p. 42–50.
40. Schmidt EM, Bak MJ, Hambrecht FT, et al. (1996), *Feasibility of a visual prosthesis for the blind based on intracortical microstimulation of the visual cortex.* Brain, **119**: p. 507–22.

41. Slavcheva E, Vitushinsky R, Mokwa W, Schnakenberg U (2004), *Sputtered iridium oxide films as charge injection material for functional electrostimulation*. J Electrochem Soc, **151**(7): p. E226–37.
42. Wang K, Fishman HA, Dai H, Harris JS (2006), *Neural stimulation with a carbon nanotube microelectrode array*. Nano Lett, **6**(9): p. 2043–8.
43. Wessling B, Mokwa W, Schnakenberg U (2006), *RF-sputtering of iridium oxide to be used as stimulation material in functional medical implants*. J Micromech Microeng, **16**(6): p. S142–8.
44. Wise KD, Angell JB, Starr A (1970), *An integrated-circuit approach to extracellular microelectrodes*. IEEE Trans Biomed Eng, **17**(3): p. 238–47.
45. Woods R (1974), *Hydrogen adsorption on platinum, iridium and rhodium electrodes at reduced temperatures and determination of real surface area*. J Electroanal Chem, **49**(2): p. 217–26.
46. Xiao Y, Cui X, Hancock JM, et al. (2004), *Electrochemical polymerization of poly(hydroxymethylated-3,4-ethylenedioxythiophene) (PEDOT-MeOH) on multichannel neural probes*. Sens Actuators B, **99**: p. 437–43.
47. Ziaie B, Nardin MD, Coghlan AR, Najafi K (1997), *A single-channel implantable microstimulator for functional neuromuscular stimulation*. IEEE Trans Biomed Eng, **44**(10): p. 909–20.

Part III
Prosthetic Visual Function: Acute and Chronic

Chapter 12
The Response of Retinal Neurons to Electrical Stimulation: A Summary of In Vitro and In Vivo Animal Studies

Shelley I. Fried and Ralph J. Jensen

Abstract The studies reviewed in this chapter are restricted to those that electrically stimulate the retina. The research studies reviewed in this chapter are further limited to those performed in animal models; the results of human clinical studies are covered in subsequent chapters.

The neural response to electrical stimulation is influenced (potentially) by a large number of stimulation-related variables (Chaps. 6–10). Stimulating electrodes can be constructed in different shapes and sizes and fabricated out of different materials. Arrays of multiple electrodes can be configured in many different arrangements and ultimately positioned on opposite sides of the retina, or even penetrate into the retina. In addition, the phase length, duration, amplitude and/or frequency of stimulus pulses can each vary, some by several orders of magnitude.

The neurobiology of the retina creates additional variables. There are five major classes of retinal neurons and each is a potential target of electrical stimulation. Each class can be subdivided into many different types; the anatomical and biophysical properties of each can vary considerably. Therefore, the response to electrical stimulation may also vary across types. Since each type is thought to convey different features of the visual world, stimulation methods that do not activate all types appropriately may not convey some or all of the important features of the visual scene.

Systematic study of the interactions between all engineering and neurobiological variables requires an extensive matrix of experiments. As a result, many basic questions remain unexplored. Here, we will focus on the experimental studies that have yielded the more important insights into either the mechanism by which retinal neurons respond to electrical stimulation or those that have led to improved

S.I. Fried (✉)
VA Boston Healthcare System, 150 South Huntington Avenue,
Boston, MA 02130, USA
and
Massachusetts General Hospital & Harvard Medical School,
429 Their, 50 Blossom Street, Boston, MA 02114, USA
e-mail: fried.shelley@mgh.harvard.edu

G. Dagnelie (ed.), *Visual Prosthetics: Physiology, Bioengineering, Rehabilitation*,
DOI 10.1007/978-1-4419-0754-7_12, © Springer Science+Business Media, LLC 2011

stimulation methods. The final section of this chapter is devoted to a discussion of some important, unanswered questions.

Abbreviations

AMD	Age-related macular degeneration
AP4	2-Amino-4-phosphonobutyric acid
CNQX	6-Cyano-7-nitroquinoxaline-2,3-dione
DS	Directionally selective
EECP	Electrically elicited cortical potentials
LED	Local edge detector
LFP	Local field potential
MK801	(+)-5-Methyl-10,11-dihydro-5H-dibenzo[a,d]cyclohepten-5,10-imine maleate
NBQX	2-3-Dioxo-6-nitro-1,2,3,4-tetrohydrobenzo[f]quinoxaline-7-sulfonamide
RCS	Royal College of Surgeons
Rd1	Retinal degeneration 1
RGC	Retinal ganglion cell
RP	Retinitis pigmentosa

12.1 Introduction

Over 10 years ago, Humayun et al. [17] demonstrated that electrical stimulation of the retina elicits light percepts, called phosphenes, in patients that had been blind for many years. This remarkable finding has since been duplicated by other research groups [11, 41, 60]. However, the size, shape, color and contrast of phosphenes vary considerably, and, despite considerable effort over the last decade, only limited improvements in quality and consistency have been reported [18, 41].

The reasons underlying the lack of consistent, high-quality percepts are not well understood. While many factors are likely to contribute, one significant factor is thought to be the disparity between the neural activity elicited by the prosthesis vs. the neural activity that arises normally in the healthy retina [42, 58]. Prosthetic elicited activity that is too non-physiological may simply be unintelligible to the brain. Although perfect replication of physiological signals is well beyond current technology, it seems intuitive that the closer elicited patterns come to matching physiological patterns, the better the elicited vision will be.

Methods to create specific patterns of neural activity will arise from a solid understanding of the basic interactions between electrical stimulation and retinal neurobiology. The challenge therefore is to improve our understanding of how and why retinal neurons respond to stimulation. An increasing number of in vitro and in vivo research studies have begun to analyze the responses of retinal ganglion cells (RGCs) to electrical stimulation and we are now beginning to understand some of the more basic mechanisms by which neural activity is generated. The focus of this chapter is to review the progress in this area.

12.2   Responses of RGCs to Electrical Stimulation in Normal Retina

The two most common retinal implant configurations are referred to as sub- and epiretinal; the primary difference being the location of the stimulating electrodes (Fig. 12.1). Epiretinal electrodes are positioned on the innermost surface of the retina; ideally, they are in close proximity to RGCs and the nerve fiber layer. Subretinal electrodes are placed in the outer retina, close to bipolar cell dendrites when used in degenerate retina (but close to photoreceptors in studies that use normal retinal tissue). In this section we will examine the studies on epiretinal and subretinal stimulation of normal retinas; later sections will focus on similar studies in degenerate retina.

12.2.1   Epiretinal Stimulation

12.2.1.1   Target of Stimulation

Surprisingly, the duration of the stimulus pulse is the key parameter that determines which class of retinal neuron is activated by epiretinal stimulation [1, 9, 12, 24, 28]; an idea first reported by Greenberg [12]. Short duration pulses, ~0.1 ms, elicit only a single action potential, typically within 1 ms after the onset of a cathodal pulse [9, 23, 49]. In some cases, the response to a short pulse consists of a spike doublet – the latency of the first spike is <1 ms while the latency of the second spike varies between 5 and 15 ms [49]. These responses persist in the presence of synaptic blockers indicating that they arise from direct activation of RGCs.

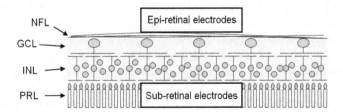

Fig. 12.1 Placement of sub- vs. epiretinal electrodes. Epiretinal electrodes are placed on or near the innermost surface of the retina while subretinal electrodes are positioned in the outer retina – approximately at the location once occupied by the photoreceptors. *NFL* nerve fiber layer, *GCL* ganglion cell layer, *INL* inner nuclear layer, *PRL* photoreceptor layer

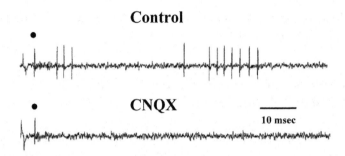

Fig. 12.2 Epiretinal stimulation elicits multiple phases of spiking. The response of a RGC before and after administration of 40 μM CNQX (a glutamate antagonist). The early phase response in each trace is indicated by a *black dot*. The late phase response (*top trace*) consists of two bursts separated by ~30 ms which was eliminated by CNQX (*bottom trace*). The deflection preceding the early phase response in both traces is an artifact of the electrical stimulus. The cell was stimulated with 2 μA (*top trace*) and 6 μA (*bottom trace*); pulse durations were 1 ms. Reprinted from [24], Fig. 1, with permission

On the other hand, long duration pulses, typically those ≥1 ms, elicit two phases of spiking (Fig. 12.2): the first phase consists of a single action potential that occurs shortly after the onset of the pulse [1, 9, 49], identical to the single action potential elicited by short duration pulses. The second phase consists of one or more bursts of action potentials [1, 9, 12, 24, 28] and does not begin until completion of the stimulus pulse [9]. Responses that contain multiple bursts can last tens [9, 23] or even hundreds of milliseconds [1, 53]. The second phase of spiking is completely eliminated by applying antagonists of glutamatergic receptors, e.g. 6-cyano-7-nitroquinoxaline-2,3-dione (CNQX) and 2-3-dioxo-6-nitro-1,2,3,4-tetrahydrobenzo[f]quinoxaline-7-sulfonamide (NBQX) [9, 24, 49] or by applying blockers of all synaptic activity, e.g. Cd^{2+} [28]. These results indicate that the second phase is mediated by glutamatergic input, most likely resulting from the activation of bipolar cells.

The use of whole cell patch clamp recordings allows for direct measurement of the synaptic input currents to RGCs and provides further support that long duration pulses result in excitatory input to RGCs [9, 28]. The bipolar origin of these currents was confirmed by Margalit and Thoreson [28] who showed that excitatory currents were eliminated in the presence of the glutamatergic blockers NBQX and (+)-5-methyl-10,11-dihydro-5H-dibenzo[a,d] cyclohepten-5,10-imine maleate (MK801). Fried et al. [9] showed that increasing the duration of the stimulus pulse (from 1 to 3 ms) resulted in larger excitatory inputs suggesting that the level of bipolar cell activation is proportional to the total amount of delivered current and/or charge.

Long duration pulses also elicit inhibitory activity in RGCs [9, 28] most likely as the result of amacrine cell activation (amacrine cells are believed to be the only source of inhibitory input to RGCs). There are two mechanisms however by which amacrine cells can become activated (Fig. 12.3). The first possibility is that long duration pulses activate amacrine cells directly. The second potential mechanism is that amacrine cells are activated secondary to the activation of bipolar cells (in the normal retina, amacrine cells are activated by glutamatergic input from bipolar cells). To distinguish between

Fig. 12.3 Long pulses activate amacrine cells secondary to bipolar cell activation. *Left*: under control conditions, long pulses elicit both excitatory and inhibitory input in RGCs. Inhibitory input to RGCs can arise either because amacrine cells are activated directly by long pulses, or as a result of excitatory input from activated bipolar cells. *Right*: application of glutamatergic blockers eliminated inhibitory input to RGCs. This indicates that amacrine cells are activated as a result of glutamatergic input from activated bipolar cells

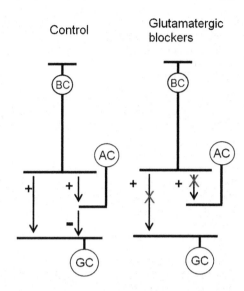

the two possibilities, Margalit and Thoreson applied glutamatergic antagonists (NBQX and MK801) in order to block the bipolar to amacrine cell pathway. Application of these blockers eliminated the inhibitory input to RGCs (Fig. 12.3, right); suggesting that the inhibitory signal to RGCs arises secondary to activation of bipolar cells.

12.2.1.2 The Site of Spike Initiation in RGCs

It is important to understand which element of the RGC (soma, axon, hillock, distal axon, etc.) is the site of spike initiation, e.g. which element has the lowest threshold for spike initiation. If spikes are initiated in the soma or a nearby element, e.g. the axon hillock, then the neurons activated by a given stimulus pulse will be restricted to those that are near the stimulating electrode. This is likely be essential for creating small, focal percepts. If, on the other hand, the axon has the lowest threshold, then elicited neural activity is not spatially restricted and the ability to create predictable percepts is significantly impeded.

Several studies have attempted to identify the RGC element with the lowest threshold. Three different modeling studies [13, 40, 47] each concluded a different region (soma, axon bend, axon initial segment, respectively) had the lowest threshold. In physiological experiments, Jensen et al. [23] found that thresholds were lowest when the stimulating electrode was positioned close to the soma although the specific anatomical site could not be identified with their methods. Sekirnjak et al. [50] inferred the location of somas and axon trajectories from multielectrode array recordings and calculated that the site of lowest threshold was probably on the proximal axon, ~13 μm from the soma. They too, were not able to correlate the low threshold region to a specific anatomical landmark. Grumet et al. [15] used multielectrode array recordings to infer that spikes were initiated in the axon (based on the

Fig. 12.4 The band of dense sodium channels is centered in the region of low threshold. (**a**) Threshold map of a DS ganglion cell. The map has been rotated ~20° in order to align it to the Ankyrin G image in (**c**). *Dark* (cool) areas are the lowest threshold, light (warm) areas are the highest. Thresholds range from 10 to 60 μA. (**b**) Overlay of the threshold map with staining for Ankyrin G, a structural protein associated with dense bands of sodium channels. The soma and axon (process extending *leftwards* from soma) are clearly visible. The *white vertical lines*, which extend up from (**c**), are used to indicate the region of high density Ankyrin G staining. (**c**) Ankyrin G staining in the same cell as in (**a**). (**d**) Low thresholds are also found in segments of the distal axon. The inset shows a threshold map that extends out along the distal axon, approximately 1.1 mm from the soma. The gaps arise from incomplete sampling along the distal axon. Similar to (**a**), each pixel of the threshold map contains an individual threshold measurement; the reduced scale makes it difficult to resolve individual pixels. Threshold at each location along the axon (*dashed line*) is plotted as a function of distance from the soma. The *circled point* indicates the threshold level when the electrode was over the soma. Scale bars: 25 μm for (**a–c**); 100 μm for the *inset* in (**d**)

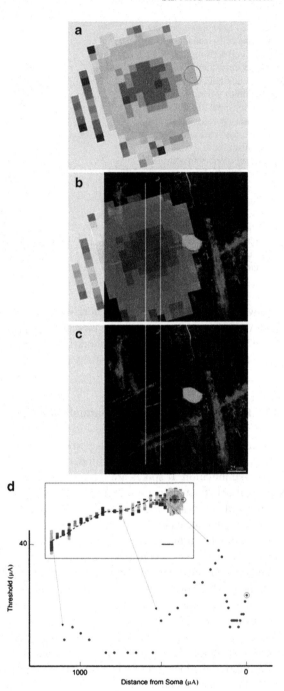

shape of the elicited spike waveform). Studies outside the retina suggest that either the initial segment or the nodes of Ranvier can be the target of stimulation [56].

To determine the site of lowest threshold, Fried et al. [10] measured threshold as a function of the position of the stimulating electrode. Measurements were made in a dense spatial grid around the soma, proximal axon and distal axon of directionally selective (DS) ganglion cells (one of the rabbit ganglion cell types). They found that thresholds were lowest in a region that was centered approximately 40 μm from the soma (Fig. 12.4a). Immunochemical staining revealed that a dense band of voltage-gated sodium channels in the proximal axon was centered at the same approximate location (Fig. 12.4c). Overlay of the two images reveals that the band of sodium channels was centered within the region of low threshold (Fig. 12.4b) suggesting that the band may be the source of low thresholds and possibly the site of spike initiation. Threshold maps were found to be qualitatively, but not quantitatively similar in other cell types.

Additional unpublished measurements from Fried et al. revealed that threshold levels in the distal axon were comparable to (and sometimes lower) than those of the proximal axon (Fig. 12.4d). These results are consistent with the Jensen et al. [23] physiological study as well as with the modeling studies mentioned earlier [13, 47]; all of which found that axonal thresholds were only slightly higher than the lowest thresholds (found in other portions of the cell).

The low activation thresholds associated with RGC axons suggests that focal percepts may be difficult to obtain via stimulation schemes that target RGCs. This further suggests that stimulation methods that avoid activation of axons are needed in order to create spatially relevant patterns of retinal activity.

12.2.1.3   Threshold vs. Stimulating Electrode Diameter

The size of the electrode used to elicit activity will ultimately determine the maximum electrode density within the array. Therefore, small diameter electrodes presumably offer the highest possible spatial resolution. To explore the effects of electrode size, Sekirnjak et al. [49] measured threshold as a function of stimulating electrode diameter and found that both the current and the charge needed to elicit activity was reduced as the stimulating electrode diameter was decreased (Fig. 12.5a, b). However, they also found that both current and charge densities increased as the stimulating electrode diameter decreased (Fig. 12.5c, d). This trade-off arises because the reduction of charge associated with a smaller electrode is less than the corresponding reduction in electrode surface area.

The findings from Sekirnjak et al. are supported by an earlier study from Jensen et al. [25], in which a 125 μm diameter electrode exhibited a reduction in threshold when compared to a 500 μm diameter electrode. The electrodes used by Jensen et al. are larger than those used by Sekirnjak et al. but support the notion that smaller diameter electrodes are associated with lower thresholds (charge and current). Whereas the above studies compare thresholds for direct activation, Ahuja et al. [1] found that 10 μm electrodes had higher thresholds than 200 μm electrodes when

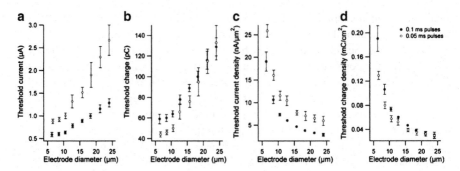

Fig. 12.5 Threshold vs. electrode size. Thresholds in response to 0.1-ms pulses (*filled circles*) and 0.05-ms pulses (*open circles*) were plotted as a function of electrode diameter for rodent RGCs. (**a**) Current, (**b**) charge, (**c**) current density, (**d**) charge density. Each is plotted for the same set of data. All responses were long-latency spikes. Reprinted from [49], Fig. 10, with permission

indirectly activating RGCs (via activation of presynaptic neurons). Ahuja et al. suggest that this may arise because the electric fields of large electrodes extend deeper into the retina and therefore can more easily activate presynaptic neurons. Further research is needed to confirm this hypothesis and also to determine whether other mechanisms are at work.

Sekirnjak et al. also compared the effects of varying the pulse duration and found that threshold increased as the pulse duration decreased from 0.1 to 0.05 ms. This finding is in agreement with an earlier study by Jensen et al. [25] which found that thresholds increased consistently as pulse duration was reduced from 50 to 0.1 ms. These studies suggest that the smallest diameter electrode that could be safely used is also a function of pulse duration and therefore, short duration pulses, which exclusively activate RGCs, may require large-diameter electrodes.

12.2.1.4 Spatial Extent of Activation

Another consideration for creating focal percepts is the spatial extent of activation arising from a single stimulating electrode. Ideally, activation should be limited to the immediate vicinity of the electrode. However, the relationship between the strength of the electric field and the extent of activation is not well known. Several studies, employing a variety of methods, have begun to explore this question.

Jensen et al. [23] measured threshold as a function of the distance between a very small stimulating electrode and the targeted cell body (Fig. 12.6a). They found that threshold was lowest when the stimulating electrode was at (or near) the soma (0.5 μA or 0.31 mC/cm²). Threshold increased as the stimulating electrode was moved away from the soma increasing by a factor of 20–30 at a distance of 100 μm. Sekirnjak et al. and Ahuja et al. [1] found similar increases in mouse and salamander RGCs respectively; Ahuja et al. showed that threshold increased with increasing distance regardless of the pulse duration (durations ranged from 60 to 1,000 μs).

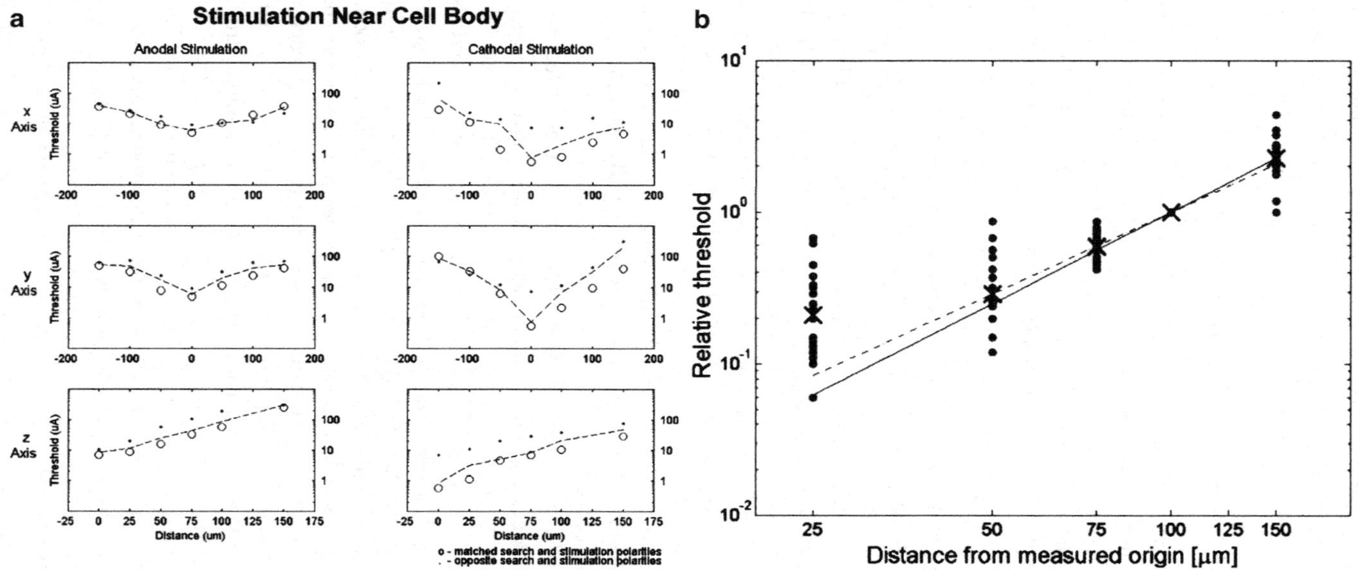

Fig. 12.6 Threshold as a function of electrode position. (**a**) Current thresholds versus electrode displacement along x-, y-, and z-axes for stimulation near the cell body. Each data point indicates a threshold on the linear scale at *left*. Thresholds obtained by anodal and cathodal current pulses are on the *left* and *right*, respectively. *Solid lines* connect median thresholds using the linear vertical axes on the *left*; *dashed lines* connect the same medians using the logarithmic axes on the *right*. (**b**) Normalized thresholds versus vertical displacement of electrode. For each axon, current thresholds at each vertical (i.e., z-axis) displacement were normalized by dividing by the threshold measured at z at 100 µm. Each *dot* represents one normalized threshold plotted on a log–log scale. (×) Median normalized threshold at each displacement. Threshold growth as a power of distance would appear as a *straight line* on these logarithmic axes. If thresholds increased with displacement *squared*, normalized thresholds would all lay along the *solid line*. *Dashed line*: best fit to the data at 75 and 150 µm of all *straight lines* passing through unity at 100 µm, representing thresholds growing as the 1.78 power of displacement. Reprinted from [23], Figs. 5 and 6, with permission

Further support comes from a study by Schanze et al. [46] who found that moving the epiretinal electrode off the retina by 50 μm resulted in a 50% decrease in the cortical response. These results are similar to studies that found the cortical signal decreased when the distance between the stimulating electrode and the retina increased [44–46, 48].

The results in the retina are consistent with a large number of previous non-retinal stimulation studies (see [56] for a review). In general, thresholds increase with the square of the distance (between stimulating electrode and targeted neuron). The equation $I = K \cdot x \cdot r^2$ can be used to describe the increase – where I is the threshold current, r is the distance between electrode and neuron and K is the excitability constant. In non-retinal neurons, the experimentally determined excitability constant was found to be small for large, myelinated neurons and large for small, unmyelinated neurons.

Jensen et al. [23] determined that the threshold for activating brisk-transient rabbit RGCs increased approximately with the square of the distance between stimulating electrode and targeted neuron (the actual threshold increase was in proportion to the distance raised to the 1.8 power). Jensen et al. also found that the rate of threshold increase was lower when the stimulating electrode was within 50 μm of the soma and higher for distances greater than 50 μm. While the reason for the different rates of increase is not known, it is possible that they arise from the long sodium channel bands described by Fried et al. (Fig. 12.4c). Thresholds are lowest when the stimulating electrode is centered directly above the sodium channel band and increase slowly as the electrode moves away from the center of the band but remains above a portion of the band. Once the stimulating electrode moves beyond the edges of the sodium channel band, threshold increases rapidly with increasing distance. Further studies are needed to confirm whether this is in fact the case.

To get a practical sense of how the increase in threshold with distance affected the activation of RGCs, Sekirnjak et al. [49] used a multielectrode array that was capable of stimulating and recording from many closely spaced electrodes. They found that threshold was always lowest when the same electrode that was used to record also delivered the stimulus pulse; if an adjacent electrode (60 μm spacing) was used to stimulate, threshold increased by a factor of three. This suggested that the use of low amplitude pulses would activate only those cells that were close to the electrode.

To confirm that low amplitude stimulation from each electrode operated independently, Sekirnjak et al. independently delivered stimulus pulses from each of seven nearby electrodes; seven distinct responses were recorded (Fig. 12.7a, spacing between neighboring electrodes: 60 μm). They then activated all seven electrodes simultaneously and measured the response in each electrode (Fig. 12.7b). They found that the response elicited by activation from all seven electrodes was nearly identical to the response elicited by activation from a single electrode. This suggests that the activation from one electrode did not interfere with that of neighboring electrodes. This is an encouraging result as it suggests that nearby electrodes can independently create activity in focal regions. At stronger stimulation levels however, Sekirnjak et al. found that stimulation from one electrode activated RGCs in the vicinity of neighboring electrodes. In some cases, activity could be detected 150 μm

Fig. 12.7 Multiple site stimulation. Rat retina was stimulated by 7 electrodes simultaneously (0.8 μA cathodal pulses). (**a**) Overlay of several trials is shown for each electrode (1–7) and evoked long-latency spikes marked with an *asterisk*. *Inset (top right)*: location of active electrodes on the array. Latencies ranged from 5 to 18 ms. (**b**) Traces from neighboring electrodes 1 (*left*) and 2 (*right*). For comparison, spikes are shown for individual stimulation at only that electrode (single) as well as when all 7 electrodes were active (all). Evoked spikes showed no difference. *Arrowhead* indicates that the large spikes seen on electrode 2 were visible on electrode 1 as small deflections. Reprinted from [49], Fig. 9, with permission

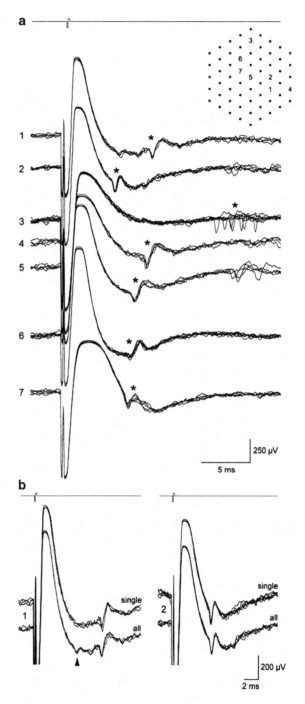

from the site of stimulation. This indicates that the radius of activation is a factor of the stimulus strength.

Several research groups have found that the electric field created by one electrode can interact with the field created by a neighboring electrode. Sekirnjak et al. found that simultaneous activation of several neighboring electrodes resulted in higher thresholds than that from a single neighboring electrode. Interestingly, these interactions were not enough to interfere with the field when the closest electrode was activated (Fig. 12.7b). Similarly, Ahuja et al. [1] measured thresholds for activating salamander RGCs from two 200 μm stimulating electrodes each positioned approximately 250 μm from the cell (center to center spacing). Thresholds for single electrode activation were approximately 13.3 nC and increased to 29.4 nC when stimulation from both electrodes was applied simultaneously. A finite element model presented in the Ahuja et al. study indicates that the threshold increase arises from a reduction in the voltage gradient caused by simultaneous stimulation from the second electrode.

More work is needed to determine under which conditions electrode interactions occur and whether there are means to reduce these interactions. One possible means would be to interleave the stimulus pulses from nearby electrodes – the slight offset in time would presumably minimize the amount of interaction between neighboring electrodes.

12.2.1.5 Selective Activation

In the normal retina, the neural activity in neighboring neurons can be quite different e.g. response duration of a "sustained" cell can last several hundred milliseconds longer than that of a "transient" cell. Similarly, ON and OFF cells typically do not generate spikes at the same time. This wide array of spatially and temporally varying neural activity is transmitted from the retina and reassembled by the visual cortex into our percept of the visual world. The concern arises that if prosthetic stimulation creates identical (or similar) activity in all neighboring neurons, the signal that arrives at the cortex is quite different from the normal signal and may not be intelligible. Methods to selectively activate specific RGC types may help to more closely re-create the signaling patterns created normally by the retina and ultimately improve the quality of the resulting percept.

A formal study of selective activation methods for RGCs has not been reported. Fried et al. [10] measured thresholds in three different types of rabbit RGCs and found that alpha cells (G11) had the lowest threshold while local edge detectors (LED, G1) had the highest (Fig. 12.8). This finding suggests that low amplitude stimulus pulses may be able to selectively activate a single type of RGC (e.g. alpha). Unfortunately, this method of selective activation would at best, apply to a single RGC type only and would not distinguish between ON and OFF cells.

In contrast to the results from Fried et al., Margalit and Thoreson [28] found no difference in thresholds between ON, OFF and ON-OFF RGCs in salamander retina. However, it is not clear whether the populations reported by Margalit and Thoreson

Fig. 12.8 Different ganglion cell types have different thresholds. Each *point ("X")* represents a threshold measurement in a different cell. Ganglion cell types were identified by the cell's light response prior to measurement of threshold. Pulses were 0.1 ms duration, cathodal with a distant ground

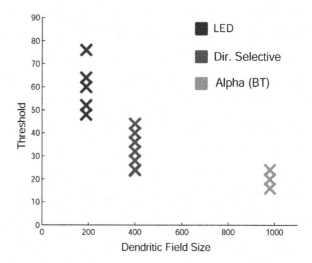

can be correlated to those from Fried et al. For example, Fried et al. found that thresholds for ON and OFF alpha cells were similar. In the Margalit and Thoreson study, it is likely that the ON-OFF cells are a different population from either the ON or OFF types and yet their thresholds were not different. Unfortunately, the limited results from Fried et al. do not preclude the possibility that some RGC types have similar thresholds. Further studies are needed to determine the threshold differences across types and if differences exist, determine whether they can be used to selectively activate specific RGC types.

It is a daunting challenge to think in terms of replicating normal light elicited patterns with a retinal prosthesis. However, there are many incremental improvements that can be realized along the way. For example, ON and OFF varieties of midget and parasol cells are the four principal types of RGCs in the human retina. Together, it is estimated that they account for >90% of all RGCs. Methods that selectively activate only one of these types, for example, would likely lead to elicited patterns of neural activity that are more physiological and therefore result in improved percepts.

12.2.1.6 Temporal Response Properties

The rates at which RGCs generate action potentials [2, 6] as well as the precise timing with which individual action potentials [30, 31] are generated are both thought to play an important role in the neural code transmitted from the retina. The upper limit on RGC spike rates can be estimated from studies by O'Brien et al. [35] and DeVries and Baylor [6]. The maximum spike rates vary for each RGC type; alpha cells have the largest maximum spike frequency (~250 Hz) which sets an approximate upper limit for the response requirements of the prosthetic.

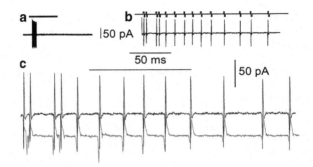

Fig. 12.9 Programmed sequences of short electrical pulse replicate light responses. (**a**) Spiking response to a 1-s light stimulus (*horizontal bar*). (**b**) *Bottom*: expanded time scale from (**a**) reveals individual spike latencies. *Top*: programmed sequence of short pulses derived from individual spike latencies: each cathodal pulse is arranged 0.5 ms before corresponding spike. (**c**) Spikes elicited by programmed sequence of short pulses (*bottom*) precisely match the light elicited spike pattern (*top*) from (**b**). Reprinted from [9], Fig. 7, with permission

As discussed in Sect. 12.2.1.1, short duration stimulus pulses (typically 100 μs) were shown to activate RGCs directly, without activating other elements of the presynaptic circuitry. Each short pulse elicits a single action potential [9, 49], typically within 0.5–1.0 ms of the pulse onset [1, 9, 24, 49]. At higher stimulation frequencies, short pulses continue to elicit one spike per pulse. This was tested originally up to 250 Hz in rabbit [9] and more recently up to 500 Hz in salamander [1]. These spike rates are comparable to the fastest rates of normal, light elicited spiking. Using the one spike per pulse paradigm, Fried et al. programmed sequences of pulses in order to precisely replicate typical RGC light responses (Fig. 12.9).

In contrast to the results from Fried et al., Sekirnjak et al. [49] found that repetitive stimulation at high frequencies resulted in a loss of the one spike per pulse response. At 50 Hz, a slight reduction (~20%) was found and a more significant reduction (~50%) was found at 100 Hz. It is not clear whether and/or how their findings impact the ability to precisely replicate light responses.

Even if the temporal properties of normal RGC signaling can be replicated using short pulses, several important obstacles must be surmounted before this paradigm can be implemented. For example, this method would presumably activate all RGCs close to the stimulating electrode with the same spike patterns resulting in patterns of retinal activation that are non-physiological. Methods for selective activation and avoiding the activation of passing axons are both needed.

The temporal response properties resulting from stimulation of bipolar cells were very different from the responses arising from stimulation of RGCs. Fried et al. [9] showed that bipolar cell input to RGCs decreased as stimulus pulse frequency increased; by 10 Hz the amplitude of the bipolar cell output was barely detectable. Ahuja et al. [1] similarly found that the RGC output was almost completely eliminated at 10 Hz. These findings suggest that temporal frequencies may be limited to <5 Hz if bipolar cells are activated (from either sub- or epiretinal stimulation).

Since activation of bipolar cells leads to activation of amacrine cells [28], the reduction in bipolar cell activity may arise from amacrine cell mediated feedback inhibition. Therefore, methods that reduce or eliminate the secondary activation of amacrine cells are likely to enhance the temporal response to stimulation. Such methods have yet to be developed.

12.2.2 Subretinal Stimulation

12.2.2.1 Target of Stimulation

Similar to epiretinal stimulation, subretinal stimulation activates many different classes of retinal neurons. Stett et al. [53] used pharmacological blockade of synaptic pathways to explore which classes of (chicken) retinal neurons were activated by subretinal stimulation. Under control conditions, 0.5 ms monophasic voltage pulses elicited RGC spiking responses whose durations lasted up to several hundred milliseconds (Fig. 12.10). Addition of magnesium (Mg^{2+}), a general blocker of neurotransmitter release, significantly reduced the RGC responses. The Mg^{2+} sensitive spiking activity in RGCs presumably results from activation of one or more presynaptic excitatory neurons; the likely candidates are photoreceptors, bipolar cells and/or starburst amacrine cells. To identify the specific neurons that were activated, synaptic blockers of the excitatory neurotransmitter glutamate were administered. Application of kynurenic acid, an AMPA/kainate receptor antagonist, greatly reduced the RGC responses to electrical stimulation. Kynurenic acid targets receptors at multiple locations [14, 57] but notably it blocks the output of bipolar cells. A more specific glutamate receptor blocker, 2-amino-4-phosphonobutyric acid (AP4), blocks the synapse between photoreceptors and ON bipolar cells [51]. Application of AP4 abolished the electrically evoked responses. Although Stett et al. [53] did not identify the physiological type of RGC shown in Fig. 12.10, the fact that AP4 abolished the evoked response suggests that this RGC cell was an ON cell. Jensen et al. [24] reported in a later study (using epiretinal stimulation) that electrically evoked responses of ON RGCs but not OFF RGCs in rabbit retina were abolished with AP4.

The AP4 results suggest that photoreceptors are the principal target of electrical stimulation in normal retina. Understanding whether photoreceptors or bipolar cells are the target of subretinal stimulation has important implications for clinical use since patients that have been blind for many years will have few or no viable photoreceptors remaining (Chap. 3). Therefore methods that target photoreceptors are not likely to be useful in clinical applications. More research is needed to determine the relative excitability between photoreceptors and bipolar cells.

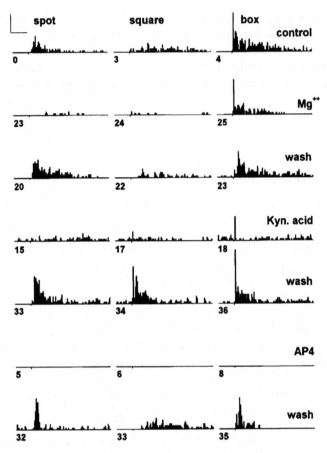

Fig. 12.10 Synaptic blockers reduce the response to subretinal stimulation. Response histograms (5 ms bin width) elicited by 20 repetitions of a single voltage pulses (2 V, 0.5 ms). Application of high [Mg²⁺], kynurenic acid or AP4 reduced the spiking activity. The number at each histogram indicates the time interval (minutes) after switching to the perfusate with the agents given at the *right* and to the standard perfusate for washing out the agents. Scale bars 100 ms, 100 spikes/s. Reprinted from [53], Fig. 6, with permission

12.2.2.2 Threshold vs. Polarity of Stimulation Pulse

It is well known that axons (including those of RGCs) are more sensitive to a cathodal current pulse than to an anodal current pulse [56]. When RGCs are activated through electrical stimulation of presynaptic cells, the situation is not so straightforward.

Jensen and Rizzo [19] reported that when the rabbit retina is stimulated with a subretinal electrode the threshold current needed to activate OFF RGCs was much lower for an anodal current pulse than for a cathodal current pulse. On the other hand, cathodal and anodal current pulses were on average equally effective for activating ON RGC cells. This is illustrated in Fig. 12.11, in which threshold measurements were made for RGC responses to stimulation of the neural network.

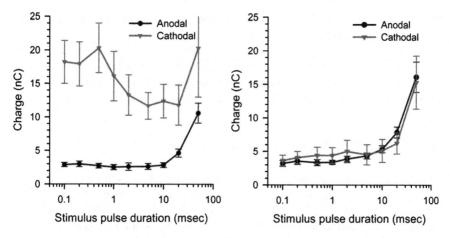

Fig. 12.11 Threshold charge as a function of stimulus pulse duration for OFF ganglion cells (*left graph*) and ON ganglion cells (*right graph*) for both cathodal and anodal stimulus pulses. For OFF ganglion cells (*left*), anodal stimulation produces a substantially lower activation threshold for all pulse durations, while as a whole ON ganglion cells (*right*) are insensitive to the polarity of stimulation. Reprinted from [19], Figs. 4 and 5, with permission

In the chicken retina, Stett et al. [54] reported that when the neural network is stimulated with a subretinal electrode, anodal voltage pulses were overall more effective than cathodal voltage pulses. They found that on average a 3.2-fold difference in thresholds. They did not distinguish between ON and OFF RGCs. Nevertheless, both studies suggest that for indirect activation of RGCs (with a subretinal electrode) an anodal stimulus is in general more effective than a cathodal stimulus. The findings of Jensen and Rizzo [19] further suggest that a cathodal current pulse may bias activation of ON cells over OFF cells. Results such as these may one day underlie methods to selectively activate ON vs. OFF pathways which would allow more physiological patterns of activity to be elicited.

In contrast to the ON-OFF selectivity found in rabbit (described above), a recent study conducted in the mouse retina [22] found that the median threshold current for cathodal stimulation of ON RGCs was only 32% lower than for OFF RGCs and this difference was not statistically significant. Thus, it would seem from the mouse experiments that a cathodal current pulse may not bias activation of ON cells over OFF cells as the findings in the rabbit would suggest. It will be of interest to examine the thresholds of ON and OFF RGCs to anodal and cathodal current pulses in the primate retina.

12.2.2.3 Spatial Extent of Activation

Stett et al. [54] examined the spatial extent of activation of RGCs in the chicken retina. They used an ultra-fine (1-μm diameter) tip electrode for stimulating the retina and a dense multielectrode array to record simultaneously from many RGCs in the retina.

They reported a half width of an "electrical point spread function" of ~100 μm. This distance on the retina corresponds to a visual angle of 21′ in the human eye. A minimum angle of resolution of 21′ corresponds to a visual acuity of ~20/400. It will be of interest to examine the electrical point spread function of RGCs in the primate fovea where the convergence of photoreceptors and bipolar cells onto RGCs is very low. The findings may indicate that a higher visual acuity is possible.

12.2.2.4 Temporal Response Properties

Fried et al. [8] showed that when rabbit RGCs are indirectly activated with an epiretinal stimulating electrode, bipolar cell output is drastically reduced by a 10 Hz stimulation frequency. The situation is not much different with a subretinal stimulating electrode. Jensen and Rizzo [20] showed that the responses of rabbit RGCs to stimulation of the neural network began to diminish in size when the retina was stimulated within ~400 ms of a preceding current pulse (Fig. 12.12). The shorter the interpulse interval, the smaller was the response to the second stimulation pulse. They also studied the responses of RGCs to trains of pulses applied at different frequencies. As expected, the responses were greatly reduced for stimulation frequencies >25 Hz. These data indicate that rapid electrical stimulation of the retina in patients with a retinal prosthesis may be counterproductive, assuming that RGCs are being activated through the neural network.

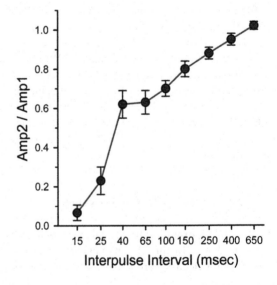

Fig. 12.12 Mean paired-pulse depression of RGC cell response amplitudes in rabbit retina. Data were collected using biphasic current pulses of 1 ms per phase. *Amp1* amplitude of first response; *Amp2* amplitude of second response. Reprinted from [20], Fig. 2, with permission

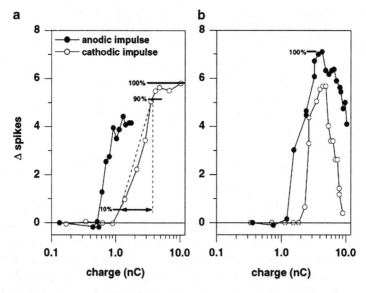

Fig. 12.13 The average number of spikes elicited in two RGCs (**a**, **b**) as a function of stimulus charge. The responses of both cells increased with increased charge injection. While the response of the cell on the *left* (**a**) appeared to plateau with high charge injection, the response of the cell on the *right* (**b**) fell with high charge injection. Both cells were more sensitive to an anodal pulse. Reprinted from [54], Fig. 3, with permission

12.2.2.5   Dynamics of the Retinal Response

Stett et al. [53] found that the number of spikes in an evoked retinal response increased with increasing voltage level. In a later study [54], they reported that the number of spikes evoked per voltage pulse was almost a logarithmic function of the charge delivered (Fig. 12.13). This finding suggests that it may be possible to influence the intensity of a visual percept in a patient with a retinal prosthesis by adjusting the amplitude of the current pulses. Unfortunately, further increases in the injected charge eventually led to a decrease in the number of spikes for some cells (Fig. 12.13b).

12.2.2.6   Comparing Sub- vs. Epiretinal Stimulation

It is difficult to draw definitive conclusions from the existing studies that compare the thresholds elicited by sub- vs. epiretinal stimulation. O'Hearn et al. [36] measured thresholds in the mouse retina using two 125 μm disk electrodes (bipolar configuration) and 1 ms duration biphasic pulses (cathodal first). Thresholds for epiretinal stimulation were 30 μA (0.61 mC/cm²) while thresholds for subretinal stimulation were 77 μA (1.57 mC/cm²) suggesting that epiretinal stimulation is more effective (Table 12.1).

Table 12.1 Comparison of sub- vs. epiretinal thresholds

O'Hearn et al. (125 μm diameter electrode, biphasic pulses)

Pulse duration	Epiretinal	Subretinal	
1.0 ms	30 μA	77 μA	

Jensen et al. (500 μm diameter electrode)

	I	II	III
	Epiretinal	Subretinal	Subretinal
Pulse duration	Cathodal	Cathodal	Anodal
0.1 ms	54 μA	180	29
2.0 ms	6.3 μA	6.7	1.3

The results of two different Jensen et al. studies [19, 25] can also be used to compare thresholds. Both studies used identical stimulation parameters, including a 500 μm stimulating electrode and pulse durations of 0.1 and 2.0 ms and were restricted to OFF RGCs of the rabbit retina. The thresholds in response to epiretinal stimulation were lower than those from subretinal stimulation for both pulse durations tested (0.1 and 2.0 ms); however the difference was small for 2.0 ms pulses. While these results are qualitatively in agreement with O'Hearn et al., Jensen and Rizzo point out that it may be more accurate to compare epiretinal cathodal pulses to subretinal anodal pulses since in both cases current flows through the retina in the same direction [19]. Under these conditions, subretinal thresholds are significantly lower with both short and long pulses (Table 12.1, bottom, compare columns I and III).

It is difficult to assess the discrepancies between the O'Hearn et al. and Jensen and Rizzo studies. Both the stimulus waveform and electrode size are different between studies and either (or both) may contribute to threshold differences. In addition, the O'Hearn study was unable to distinguish between the early and late phase responses and therefore comparison with the Jensen et al. studies may not be appropriate. In addition, neither group was able to ascertain whether photoreceptors were activated by stimulation raising the possibility that the differences arise because different neurons were activated in each study. Further research is needed to better understand the relative thresholds and underlying mechanism differences between sub- and epiretinal stimulation.

12.3   Electrophysiological Properties of RGCs in Degenerate Retina

Electrophysiological studies on RGCs in degenerate retina have been made in the *rd1* mouse and the dystrophic Royal College of Surgeons (RCS) rat. Similar to a form of retinitis pigmentosa (RP) in humans, the *rd1* mouse has a mutation in the gene for the β-subunit of cGMP phosphodiesterase-6 [7]. As a consequence, rapid rod photoreceptor degeneration begins at approximately postnatal day (P)10, with nearly all photoreceptors lost by P36 [3]. The RCS rat has a mutation in the receptor tyrosine kinase gene Mertk, which results in the failure of the outer segments of photoreceptors to be phagocytosed and eventually causing the death of photoreceptors [5, 34]. Degeneration of photoreceptors in RCS rat is slower than that in *rd1* mouse. Only ~1/3 of photoreceptors are lost by P30. However, by P75 only scattered photoreceptors are present [37].

Pu et al. [38] reported that RGCs in RCS rats show an increased level of baseline spiking and by P47 (before complete degeneration of photoreceptors) were predominately OFF cells. Similar findings were reported in the *rd1* mouse retina [52] where baseline spiking levels went from <1 Hz in RGCs of normal, wild type mice to as high as 20 Hz in retinas from *rd1* mice. Furthermore, Stasheff [52] found that ~2/3 of *rd1* RGCs exhibited rhythmic bursts (peaks at ~6 and ~12 Hz) of activity. As in the RCS rat retina, Stasheff [52] reported that ON and OFF responses are differentially affected in early stages of retinal degeneration *rd1* mouse retina. Light-evoked ON responses in *rd1* RGCs at P14-P15 were reduced more than OFF responses, and unlike OFF responses many of the ON responses showed an increased latency.

From patch-clamp recordings, Margolis et al. [29] reported rhythmic spike activity (frequency of ~10 Hz) in *rd1* mouse retinas of P36-P50. The spike bursting was present in older mice as well, although the frequency of the bursts decreased 2-3-fold. Margolis et al. further showed that the rhythmic bursting was not generated intrinsically in RGCs but was due to strong, aberrant synaptic input. Intrinsic electrical properties of (morphologically identified) ON and OFF RGCs in *rd1* mouse retinas were similar to those in wild-type mouse retinas.

The hyperactivity and bursting activity in RGCs indicates that new strategies for prosthetic stimulation of the retina may be necessary. For example, higher spiking levels may be needed to obtain an adequate signal to noise ratio. Alternatively, it may be necessary to develop stimulation strategies that reduce the level of baseline activity in order for the brain to be able to understand the signal leaving the retina. However, it is important to first determine whether increased spiking levels and bursts of activity are present clinically (in patients) before new stimulation strategies are developed.

12.4   Responses of RGCs to Electrical Stimulation in Degenerate Retina

The prosthetic must ultimately function in patients with retinal degenerative disease and therefore stimulation methods for clinical use must be tailored to these types of retinas. While many studies have examined the structural changes that occur as part of the degenerative process (Chap. 3), the corresponding physiological changes are not as well known. It is important to understand how the anatomical and physiological changes in degenerate retina will affect the response to electrical stimulation. In this section, we explore the few studies that have looked at the electrophysiological properties of RGCs in degenerate retina as well as those that have looked at the responses of RGCs in these retinas to electrical stimulation.

12.4.1 Epiretinal Stimulation

Suzuki et al. [55] measured the thresholds of RGCs in wild-type and 16-week old *rd1* mouse retinas to biphasic current pulses delivered through a 125-μm electrode positioned on the epiretinal surface. The thresholds were on average 20–50% higher in *rd1* mouse retinas, depending on pulse duration. A description of the evoked responses was not provided so it is unclear whether the thresholds were from direct or indirect activation of the RGCs.

O'Hearn et al. [36] examined the thresholds for activation of RGCs in 8–12 week old wild-type and *rd1* mouse retinas. Retinas were stimulated with a pair of electrodes (125-μm diameter) that were positioned either epiretinally or subretinally. With epiretinal stimulation, the thresholds for activation of RGCs in *rd1* mouse retinas were 1.8-fold higher. The short-latency responses suggest that the RGCs were directly activated although this was not discussed by the authors. If so, then perhaps the properties of the sodium channel bands that presumably underlie RGC activation thresholds (described in Sect. 12.2.1.2) change during the degenerative process and the change results in a higher threshold.

12.4.2 Subretinal Stimulation

12.4.2.1 Response Properties of RGCs

Jensen and Rizzo [21] compared the electrically evoked responses of RGCs in wild-type and *rd1* mouse retinas to stimulation of the neural network. Using biphasic current pulses, they found that RGCs in *rd1* mouse retinas respond similarly to wild-type RGCs. In both wild-type and *rd1* mouse retinas, three types of electrically evoked responses were observed (Fig. 12.14). Type I cells

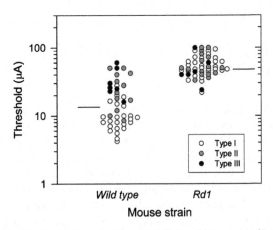

Fig. 12.14 *Dot plot* of thresholds for activation of RGCs in wild-type and *rd1* mouse retinas with biphasic current pulses. See Sect. 12.4.2.1 for a description of the three types of RGCs. Reprinted from [22], Fig. 4, with permission

elicited a single burst of spikes within 20 ms following application of the electrical stimulus, type II cells elicited a single burst of spikes with a latency greater than 37 ms, and type III cells elicited two or more bursts of spikes. The similarity of electrically evoked responses in wild-type and *rd1* mouse retinas led Jensen and Rizzo to suggest that postreceptoral neurons (not photoreceptors themselves) determine the response properties of RGCs to electrical stimuli.

12.4.2.2   Activation Thresholds of RGCs

As noted above (Sect. 12.4.1), O'Hearn et al. [36] examined the thresholds for activation of RGCs in wild-type and *rd1* mouse retinas with a pair of electrodes that were positioned either epiretinally or subretinally. With subretinal stimulation, the thresholds for activation of RGCs in wild-type and *rd1* retinas were not statistically different. In contrast, Jensen and Rizzo [21] reported that the thresholds of RGCs in *rd1* mouse retinas were 3.6-fold higher than the thresholds of RGCs in wild-type mouse retinas. The elevated thresholds occurred for all ages examined, ranging from postnatal day (P) 25 to P186. Of the three types of RGCs identified (see Sect. 12.4.2.1), type I RGC cells appeared to be particularly affected.

The discrepancy in the findings between Jensen and Rizzo [21] and O'Hearn et al. [36] may be because Jensen and Rizzo were examining the thresholds of RGCs due to indirect activation (through the network), whereas O'Hearn et al might have been examining thresholds of RGCs to direct activation, judging from the short-latency (<3 ms) responses they reported in their study.

Although Jensen and Rizzo [22] proposed that photoreceptors lower the thresholds for activation of RGCs, the reason for the higher thresholds in *rd1* mouse retinas is not yet fully understood. Bipolar cells provide most of the excitatory input to RGCs. Perhaps, a greater amount of depolarizing current is needed to stimulate neurotransmitter release from these cells, either from the absence of facilitatory effects from photoreceptors or due to early structural changes in the bipolar cells themselves.

12.5   Cortical Responses to Retinal Stimulation

12.5.1   Spatial Properties Revealed by Cortical Measurements

Several studies have explored the cortical response that result from electrical stimulation of the retina. These types of studies not only confirm that elicited retinal activity is transmitted to higher visual centers, they also provide a more global view of the results of stimulation. The position of cortical electrodes can vary from surface (scalp) to penetrating: less invasive methods typically record electrically elicited cortical potentials (EECPs) – slow potentials that reflect activity of large populations of neurons. On the other hand, more invasive methods (e.g. penetrating electrodes) record brief activity (action potentials) from only a few neurons (typically one or two).

12.5.2 Local Field Potentials

Penetrating electrodes that are appropriately positioned and configured can also record local field potentials (LFPs). Whereas spiking measures neuronal output, LFPs are thought to reflect input signals to the neuron and arise from the electric currents associated with dendritic activity of a relatively large number of nearby neurons [26, 27, 32, 33]. The time course of the LFP signal contains one early and several late components; the early component reflects direct synaptic input while later components reflect later processing most likely arising from intracortical connections [7, 59]. By carefully positioning the recording electrodes at layer IV of visual cortex (the input layer), and looking at only the early LFP components, a measure of the input signal to the visual cortex can be obtained. Presumably this is a direct assessment of the retinal output arising from electrical stimulation.

12.5.3 Elicited Responses Are Focal

Similar to the findings of Sekirnjak et al. [49], Schanze, Wilms, Eger and colleagues [8, 43, 45, 46, 59] found that retinal stimulation from a multielectrode array elicited small focal and spatially discrete regions of cortical activity (LFPs). The average size was approximately 1.3 mm of cortex corresponding to a visual angle of 1.5° although nearly 10% of the regions were much smaller (~0.4 mm cortex/0.5° visual angle). Similar sizes (0.68° visual angle) were obtained by Cottaris and Elfar [4] in a recent study that used similar methodology. Activated regions elicited by individual stimulating electrode were topographically arranged [46, 59] confirming that electrical stimulation is transmitted retinotopically. The activated cortical regions and corresponding stimulating electrodes were not perfectly aligned however [46, 59], possibly because of limited resolution from the cortical array or possibly because electrical stimulation activated axons of distal neurons.

The width of the activated cortical region increased with the amplitude of the stimulus pulse [59]. At threshold, the width of the activated region was smallest (average ~1°) increasing by 10% in response to doubling the amplitude and by 67% in response to a tenfold increase in amplitude. These increases are smaller than might be expected since a tenfold increase in amplitude would more than double the retinal area containing activated neurons. In contrast to the *increases* observed by Schanze and colleagues, Cottaris and Elfar [4] observed a *reduction* in width with increases in stimulus strength. Cottaris and Elfar speculate that the differences may arise from the differences in pulse durations used across the two studies, perhaps because different pulse durations target different classes of retinal neurons. Further study is needed to resolve the mechanism(s) by which patches of retinal activity are transformed into patches of cortical activity.

12.5.4   Cortical Measurements Reveal Electrode Interactions

A series of papers by Schanze, Wilms, Eger and colleagues [8, 43, 45, 46, 59] found evidence of interactions between nearby electrodes. Schanze et al. [45] found that both excitatory and inhibitory interactions occur and that the sign of the interaction depended on the distance between the retinal stimulating electrodes. Specifically, they found that electrode separations of 3–4° generated excitatory interactions resulting in cortical activation regions that were larger than the sum of the two individual regions. Smaller or larger electrode separations generated the opposite effect: inhibitory interactions where the size of the activation regions was less than the sum of the individual regions.

These studies are seemingly in conflict with those of Sekirnjak et al. [49] however, there are considerable methodological differences between the studies that might underlie the observed differences. For example, while the Sekirnjak et al. study reflects the activity of a limited number of RGCs, the cortical measurements presumably reflect the activity of a large population of retinal neurons. In addition, the input activity to cortex reflects synaptic processing that occurs in the thalamus. Any of these factors might account for the electrode interactions observed in the cortical studies. Understanding these interactions will be important for effectively constructing complex spatial images from individual percepts (phosphenes).

12.5.5   Temporal Responsiveness in Cortex

The maximum spiking rate in cortical neurons is typically around 50 Hz [16, 39]; considerably slower than the maximum rate of 250 Hz generated by RGCs (Sect. 12.2.1.6). This reduction in maximum spike rate is thought to occur at the synapse between RGCs and lateral geniculate neurons. Therefore, it is important to understand how the temporal parameters of retinal stimulation translate into the temporal response patterns of cortical neurons.

To study this, Wilms et al. [59] measured the temporal properties of the LFP early component. As mentioned earlier the early component of the LFP is thought provide a representative measure of the retinal output. Rise times (10–90%) ranged from 4.8 to 8.3 ms with lower values arising from higher amplitude stimuli. A sine curve was best fit to each rise time; the resultant frequencies of 40–70 Hz were obtained by multiplying each rise time by three. More sophisticated methods of estimating the temporal responsiveness yielded similar results. This suggests that that adequate temporal resolution may be achievable with electrical stimulation of the retina. Additional findings from Wilms et al. [59] suggest that larger amplitude pulses create better temporal resolution.

Unfortunately, the large amplitude stimulus pulses that increase temporal resolution may simultaneously lower the spatial resolution. The authors suggest that in clinical use these parameters could be varied depending on whether the patient has a more pressing need for spatial or temporal information. The study from Cottaris and Elfar [4] however suggests that such a tradeoff may not be necessary since they did not see a decrease in spatial resolution; additional studies are needed to resolve this.

12.6 Suggestions for Future Studies

Several important research goals emerge from a review of the studies to date. First, the thresholds associated with eliciting activity in RGCs during animal studies are consistently low – well below the charge density limits considered to be safe for electrodes. These low levels are in contrast to the much higher thresholds required to elicit percepts during clinical trials [17, 18, 41]. Elevated thresholds raise many concerns. For example, larger power supplies that generate more heat will be required. Also, the diameter of the stimulating electrodes will need to be increased in order to maintain charge densities below established safety levels. Unfortunately, increasing the size of the electrode reduces the potential resolution of these devices.

The reasons underlying the threshold differences are not well understood. Possibilities include structural and functional alterations in the diseased retina, variability in the distance between electrode and neurons (smaller and more controllable during animal studies), uncertainty in the intactness of the ascending visual pathways in patients that have been blind for many years, and appropriateness of the stimulation methods used to elicit clinical percepts. Studies that systematically elucidate which factors contribute most to threshold differences are needed. Presumably, an improved understanding of the factors that influence thresholds will lead to more efficient stimulation methods for eliciting clinical percepts.

A second research goal that emerges is to develop better methods of stimulation. Existing retinal prosthetics typically use stimulating electrodes with diameters that are 10–20 times larger than the diameter of RGC somata. Stimulation from these larger electrodes presumably results in similar patterns of elicited neural activity in large numbers of RGCs situated in and around the electrode region. Light responses from neighboring RGCs (of different types) utilize different patterns of spike activity – differences can include variations in both spike frequency and total spike count. Thus, the prosthetic elicited patterns of activity are considerably different than normal physiological patterns. Stimulation methods that bring the elicited neural activity closer to physiological patterns are likely to improve the quality of elicited vision, even if every aspect of normal signaling patterns cannot be replicated by existing devices.

The need for improved stimulation methods becomes even more necessary after considering the likely clinical applications for retinal prosthetics. The most common form of retinal blindness arises from age-related macular degeneration (AMD). However many patients with AMD retain some useful peripheral vision. Therefore, surgical implantation of the prosthetic needs to provide clinical benefit that outweighs the risk of damage to existing vision. Sophisticated methods of activation will be to needed to achieve these high levels of vision. For patients that are completely blind, the criteria for success may be lower but will still need to meet or exceed the information provided from a white cane or guide dog in order to justify the risk and the costs associated with implantation. Once again, fairly complex methods of stimulation are likely to be needed.

The third research goal is to develop a better understanding of the changes that occur in the retina as part of the degenerative process. For example, studies that show that background spiking levels increase in RGCs of the degenerate retina suggest that the prosthetic may not only need to create spiking activity, it may also need to suppress activity as well. Changes in baseline activity must be fully understood before appropriate stimulation schemes can be developed. In addition, several genetic models of retinal degeneration exhibit drastic changes in both cell structure and synaptic connections. If portions of the inner retina are destroyed, stimulation schemes that target neurons presynaptic to RGCs may not be effective. It is necessary to fully understand these changes before appropriate stimulation methods can rationally be developed. Interestingly, since clinical trials (in blind subjects) indicate that large areas of the human retina remain viable and that retinotopic wiring persists, it is possible that the degenerative process in humans may be less severe than those reported in laboratory animals.

Finally, while methods for selective activation of individual types of RGCs await development, they promise great insight to our understanding of visual processing. For example, if it were possible to selectively activate a single population of RGCs (e.g. midget or parasol), its role in visual perception could be explored. If more than one population could be independently activated, knowledge of how multiple RGC types act in concert could also be explored (e.g. do spikes from the two types need to be generated synchronously?). Similar studies could be performed to elucidate the roles of the ON and OFF systems – a possibility arising from work that indicates these two types may have different thresholds in response to subretinal stimulation. Questions such as these have been difficult to explore using more conventional research tools but offer tremendous insight into the function of the visual system once appropriate methods are developed.

Acknowledgments Support provided by Department of Veterans Affairs, Rehabilitation Research and Development Service.

References

1. Ahuja AK, Behrend MR, Kuroda M, et al. (2008), *An in vitro model of a retinal prosthesis.* IEEE Trans Biomed Eng, **55**(6): p. 1744–53.
2. Caldwell JH, Daw NW (1978), *New properties of rabbit retinal ganglion cells.* J Physiol, **276**: p. 257–76.
3. Carter-Dawson LD, LaVail MM, Sidman RL (1978), *Differential effect of the rd mutation on rods and cones in the mouse retina.* Invest Ophthalmol Vis Sci, **17**(6): p. 489–98.
4. Cottaris NP, Elfar SD (2009), *Assessing the efficacy of visual prostheses by decoding ms-LFPs: application to retinal implants.* J Neural Eng, **6**(2): 026007.

5. D'Cruz PM, Yasumura D, Weir J, et al. (2000), *Mutation of the receptor tyrosine kinase gene Mertk in the retinal dystrophic RCS rat.* Hum Mol Genet, **9**(4): p. 645–51.
6. DeVries SH, Baylor DA (1997), *Mosaic arrangement of ganglion cell receptive fields in rabbit retina.* J Neurophysiol, **78**(4): p. 2048–60.
7. Eckhorn R, Wilms M, Schanze T, et al. (2006), *Visual resolution with retinal implants estimated from recordings in cat visual cortex.* Vision Res, **46**(17): p. 2675–90.
8. Eger M, Wilms M, Eckhorn R, et al. (2005), *Retino-cortical information transmission achievable with a retina implant.* Biosystems, **79**(1–3): p. 133–42.
9. Fried SI, Hsueh HA, Werblin FS (2006), *A method for generating precise temporal patterns of retinal spiking using prosthetic stimulation.* J Neurophysiol, **95**(2): p. 970–8.
10. Fried SI, Lasker AC, Desai NJ, et al. (2009), *Axonal sodium-channel bands shape the response to electric stimulation in retinal ganglion cells.* J Neurophysiol, **101**(4): p. 1972–87.
11. Gekeler F, Messias A, Ottinger M, et al. (2006), *Phosphenes electrically evoked with DTL electrodes: a study in patients with retinitis pigmentosa, glaucoma, and homonymous visual field loss and normal subjects.* Invest Ophthalmol Vis Sci, **47**(11): p. 4966–74.
12. Greenberg R (1998). *Analysis of electrical stimulation of the vertebrate retina – work towards a retinal prosthesis.* Thesis, Biomedical Engineering, Johns Hopkins, Baltimore.
13. Greenberg RJ, Velte TJ, Humayun MS, et al. (1999), *A computational model of electrical stimulation of the retinal ganglion cell.* IEEE Trans Biomed Eng, **46**(5): p. 505–14.
14. Grilli M, Raiteri L, Patti L, et al. (2006), *Modulation of the function of presynaptic alpha7 and non-alpha7 nicotinic receptors by the tryptophan metabolites, 5-hydroxyindole and kynurenate in mouse brain.* Br J Pharmacol, **149**(6): p. 724–32.
15. Grumet AE, Wyatt JL, Jr., Rizzo JF, 3rd (2000), *Multi-electrode stimulation and recording in the isolated retina.* J Neurosci Methods, **101**(1): p. 31–42.
16. Grusser OJ, Creutzfeldt O (1957), *[Neurophysiological basis of Brucke-Bartley effect; maxima of impulse frequency of retinal and cortical neurons in flickering light of middle frequency.].* Pflugers Arch, **263**(6): p. 668–81.
17. Humayun MS, de Juan E, Jr., Dagnelie G, et al. (1996), *Visual perception elicited by electrical stimulation of retina in blind humans.* Arch Ophthalmol, **114**(1): p. 40–6.
18. Humayun MS, de Juan E, Jr., Weiland JD, et al. (1999), *Pattern electrical stimulation of the human retina.* Vision Res, **39**(15): p. 2569–76.
19. Jensen RJ, Rizzo JF, 3rd (2006), *Thresholds for activation of rabbit retinal ganglion cells with a subretinal electrode.* Exp Eye Res, **83**(2): p. 367–73.
20. Jensen RJ, Rizzo JF, 3rd (2007), *Responses of ganglion cells to repetitive electrical stimulation of the retina.* J Neural Eng, **4**(1): p. S1–6.
21. Jensen RJ, Rizzo JF, 3rd (2008), *Activation of retinal ganglion cells in wild-type and rd1 mice through electrical stimulation of the retinal neural network.* Vision Res, **48**(14): p. 1562–8.
22. Jensen RJ, Rizzo JF (2009), *Activation of ganglion cells in wild-type and rd1 mouse retinas with monophasic and biphasic current pulses.* J Neural Eng, **6**(3): 035004.
23. Jensen RJ, Rizzo JF, 3rd, Ziv OR, et al. (2003), *Thresholds for activation of rabbit retinal ganglion cells with an ultrafine, extracellular microelectrode.* Invest Ophthalmol Vis Sci, **44**(8): p. 3533–43.
24. Jensen RJ, Ziv OR, Rizzo JF (2005), *Responses of rabbit retinal ganglion cells to electrical stimulation with an epiretinal electrode.* J Neural Eng, **2**(1): p. S16–21.
25. Jensen RJ, Ziv OR, Rizzo JF, 3rd (2005), *Thresholds for activation of rabbit retinal ganglion cells with relatively large, extracellular microelectrodes.* Invest Ophthalmol Vis Sci, **46**(4): p. 1486–96.
26. Kruse W, Eckhorn R (1996), *Inhibition of sustained gamma oscillations (35-80 Hz) by fast transient responses in cat visual cortex.* Proc Natl Acad Sci USA, **93**(12): p. 6112–7.
27. Logothetis NK (2002), *The neural basis of the blood-oxygen-level-dependent functional magnetic resonance imaging signal.* Philos Trans R Soc Lond B Biol Sci, **357**(1424): p. 1003–37.
28. Margalit E, Thoreson WB (2006), *Inner retinal mechanisms engaged by retinal electrical stimulation.* Invest Ophthalmol Vis Sci, **47**(6): p. 2606–12.

29. Margolis DJ, Newkirk G, Euler T, Detwiler PB (2008), *Functional stability of retinal ganglion cells after degeneration-induced changes in synaptic input.* J Neurosci, **28**(25): p. 6526–36.
30. Mastronarde DN (1989), *Correlated firing of retinal ganglion cells.* Trends Neurosci, **12**: p. 75–80.
31. Meister M, Lagnado L, Baylor DA (1995), *Concerted signaling by retinal ganglion-cells.* Science, **270**(5239): p. 1207–10.
32. Mitzdorf U (1985), *Current source-density method and application in cat cerebral cortex: investigation of evoked potentials and EEG phenomena.* Physiol Rev, **65**(1): p. 37–100.
33. Mitzdorf U (1987), *Properties of the evoked potential generators: current source-density analysis of visually evoked potentials in the cat cortex.* Int J Neurosci, **33**(1–2): p. 33–59.
34. Nandrot E, Dufour EM, Provost AC, et al. (2000), *Homozygous deletion in the coding sequence of the c-mer gene in RCS rats unravels general mechanisms of physiological cell adhesion and apoptosis.* Neurobiol Dis, **7**(6 Pt B): p. 586–99.
35. O'Brien BJ, Isayama T, Richardson R, Berson DM (2002), *Intrinsic physiological properties of cat retinal ganglion cells.* J Physiol, **538**(Pt 3): p. 787–802.
36. O'Hearn TM, Sadda SR, Weiland JD, et al. (2006), *Electrical stimulation in normal and retinal degeneration (rd1) isolated mouse retina.* Vision Res, **46**(19): p. 3198–204.
37. Pennesi ME, Nishikawa S, Matthes MT, et al. (2008), *The relationship of photoreceptor degeneration to retinal vascular development and loss in mutant rhodopsin transgenic and RCS rats.* Exp Eye Res, **87**(6): p. 561–70.
38. Pu M, Xu L, Zhang H (2006), *Visual response properties of retinal ganglion cells in the royal college of surgeons dystrophic rat.* Invest Ophthalmol Vis Sci, **47**(8): p. 3579–85.
39. Rager G, Singer W (1998), *The response of cat visual cortex to flicker stimuli of variable frequency.* Eur J Neurosci, **10**(5): p. 1856–77.
40. Resatz S, Rattay F (2004), *A model for the electrically stimulated retina.* Math Comput Model Dyn Syst, **10**(2): p. 93–106.
41. Rizzo JF, 3rd, Wyatt J, Loewenstein J, et al. (2003), *Perceptual efficacy of electrical stimulation of human retina with a microelectrode array during short-term surgical trials.* Invest Ophthalmol Vis Sci, **44**(12): p. 5362–9.
42. Roska B, Werblin F (2001), *Vertical interactions across ten parallel, stacked representations in the mammalian retina.* Nature, **410**(6828): p. 583–7.
43. Sachs HG, Gekeler F, Schwahn H, et al. (2005), *Implantation of stimulation electrodes in the subretinal space to demonstrate cortical responses in Yucatan minipig in the course of visual prosthesis development.* Eur J Ophthalmol, **15**(4): p. 493–9.
44. Salzmann J, Linderholm OP, Guyomard JL, et al. (2006), *Subretinal electrode implantation in the P23H rat for chronic stimulations.* Br J Ophthalmol, **90**(9): p. 1183–7.
45. Schanze T, Greve N, Hesse L (2003), *Towards the cortical representation of form and motion stimuli generated by a retina implant.* Graefes Arch Clin Exp Ophthalmol, **241**(8): p. 685–93.
46. Schanze T, Wilms M, Eger M, et al. (2002), *Activation zones in cat visual cortex evoked by electrical retina stimulation.* Graefes Arch Clin Exp Ophthalmol, **240**(11): p. 947–54.
47. Schiefer MA, Grill WM (2006), *Sites of neuronal excitation by epiretinal electrical stimulation.* IEEE Trans Neural Syst Rehabil Eng, **14**(1): p. 5–13.
48. Schwahn HN, Gekeler F, Kohler K, et al. (2001), *Studies on the feasibility of a subretinal visual prosthesis: data from Yucatan micropig and rabbit.* Graefes Arch Clin Exp Ophthalmol, **239**(12): p. 961–7.
49. Sekirnjak C, Hottowy P, Sher A, et al. (2006), *Electrical stimulation of mammalian retinal ganglion cells with multielectrode arrays.* J Neurophysiol, **95**(6): p. 3311–27.
50. Sekirnjak C, Hottowy P, Sher A, et al. (2008), *High-resolution electrical stimulation of primate retina for epiretinal implant design.* J Neurosci, **28**(17): p. 4446–56.
51. Slaughter MM, Miller RF (1981), *2-amino-4-phosphonobutyric acid: a new pharmacological tool for retina research.* Science, **211**(4478): p. 182–5.
52. Stasheff SF (2008), *Emergence of sustained spontaneous hyperactivity and temporary preservation of OFF responses in ganglion cells of the retinal degeneration (rd1) mouse.* J Neurophysiol, **99**(3): p. 1408–21.

53. Stett A, Barth W, Weiss S, et al. (2000), *Electrical multisite stimulation of the isolated chicken retina.* Vision Res, **40**(13): p. 1785–95.
54. Stett A, Mai A, Herrmann T (2007), *Retinal charge sensitivity and spatial discrimination obtainable by subretinal implants: key lessons learned from isolated chicken retina.* J Neural Eng, **4**(1): p. S7–16.
55. Suzuki S, Humayun MS, Weiland JD, et al. (2004), *Comparison of electrical stimulation thresholds in normal and retinal degenerated mouse retina.* Jpn J Ophthalmol, **48**(4): p. 345–9.
56. Tehovnik EJ, Tolias AS, Sultan F, et al. (2006), *Direct and indirect activation of cortical neurons by electrical microstimulation.* J Neurophysiol, **96**(2): p. 512–21.
57. Wang J, Simonavicius N, Wu X, et al. (2006), *Kynurenic acid as a ligand for orphan G protein-coupled receptor GPR35.* J Biol Chem, **281**(31): p. 22021–8.
58. Wassle H (2004), *Parallel processing in the mammalian retina.* Nat Rev Neurosci, **5**(10): p. 747–57.
59. Wilms M, Eger M, Schanze T, Eckhorn R (2003), *Visual resolution with epi-retinal electrical stimulation estimated from activation profiles in cat visual cortex.* Vis Neurosci, **20**(5): p. 543–55.
60. Zrenner E, Besch D, Bartz-Schmidt KU, et al. (2006), *Subretinal chronic multielectrode arrays implanted in blind patients.* Invest Ophthalmol Vis Sci, **47**(1538).

Chapter 13
Findings from Acute Retinal Stimulation in Blind Patients

Peter Walter and Gernot Roessler

Abstract In acute retinal stimulation experiments retinal stimulators are inserted into the eye, activated, and responses from patients to electrical stimulation are recorded. These tests were done to obtain evidence that the principle of electrical stimulation of the retina works in terms of elicitation of phosphenes or visual perception, respectively. These tests were also done to narrow the parameter range for electrode size and stimulation energy before efforts were undertaken to fabricate a device for chronic stimulation. Results from such tests were also helpful to describe possible perception patterns of patients and also to estimate possible visual acuities after implantation. Usually these tests were done in local anaesthesia so that the patient can respond verbally or by means of an interface to the stimulation. In different experiments rheobase and chronaxie data were reported showing a large variation depending on the device and on individual factors such as the disease state or the proximity between the electrode and the retina. Possible spatial and temporal resolution data were calculated from such experiments demonstrating that the concept of retinal stimulation in blind RP subjects can really help to restore some useful visual function in such patients.

Abbreviations

DTL	Dawson Trick, Litzkow
I	Intensity e.g. current for stimulation
LP	Light probe
MUX	Multiplexer
PC	Computer system
PS	Power source
RCS	Royal College of Surgeons

P. Walter (✉)
Department of Ophthalmology, RWTH Aachen University, Pauwelsstr. 30, 52074 Aachen, Germany
e-mail: pwalter@ukaachen.de

G. Dagnelie (ed.), *Visual Prosthetics: Physiology, Bioengineering, Rehabilitation*, DOI 10.1007/978-1-4419-0754-7_13, © Springer Science+Business Media, LLC 2011

RI   Response interface
RP   Retinitis pigmentosa
SIU   Stimulus isolation unit
STIM   Stimulator
T   Time
VD   Video documentation

13.1   Introduction

The basic assumption behind the development of implantable devices for retinal stimulation is that electrical stimulation of the retina may provide useful vision in patients suffering from advanced forms of degenerative diseases of the retina. Either from theoretical considerations but also from early experiments in blind human subjects one may conclude that this assumption should be correct. An early example of a human experiment was the implantation of stimulation electrodes across the visual cortex as reported by Brindley and associates. A blind RP patient reported phosphenes upon electrical stimulation of the visual cortex [1]. Dobelle and his group continued the work of Brindley and they were also able to demonstrate that blind subjects do have visual sensations when the posterior parts of the visual system are electrically stimulated [5]. The application of electrodes onto, underneath, or within the retina was limited to basic research approaches and did not extend to therapeutic efforts. Not earlier than 1991 devices and surgical techniques became available with which in patients suffering from retinitis pigmentosa (RP) experiments for retinal stimulation could be performed in the operating room without considerable risk to the patients. The rationale to do these experiments was that only data was available from retinal stimulation experiments in animals with a normal retina using preliminary electrode arrays or in tissue preparations of RCS rat retina using multielectrode array devices but not implantable electrodes. From these animal experiments only some information was known about the range of stimulation currents and about the timing of the stimulation pulses. It was not known to what extent the stimulation parameters would have to be changed to achieve visual percepts in blind humans suffering from such a disease. Three major questions should be answered by acute retinal stimulation experiments in humans. (a) Is it possible to elicit visual percepts when stimulation pulses are emitted by electrodes placed near the degenerated retina? (b) What charge delivery is necessary to obtain such responses? (c) Is it possible to elicit several percepts when several electrodes are activated and what is the two-point discrimination? All three questions were crucial. If it was found that the energy required to obtain visual percepts in RP patients was above the maximum charge delivery capacity of the electrode material or beyond a level indicating toxic tissue reactions then it would not have been possible to further pursue these research projects. If only unpatterned chaotic percepts were registered than there would also be no chance to establish artificial vision in terms of useful vision.

13.2 General Considerations for Acute Retinal Stimulation Experiments

Acute experiments for retinal stimulation in blind humans require the possibility to measure more or less quantitatively the visual response. Objective measurements are not possible in a clinical setting because the obtained local responses are too small to detect them with surface electrodes attached to the skull, although Chen reported one blind patient in which he recorded evoked cortical potentials with scalp electrodes upon electrical stimulation of the retina with eight electrodes simultaneously and 10% above threshold [3]. Due to obvious reasons microelectrodes inserted in the visual cortex to record local field potentials or functional imaging experiments in humans were not performed in contrast to such experiments which have been reported for animal studies [9, 18].

Acute tests for electrical stimulation of the retina have to be performed under local anesthesia. Only superficial anesthesia techniques such as subconjunctival or subtenon injections are recommended because any effect of the anesthetic drug on the optic nerve must be excluded. Sedative drugs should also be avoided because the patient has to indicate the visual response either by voice but more reliable by a response interface such as a set of buttons which he is asked to press to indicate whether he sees something or not. All patient responses must be recorded using such response interfaces to correlate them afterwards with the stimulus parameters. When using single electrodes, stimulus thresholds can be recorded by a two-alternative forced choice method at several points of the retinal surface.

When using electrode arrays, stimulus threshold data can be determined for each electrode or electrode pattern. Electrode arrays could also be used to estimate if two points or lines can be differentiated by the patient when two electrodes or two clusters of electrodes are stimulated simultaneously. Information on the distance of distinguishable electrodes or angles should give some information on the possible visual acuity that can be achieved with such systems. Important aspects of the neurophysiology of the target tissue can also be investigated, such as the determination of rheobase, which is the minimum stimulus intensity necessary to elicit a response at very long stimulus durations, and chronaxie which is the stimulus duration necessary to elicit a response at twice the rheobase level of stimulus strength. These data are characteristic for certain elements of nervous tissue.

The main limitation of acute retinal stimulation experiments in humans is that the time to perform these experiments is limited. Usually 1 h of experimentation is possible. Within this time all the possible combinations of stimulus intensity and time at all electrode positions cannot be included in the experimental setup. Another limitation is that the patient's response is not a uniform standardized yes or no. The answer sometimes also contains information on shape or color or maybe on temporal aspects of phosphenes. This information can usually not be interpreted systematically.

It should also be pointed out that acute tests for retinal stimulation have been performed in two types of patients: blind patients with RP and patients in which the eye has to be removed because of cancer. In the latter the retina itself usually was normal [12–14, 19].

13.3   Surgical Technique

Full pupil dilation should be obtained and then the patient is prepared for vitrectomy. Sclerotomies are made 3–4 mm behind the limbus. A vitrectomy is performed to avoid any traction at the entry sites or elsewhere to the retina. Wide angle viewing systems are indispensable. The size of the sclerotomy depends on the size of the implant. Usually handheld devices are used for acute retinal stimulation experiments. These devices are held onto or above the retinal surface. They are usually connected via a cable with a programmable power unit providing the requested pulse sequences to each electrode (Fig. 13.1). The precision with which such devices are held to the retinal surface is usually not constant throughout the experimental procedure. In such approaches eye movements may be a problem. Therefore some authors suggest the use of botulinum toxin to achieve akinesia [14, 15]. Movement of the device should be avoided during the stimulation procedure for several obvious reasons. Threshold determination may vary significantly depending on the force with which an electrode is pushed towards the retinal surface

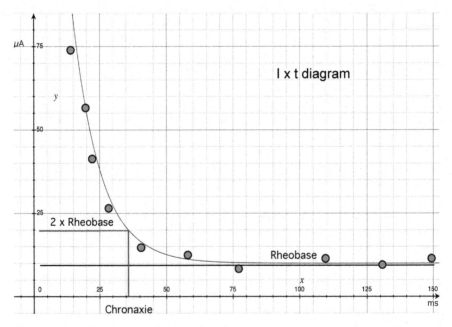

Fig. 13.1 Typical I×t diagramme for the electric stimulation of neural tissue. The I×t diagramme is determined by finding the stimulus current for a given stimulus duration or by finding the stimulus duration for a given stimulus current necessary to evoke a certain response, usually the threshold response. The minimal current to evoke a response with very long stimulus durations is called rheobase. The stimulus duration at the twofold rheobase intensity is called chronaxie. Rheobase and chronaxie are values characteristic for certain tissues and stimulation settings. The data points are experimental data fitted by a mathematical model

and also depending on the location of the electrode. For animal experiments therefore devices were used which were placed onto the retinal surface and held here in place with heavy liquids such as Perfluorodekaline [17]. Rizzo and coworkers used gold weights to apply pressure to the devices in a series of human experiments [14].

Quantification of the precision in terms of distance between electrodes and retina or pressure between retina and electrode and the constancy of the position is difficult. Even with such tools movement of the array may occur during the experiment as mentioned by Rizzo [15]. Therefore, the conclusions drawn from such experiments should be regarded cautiously. It is important when such experiments are performed and their results interpreted that the position of the array on the retinal surface is known. Much better information could be gained with experiments where the electrode array is chronically mounted onto the retina. Weiland and coworkers found in acute retinal stimulation tests in normal eyes that lifting an epiretinal electrode more than 0.5 mm off the retina resulted in loss of the electrically evoked percept [19] (Figs. 13.2 and 13.3).

After removal of the implant the sclerotomies are closed. Clinically, the patients showed adverse events in rare cases only. As in every vitrectomy the patient should be informed that a retinal detachment may occur in up to 5% of cases, as may cataract formation or in rare cases endophthalmitis. Such adverse events may require secondary interventions.

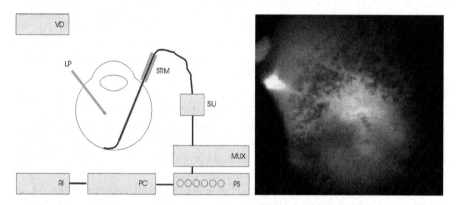

Fig. 13.2 *Left*; General setup for acute experiments on retinal stimulation. Under vitrectomy conditions the stimulator (STIM) is handheld at the desired position. A light probe (LP) is also inserted to allow visualization of the stimulator position onto the retinal surface. The electrodes are connected to a power source (PS) controlled by a computer system (PC) and possibly using a multiplexer (MUX) if several electrodes are desired. The electrodes are physically isolated from the high voltage devices using stimulus isolation units (SIU). The patient's responses are registered using response interfaces (RI) and the whole procedure is usually video documented (VD). *Right*; Intraoperative situation during an acute experiment for retinal stimulation – surgeon's view, inferior retina is in the *upper* part of the picture. The handheld microelectrode device is placed onto the retinal centre with the active electrodes near the superior arcade

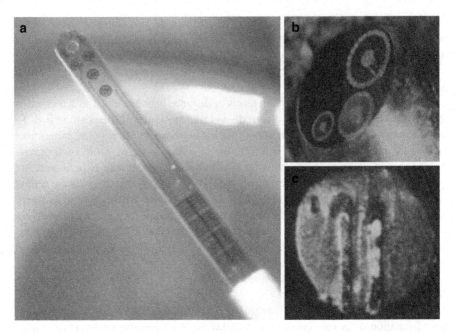

Fig. 13.3 Examples of probes for acute retinal stimulation experiments in human. (**a**) Polyimide stimulator used by Hornig et al. [10]. The thin and flexible polyimide device has a width of 1.6 mm. Active electrodes with diameters between 50 and 360 μm are placed at the tip of the device, the return electrodes are at the base of the device adjacent to the metal housing. (**b**) 17.5 Gauge steel cannula with several stimulation electrodes at the tip as used by Humayun and co-authors [12]. (**c**) 125 μm thin curved platinum wires for direct electrical stimulation of the retinal surface as used by Humayun and co-authors [12]

13.4 Threshold Measurements

The first approach in such settings is always the determination of thresholds at which a stimulation pulse or pulse series yields a phosphene. A two alternative forced choice method is usually applied. In such experiments threshold is commonly defined as the lowest stimulus intensity at which on 75% or more of the test repetitions the patient correctly reports a visual percept.

Also catch trials are usually employed in acute retinal stimulation testing, i.e. a stimulus is indicated, e.g., by a warning tone, but no electric pulse is given. If the subject gives a positive answer the answer is classified as false positive. Such catch trials are needed to test the reliability of a subject in this very demanding situation. Only those experiments should be analyzed in which patients do not give too many false positive responses. The criterion at which tests should not be used because of too many false positive responses should be defined for each experiment [10].

Threshold measurements were reported by Rizzo [14, 15]. In his series visual responses in patients with advanced RP could not be obtained with needle type electrodes because the charge density would exceed toxicity limits. They used oxidized iridium as electrode material based on a polyimide substrate. With 250 μs per phase stimulus duration threshold currents were around 1.5 mA for 400 μm electrodes in diameter. For 1 ms per phase stimulus duration the stimulus current at threshold was between 0.8 and 0.4 mA and for 16 ms per pulse phase stimulus currents at threshold were around 200 μA. Based on these experiments the rheobase was calculated as 125 μA with a chronaxie of 2.3 ms per pulse phase. When charge densities were calculated they found that for 100 μm electrodes the charge densities at threshold were between 4 and 10 mC/cm^2, for 400 μm between 0.28 and 2.8 mC/cm^2 which was larger than their own safety limit (1 and 0.252 mC/cm^2, resp.).

In Humayun's series nine blind RP patients were acutely stimulated with either platinum wires (25–125 μm diameter) used as electrodes or disc electrodes each 400 μm in diameter. Charges et threshold were reported between 0.2 and 2.4 μC which gives charge densities between 1 and 96 mC/cm^2. The higher values were obtained for needle type electrodes [13].

13.5   Spatial Resolution and Pattern Perception

The main prerequisite for the recognition of forms and patterns is a correct retino-topical representation of the stimulus within the visual field. Humayun was able to show in his series, that stimuli were correctly identified in terms of their location within the retina resp. within the visual field. This was also confirmed by animal experiments. Spatial resolution can only be tested with electrode arrays when two electrodes or electrode clusters are stimulated simultaneously. The patient has to be asked if he sees two spots of light or two distinct patterns. In Humayun's series he calculated a spatial resolution of 1.75° which could be achieved with epiretinal stimulation [11, 12]. These authors calculated based on simulations by Cha et al. [2] that placing a 32 × 32 array over a central field of 0.5 × 0.5 mm onto the macula surface would result in a 20/26 Snellen visual acuity. A spacing of 90 μm between each electrode would then reduce the visual acuity to 20/200. However, such electrode arrays are currently not available. Such devices are desirable when AMD treatment is supposed to be performed with such implants.

In a series of experiments in four blind subjects Rizzo and coworkers answered the question if blind patients are able to identify a stimulus pattern [15]. Only in one out of three patients more than 50% of the given patterns could be identified. Two-point discrimination was also tested in this series and only in very few experiments Rizzo's group was able to find patient responses suggesting the perception of two separate objects. In these experiments the electrodes were 1,860 μm apart.

Acute stimulation experiments were not performed with any camera picture input. However, electrodes can be activated as if a certain pattern should be seen such as a large letter H. Humayun did such experiments in his series and he found that patients were able to detect patterned phosphenes. They used a 25 electrode array with a pattern "U" and patient reported a "H" type pattern which the authors thought to be the result of a blurring effect due to unstable positioning [13].

13.6   Temporal Resolution

Experiments to systematically determine temporal resolution in acute tests for retinal stimulation have not been reported in detail so far for human subjects. Early results indicated that the flicker fusion frequency may be similar for electrical stimulation and for normal vision [12]. From animal experiments it should be expected that 25 frames per second could be transferred with a retinal implant [6]. It can be expected that data on temporal resolution will be extracted from the clinical trials on retinal prostheses, which are now being performed. This information is important, because useful vision elicited with prosthetic devices does not only depend on spatial characteristics but also on the time necessary to transmit a pattern from the implant to the primary visual cortex and to identify it. The type of visual sensations that can be obtained.

If patients are asked after the procedure what they have seen, the descriptions vary considerably between individuals. However, a few aspects are common. Usually the patients did not experience any unpleasant sensations. With suprathreshold stimulation patients reported dots, arcs, circles and lines of different intensity, color, and orientation. Stimulation by one electrode may not necessarily result in the perception of one single percept or phosphene but also in multiple phosphenes as reported in one volunteer by Rizzo [15]. The size of the objects does also vary with respect to the electrode size, the stimulus intensity, and duration. The characteristics of the phosphenes were reproducible; i.e. the stimulus pattern X elicits the same phosphene when it was repeated at the same retinal area. Weiland and colleagues tested two subjects before removal of the eye due to cancer. In these patients they destroyed part of the outer retina with argon green and krypton red laser photocoagulation. They stimulated normal retina and the laser treated areas with a 125 μm platinum wire electrode. The percept after stimulation of the normal retina was a dark oval shaped phosphene whereas stimulation in the krypton red treated area revealed a small white spot. Stimulation of the argon green treated area resulted in a line type of percept. Stimulus threshold at the normal retinal area in their experiments were 0.8 and 4.8 mC/cm^2 respectively. The percepts being described by blind RP patients were similar to those obtained over the krypton red treated normal retinas. From a histological work-up of the stimulated retina Weiland concluded that the target for electrical stimulation with an epiretinal electrode is not the ganglion cell layer but the inner nuclear layer [19].

The shape of the percepts varied significantly with the stimulus pattern or with the orientation of the activated electrodes of the array. In Rizzo's series circles were seen when columnar electrodes along the axon orientation were activated but also curved lines [15].

13.7   Subretinal Versus Epiretinal Stimulation

Acute retinal stimulation experiments in blind humans were only reported for epiretinal stimulation, not for subretinal stimulation. Acute tests on subretinal stimulation so far have only been reported for rabbits and pigs but not for blind humans [7, 16]. Therefore major assumptions as confirmed for epiretinal stimulation were so far not confirmed for subretinal stimulation scenarios.

13.8 Less Invasive Stimulation Procedures

Electrical retinal stimulation can also be performed placing electrodes outside the eye. It has been shown that even by using corneal electrodes or electrodes placed onto the scleral surface it is possible to elicit visual phosphenes. Gekeler and colleagues investigated phosphenes upon electrical stimulation via DTL electrodes placed in the fornix. They found that in some RP patients such phosphenes could not be elicited. The rheobase for RP patients was $0.69 + 0.10$ mA which is 18 times higher than for normal individuals [8]. Similar experiments were performed by Delbeke and colleagues. They used corneal surface electrodes with large periorbital reference electrodes. Their estimation for rheobase varied between 2.14 and 8.16 mA and for chronaxie they found values between 0.45 and 0.87 ms whereas for healthy volunteers rheobase was 0.28 mA and the mean chronaxie was 3.07 ms [4]. In our own experience in patients with advanced stages of retinitis pigmentosa the currents necessary to obtain visual percepts with such approaches are very high and close to current with which other subjective sensations such as pain or muscle tics were evoked. Such approaches are proposed to identify patients prior to surgery who may benefit from a chronic implant. Only those patients would be selected in which phosphenes can be obtained with such non-invasive techniques. However, we learned that in RP patients even when they do not respond to external stimulation or the stimulation has to be stopped because of unpleasant somatosensory sensations the same patient may have visual sensations with a chronic implant.[1] Therefore, we feel that such a test is not useful to identify good candidates for retinal implants.

13.9 Conclusions and Outlook

The available data from acute trials in electrical retinal stimulation showed that at least with epiretinal stimulation visual responses can be obtained and that the energies necessary to achieve such responses are dependent on the material of the electrodes and on their sizes and shapes. A close contact between the electrode and the retina is desirable to elicit visual responses with low stimulus intensities. A deeper knowledge of the mechanisms of retinal stimulation as well as of the phosphenes which are elicited and their role in providing vision for a blind patient is necessary. Such information can only be obtained by chronic stimulation experiments in which enough time is available to study more stimulus parameters, more electrode positions over a longer time period and to allow the patients to learn how to interpret the percepts

---

[1]Walter (2005) unpublished observation.

induced with a visual prosthesis and to use the potential of visual system plasticity. Such data will be provided in the future with semichronic or chronic implants.

The three crucial questions asked at the beginning of the chapter can now been answered: (a) Yes, visual responses can be obtained through electrical stimulation of the retina. (b) The stimulus intensity necessary to elicit visual responses depends on several factors. Visual responses can be obtained within safe stimulation levels in certain scenarios. (c) The two-point discrimination and the perception of patterns using such an approach is under debate. In many acute trials there was not enough time for retinal stimulation to collect enough data to really answer this question. However, the chronic trials, that have already been initiated will answer the question in the near future.

References

1. Brindley GS, Lewin WS (1968), *The sensations produced by electrical stimulation of the visual cortex.* J Physiol 196: p. 479–493.
2. Cha K, Horch KW, Norman RA (1992), *Simulation of a phosphene based visual field: visual acuity in a pixelized vision system.* Ann Biomed Eng 20: p. 439–449.
3. Chen SJ, Mahadeveppa M, Roizenblatt R et al (2006), *Neural responses elicited by electrical stimulation of the retina.* Trans Am Ophthalmol Soc 104: p. 252–259.
4. Delbeke J, Pins D, Michaux G et al (2001), *Electrical stimulation of anterior visual pathways in retinitis pigmentosa.* Invest Ophthalmol Vis Sci 42: p. 291–297.
5. Dobelle WH, Mladejovsky MG (1974), *Phosphenes produced by electrical stimulation of human occipital cortex, and their application to the development of a prosthesis for the blind.* J Physiol 243(2): p. 553–576.
6. Eckhorn R, Wilms H, Schanze T et al (2006), *Visual resolution with retinal implants estimated from recordings in cat visual cortex.* Vision Res 46: p. 2675–2690.
7. Gekeler F, Kobuch G, Schwahn HN et al (2004), *Subretinal electrical stimulation of the rabbit retina with acutely implanted electrode arrays.* Graefes Arch Clin Exp Ophthalmol 242(7): p. 587–596.
8. Gekeler F, Messias A, Ottinger M et al (2006), *Phosphenes electrically evoked with DTL electrodes: a study in patients with retinitis pigmentosa, glaucoma, and homonymous visual field loss and normal subjects.* Invest Ophthalmol Vis Sci 47: p. 4966–4974.
9. Hesse L, Schanze T, Wilms H et al (2000), *Implantation of retina stimulation electrodes and recording of electrical stimulation responses in the visual cortex of the cat.* Graefes Arch Clin Exp Ophthalmol 238: p. 840–845.
10. Hornig R, Laube T, Walter P et al (2005), *A method and technical equipment for an acute human trial to evaluate retinal implant technology.* J Neural Eng 2: p. S129–S134.
11. Humayun MS, deJuan E (1998), *Artificial vision.* Eye 12: p. 605–607.
12. Humayun MS, deJuan E, Dagnelie G et al (1996), *Visual perception elicited by electrical stimulation of the retina in blind humans.* Arch Ophthalmol 114: p. 40–46.
13. Humayun MS, deJuan E, Weiland JD et al (1999), *Pattern electrical stimulation of the human retina.* Vision Res 39: p. 2569–2576.
14. Rizzo JF, Wyatt J, Loewenstein J et al (2003), *Methods and perceptual thresholds for short term electrical stimulation of human retina with microelectrode arrays.* Invest Ophthalmol Vis Sci 44: p. 5355–5361.
15. Rizzo JF, Wyatt J, Loewenstein J et al (2003), *Perceptual efficacy of electrical stimulation of human retina with a microelectrode array during short term surgical trials.* Invest Ophthalmol Vis Sci 44: p. 5362–5369.

16. Sachs HG, Gekeler F, Schwahn HN et al (2005), *Implantation of stimulation electrodes in the subretinal space to demonstrate cortical responses in Yucatan minipig in the course of visual prosthetics development.* Eur J Ophthalmol 4: p. 493–499.
17. Walter P, Heimann K (2000), *Evoked cortical potentials after electrical stimulation of the inner retina in rabbits.* Graefes Arch Clin Exp Ophthalmol 238: p. 315–318.
18. Walter P, Kisvarday Z, Goertz M et al (2005), *Cortical activation via an implanted wireless retinal prosthesis.* Invest Ophthalmol Vis Sci 46: p. 1780–1785.
19. Weiland JD, Humayun MS, Dagnelie G et al (1999), *Understanding the origin of visual percepts elicited by electrical stimulation of the human retina.* Graefes Arch Clin Exp Ophthalmol 237: p. 1007–1013.

Chapter 14
The Perceptual Effects of Chronic Retinal Stimulation

Alan Horsager and Ione Fine

Abstract Can functional vision be restored in blind human subjects using a microelectronic retinal prosthesis? The initial indications suggest that, yes, it is possible. However, the visual experience of these subjects is nothing like a digital scoreboard-like movie, with each electrode acting as an independent pixel. The work described here in this chapter suggests that there are interactions between pulses and across electrodes, at the electrical, retinal, or even cortical level that influence the quality of the percept. In particular, this work addresses the question, "how does the percept change as a function of pulse timing on single and multiple electrodes"? The motivation for the work described here is that these interactions must be understood and predictable if we are to develop a functional tool for blind human patients. In this chapter, we review work evaluating perceptual effects using chronic electric stimulation in three different implantable systems.

Abbreviations

AltFC	Alternative forced-choice
AMD	Age-related macular degeneration
BLP	Bare light perception
ChR2	Channelrhodopsin-2
DS	Direct stimulation
IMI	Intelligent medical implants
IRI	Intelligent retinal implant
LGN	Lateral geniculate nucleus
LP	Light perception

A. Horsager (✉)
Eos Neuroscience, Inc., 2100 3rd Street, 3rd floor, Los Angeles, CA 90057, USA
and
Department of Ophthalmology, University of Southern California,
Los Angeles, CA 90089, USA
e-mail: horsager@usc.edu

G. Dagnelie (ed.), *Visual Prosthetics: Physiology, Bioengineering, Rehabilitation*,
DOI 10.1007/978-1-4419-0754-7_14, © Springer Science+Business Media, LLC 2011

MPDA   Microphotodiode array
NLP   No light perception
OCT   Optical coherence tomography
rd1   Retinal degeneration 1
RP   Retinitis pigmentosa
RPE65   Retinal pigment epithelium-specific 65 kDa protein
SSMP   Second Sight Medical Products, Inc.
V1   Primary visual cortex
VPU   Visual processing unit

14.1 Introduction

Visual impairment is one of the most common disabilities: at the most recent estimate, 110 million people worldwide have low vision and 40 million are blind [69]. Photoreceptor diseases such as retinitis pigmentosa (RP) and age-related macular degeneration (AMD) are responsible for blindness in approximately 15 million of those people [15], a number that continues to increase with the aging population [18]. Currently, there are no FDA approved treatments for blindness due to photoreceptor disease.

Although a number of highly promising treatments are being developed, each suffers from its own set of difficulties. For example, gene replacement therapy efforts have made great progress in treating one form of Leber's Congenital Amaurosis (an RPE65 mutation) in humans [1, 2, 4, 7, 8, 52]; however, this form of RP is relatively rare, and photoreceptor diseases are genetically heterogeneous, with single and multi-gene mutations occurring in over 180 different genes responsible for photoreceptor function [19]. For gene replacement therapy to broadly cure photoreceptor disease would require at least as many (and, most likely, many more) treatments as there are mutations. Another genetic approach uses optical neuromodulators such as channelrhodopsin-2 (ChR2) that can be genetically targeted to retinal bipolar [41] or ganglion cells [10, 43] to restore visual responsiveness in a mouse model of blindness (*rd1*). However ChR2 activation requires light stimulation levels that are 5 orders of magnitude greater than the threshold of cone photoreceptors [63], and the induced light responses have a substantially limited dynamic range (2 log units) [72]. An ideal therapy would be able to treat blindness independently of the genetic mutation, in the absence of photoreceptors, and with reasonable response sensitivity and range.

Therapies employing direct electrical stimulation of the retina have the potential to fulfill those two particular constraints. However, electrical stimulation suffers from its own set of limitations. There are a number of engineering concerns such as charge density safety limits which limit the miniaturization of implanted electrodes, difficulties in placing the electrode array close to the target retinal cells, and limitations is in the available power supply that make prosthesis design extremely challenging. Electric current fields from relatively large electrodes indiscriminately

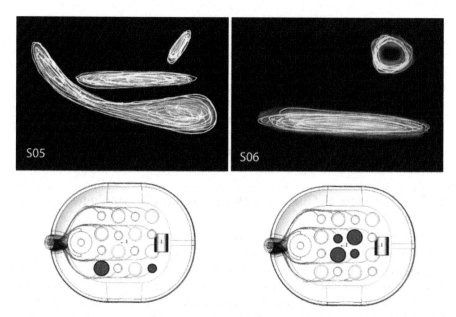

Fig. 14.1 Patient percepts. Example percepts generated by retinal electrical pulse train stimulation in two blind subjects, S05 and S06, respectively, using the Second Sight Medical Products, Inc. A16 epiretinal prosthesis. Percepts (*top*) were hand drawn by experimenter based upon patient report. The electrodes that were stimulated for each condition are shown with *solid dots*. Stimulation patterns were 50 Hz pulse trains on each of the electrodes

drive local retinal circuits in an unnatural way, leading to complex retinal responses. Although electrically-driven retinal activation produces phosphenes in blind human subjects, these percepts are complex and cannot be simply thought of as a one to one, electrode to pixel, scoreboard-like experience with punctate individual phosphenes (Fig. 14.1).

There is a substantial literature evaluating the use of electrical stimulation to generate visual percepts in both sighted and blind human subjects [11, 22, 24, 35, 37, 42, 44, 45, 50, 53, 54, 61, 62, 75, 77]. However, partly because many of these studies were carried out acutely, there has not yet been a been a thorough quantitative and systematic analysis of how these electric pulses interact within the network of retinal neurons in time and space to form the visual image the subject sees. With the goal of creating a visual prosthesis that is capable of restoring functional vision in blind human patients, much needs to be learned about how the timing of pulses (within single electrodes and across multiple electrodes) interact at the electrical, retinal, and cortical level to form a percept.

There is relatively little published data quantitatively examining chronic retinal stimulation in human subjects. To date, only three commercial groups have been able to collect chronic data: Retinal Implant AG, Intelligent Medical Implants GmbH, and Second Sight Medical Products, Inc. To summarize their collective findings: (1) an electrode array can be safely and chronically implanted in human patients (more than 5 years as of this writing), (2) stimulation via these electrodes consistently produces visual percepts, (3) provided the array is stable on the retinal surface, the current-brightness relationship is stable, repeatable, and monotonic, (4) the brightness of the percept can be controlled through both amplitude and frequency modulation, and (5) signal integration during single or multi-electrode stimulation can be approximated using very simple models. In the last part of the chapter we discuss the use of these implants during "real world" and mobility tasks. While findings are promising, it is still not demonstrated that the devices that are currently available can provide useful function vision outside of the laboratory setting.

14.2   Overview of Chronic Retinal Implant Technologies

The earliest documented electrically generated percept in a blind human patient was in 1755, when Charles LeRoy, a French chemist and physician, discharged a Layden jar and supplied electrical current to a brass coil that wrapped around the head of a blind man [42]. In addition to "provoking terrible cries [47]", the young patient perceived a flame that rapidly descended before his eyes. This is, more than likely, the first documented visual phosphene perceived by a blind subject via electrical stimulation.

Despite this somewhat unpromising beginning, restoring functional vision using electrical stimulation has been a goal of ophthalmologists and vision scientists for more than a century. The inspiration for these studies comes from very early (and probably inadvertent) electrical activation of visual cortex during neurosurgery. However, it wasn't until the middle of the twentieth century that scientists and clinicians began to investigate, more deliberately and rigorously, the relationship between electrical stimulation of neural tissue and visual perception.

In recent years much of the effort in developing a visual prosthesis for the blind has focused on electrical stimulation of the retina. There is a substantial amount of neural processing within the LGN [5, 20, 73] and V1 [34] that transforms the visual signal in ways that are complex, nonlinear, and poorly understood. Targeting stimulation as early as possible in the visual pathway allows one to maximize the use of the innate computational processing of the visual system.

Even within the retina, a variety of approaches have received attention. Retinal stimulation devices have been developed for both subretinal (between outer retinal layers and the choroid) and epiretinal (between inner retinal layers and the vitreous humor) activation. Here, we provide a brief overview of the basic technology and the types of psychophysical and behavioral studies that have been conducted with subretinal and epiretinal devices.

14.2.1   The Retinal Implant AG Microphotodiode Prosthesis

The Retinal Implant AG device has two subretinally implanted components. The first is a wire-bound microphotodiode array (MPDA), consisting of approximately

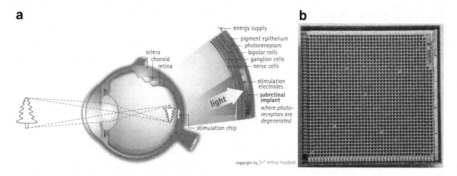

Fig. 14.2 (**a**) General schematic of Retinal Implant AG device. Note the placement of the device in the subretinal space. (**b**) Close-up of the microphotodiode array (MPDA). Permission for reproduction provided by Retina Implant AG

1,500 photosensitive cells on a $3\,mm^2$ surface (Fig. 14.2). Each cell unit contains an amplifier and electrode, spaced $70\,\mu m$ apart. The amplitude of the electrical signal across each electrode is proportional to the overall illumination of the specific photosensitive cell. The second implanted component consists of a 4×4 array of $50\,\mu m$ electrodes (with $280\,\mu m$ spacing) that can be used for direct stimulation (DS). The stimulation presented on the electrodes of the DS array can also be independently controlled, and each of the parameters (e.g., pulse width and amplitude) can be independently modulated. The MPDA and DS arrays are positioned on a small polyimide foil surface and are powered via a transchoroidal, transdermal line.

This MPDA device was implanted in the fovea of one eye of seven patients blind from RP (all seven patients had the MPDA array and six had, in addition, the DS array). Devices were chronically implanted for a total of 4 weeks [78]. With the exception of one, patients were explanted at the end of this time. Visual perception and performance was evaluated in the following ways using DS: (1) the brightness of a biphasic pulse (1–2.5 Volts (V), 3 milliseconds (ms) per phase) was assessed using a rating scale from 5 (very bright) to 0 (no perception), (2) subjects were asked to discriminate stimulation of rows and columns of electrodes in a vertical vs. horizontal discrimination task, (3) motion discrimination for sequential stimulation of electrodes, (4) and subjective reporting of the apparent size of percepts. Three additional patients were implanted at a later date with a similar device (for this device the stimulating electrodes were 100 μm). With these three patients, the researchers conducted more complex visual perception tasks such as letter recognition and orientation discrimination [81]. Data collected on all ten patients using this device are described in Sects. 14.4.1 and 14.5.1.

14.2.2   The Intelligent Retinal Implant System

The Intelligent Retinal Implant System™ has two external components (a Visual Interface and the Pocket Processor), and a subretinally-implanted Retinal Stimulator

Fig. 14.3 Schematic of the
Intelligent Retinal Implant
System. Illustration kindly
provided by IMI

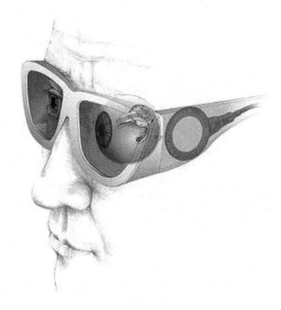

(Fig. 14.3) designed by IMI Intelligent Medical Implants. The Visual Interface consists of a pair of glasses with a camera to capture the visual image and other components for data communication with the Pocket Processor and Retinal Stimulator. Communication with the Retinal Stimulator is conducted via wireless transmission. The Pocket Processor supplies power to the entire system and contains a microcomputer that translates the image data into the stimulation protocol for the Retinal Stimulator. The internal flexible Retinal Stimulator consists of a 49 electrode array and is attached using a silicon ring to a titanium tack which had been placed in the sclera.

Four patients (56–66 years old) with visual acuity ranging from no light perception (NLP) to hand movement were implanted. Approximately 20 different testing sessions were conducted with the patients over a 12 month period. Each testing session consisted of single or multi-electrode pulse train stimulation. In these patients, performance on absolute threshold, point discrimination, and pattern recognition tasks was measured. See Sects. 14.4.2 and 14.5.2 for details regarding psychophysical data collected using this system.

14.2.3   Second Sight Medical Products, Inc. A16 System

The Second Sight Medical Products, Inc. (SSMP) A16 epiretinal prosthesis contains similar intraocular (electrode array) and extraocular (e.g., glasses, Visual Processing Unit) components as the Intelligent Retinal Implant System™. The intraocular array consists of 16 platinum electrodes in a 4×4 arrangement, held in place within a clear silicone rubber platform [37, 45]. The electrode array is implanted epiretinally in the macular region and held in place using a retinal tack. Electrodes are either 260 or 520 μm in diameter (subtending 0.9° and 1.8° of visual angle, respectively). Electrodes are spaced 800 μm apart, center to center.

Pulse train signals are generated and sent to an external Visual Processing Unit (VPU) using custom software run on a PC laptop. Power and signal information are sent from this processor through a wire to an external transmitter coil that attaches magnetically, and communicates inductively, to a secondary coil that is implanted subdermally in the patient's temporal skull. From this secondary coil, power and signal information are sent through a subdermally implanted wire that traverses the sclera to the array of electrodes (Fig. 14.4). Stimulation can be presented using two different protocols: (1) camera mode – real-time video captured by a miniature video camera mounted on the subject's glasses is continuously sampled by the VPU and a monotonic transform determines the stimulation current amplitude in each electrode based on the (normalized) luminance at the corresponding area of the scene and (2) direct stimulation mode – the stimulation signal sent to each electrode is independently controlled by the VPU or an external computer.

Six patients have been examined that were chronically implanted with the A16 retinal prosthesis. See Table 14.1 for details regarding these subjects. Testing sessions lasted a maximum of 4 h with frequent rest periods. Testing sessions included threshold and impedance measurements as well as other measures of visual performance reported elsewhere [77]. When performed, threshold measurements were usually carried out at the beginning of a given testing session. The frequency of testing sessions was limited by the subjects' availability and the clinical trial protocol. In general, testing was carried out 1–2 sessions/week for each subject. The protocol

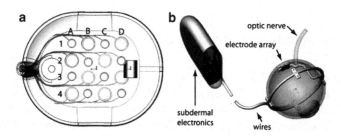

Fig. 14.4 (a) Electrode array. The electrode array consisted of 260 or 520 micrometers (μm) electrodes arranged in a checkerboard pattern, with center-to-center separation of 800 μm. The entire array covered ~2.9 mm by 2.9 mm of retinal space, subtending ~10° of visual angle. (b) Prosthesis system schematic. The stimulus sets are programmed using Matlab® on a PC, which then communicated the stimulus parameters to an external Visual Processing Unit (not shown). Signal and power information was then passed through an external inductive coupling device (not shown) that attaches magnetically to a subdermal coil implanted in the patient's temporal skull. This signal is then sent through a parallel system of wires to the epiretinally implanted electrode array. Note that the power and signal information can be independently controlled for each electrode. Reprinted from [33], with permission

Table 14.1 A16 subjects' age at implantation, eye of implantation, preoperative acuity in the implanted and non-implanted eyes, and electrode sizes

	Age	Eye	VA (implanted)	VA (non-implanted)	Electrode size (μm)
S01	76	R	NLP	LP	520
S02	56	L	LP	LP	520
S03	74	R	NLP	NLP	260
S04	60	L	LP	LP	260/520
S05	59	R	LP	LP	260/520
S06	55	R	NLP	NLP	260/520

Where there was a difference in pre-operative vision between the two eyes, implantation was carried out in the eye with worse vision. One subject (S02) had two operations in the same eye since her electrode array detached from the retina after 11 months due to the subject falling and bumping her head (no retinal detachment occurred). In the second surgery, the electrode array was reattached in a nearby macular area no more than 500 μm distant from the position of the original implant. Testing for S01 was limited in duration due to geographical constraints. Testing for S03 ended due to medical reasons unrelated to the implant. Testing in S04 was ended after microperforation of her conjunctiva which led to cable exposure. Because her cardiac status had deteriorated since the initial implantation she could not undergo anesthesia. This prevented the use of a scleral patch graft to repair the microperforation. To minimize the risk of possible infection, the multi-wire cable connecting the electrode array to the extraocular stimulator was cut and the electrode array was left in place

specified that optical coherence tomography (OCT) measurements were only carried out on the subjects for clinical reasons, and as a result OCT data were only collected at irregular intervals.

The bulk of the data described in this chapter were collected using this epiretinal device. As a result more detailed information is given about the patients implanted with this device (Sects. 14.3, 14.4.3, and 14.5.3).

14.3 Thresholds on Individual Electrodes

One major concern in the field of retinal prostheses is that the current amplitude required to elicit percepts may fluctuate unpredictably over time, due to neurophysiological changes of the retina due to reorganization [40], electrochemical changes on the electrode surface, or instability of position of the electrode array on the retinal surface [21, 77]. Previous acute studies found that localized retinal electrical stimulation of blind subjects resulted in discrete visual percepts; however, the amount of electrical current required to elicit visual responses was relatively large compared to animal in vitro retinal studies examining responses to electrical stimulation [36, 61].

The most exhaustive examination of thresholds and brightness reported to date has been within the six subjects implanted with the A16 epiretinal prosthesis (Second Sight Medical Products, Inc.). Over the course of several years, we measured the distance of the electrodes from the retinal surface (using Optical Coherence

Tomography, OCT), retinal thickness, electrode impedance, and perceptual thresholds for both single pulse and pulse train stimuli [21, 33]. These data allowed us to examine the relationship of perceptual threshold to electrode size, electrode impedance, distance of the electrodes from the retinal surface, and retinal thickness.

14.3.1   Single Pulse Thresholds Using the SSMP System

Thresholds were measured on single electrodes using a single interval, yes-no procedure. On each trial, subjects were asked to judge whether or not a stimulus was present. This reporting procedure meant that subjects were likely to report stimulation for either a light or dark spot; subjects were explicitly instructed to include either type of percept in making their decision. Half of the trials were stimulus-absent catch trials. Current amplitude was varied using a three-up-one-down staircase procedure to find the threshold current amplitude needed for the subjects to see the stimulus on 50% of stimulus-present trials, corrected for the false alarm rate. During each staircase, only amplitude varied. All other parameters (frequency, pulse width, pulse train duration, and the number of pulses) were held constant. Thresholds were measured for each of the 16 electrodes using a single "standard pulse" consisting of a 0.975 ms cathodic pulse followed by a 0.975 ms anodic pulse.

Thresholds measurements for each subject are shown in Fig. 14.5. Differences in threshold did not change systematically as a function of patient age or pre-operative vision [76]. However, thresholds did appear to decrease as a function of successive surgeries. Generally, thresholds decreased across subject implantations. Indeed, subject S01 had the highest threshold overall. This improvement across surgeries was perhaps due to the overall improvement of the surgical procedure, leading to the electrode array lying successively closer to the retinal surface. For subjects S05 and S06, most of the measured single pulse thresholds were well below 100 µA and charge density limits of 0.35 mC/cm². It should also be noted that these thresholds are for a single pulse, whereas functional electrical stimulation is likely to be mediated by pulse trains, which generally require lower stimulation thresholds (see Sect. 14.3.2).

Mean thresholds for subjects S04–S06 across the 260 and 520 µm electrodes were compared to determine if electrode size had any effect on the measured values. Interestingly, there was no noticeable difference in threshold between 260 and 520 µm electrodes, either within or across subjects (two-factor, subject × electrode size, ANOVA, $p > 0.05$ F = 0.367), see Fig. 14.6. This was in direct contrast to a recent literature review by Sekirnjak et al. who found, across a wide range of in vitro and in vivo studies, that log thresholds increase linearly with log electrode area [64]. However, only two electrode sizes are evaluated here, and it is possible that a wider range of sizes would make threshold differences, as a function of electrode size, more apparent.

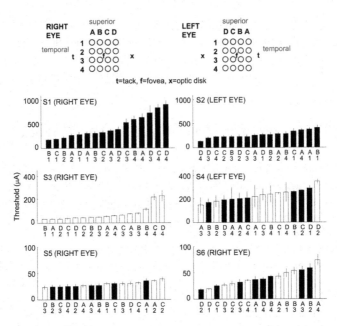

Fig. 14.5 Mean thresholds across the entire time period of implantation (μA) for all 16 electrodes for each subject. The *upper panels* show the labeling scheme used to identify electrodes, as viewed through the pupil. For each subject, electrodes are ordered from most to least sensitive along the x-axis. *White* and *black bars* represent electrode diameters of 260 and 520 μm respectively. Threshold current is shown along the y-axis. Note the dramatic change of scale along the y-axis across subjects. Error bars are +/− one standard error of the mean. Reprinted from [21], with permission

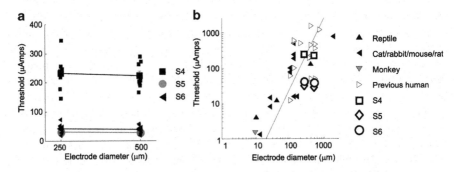

Fig. 14.6 In both (**a**) and (**b**), the x-axis represents electrode diameter and the y-axis represents the current needed to reach perceptual threshold. (**a**) is the data are taken from subjects S04–S06. The *large symbols connected by lines* represent the mean threshold across each of the eight electrodes of a given diameter for each subject. Error bars are +/− one standard error of the mean. In many cases error bars are smaller than the symbols. Individual electrodes are shown with *small symbols*. (**b**) compares our measured thresholds in S04-06 (*large open shapes*) and those reported in the literature [64]. Reprinted from [21], with permission

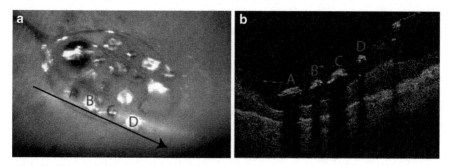

Fig. 14.7 OCT imaging of the array. (**a**) Fundus photograph of an intraocular electrode array viewed through a dilated pupil, imaged by the OCT machine (OCT, STRATUSOCT™; Carl Zeiss Meditec AG) just previous to the cross-sectional OCT image shown in (**b**). (**b**) Cross-sectional OCT image. Location of electrodes that were imaged are denoted by the *letters* A, B, C, and D. Reprinted from [21], with permission

The distance from the top of each electrode to the internal limiting membrane of the retina varied both within and across subjects, as measured using optical coherence tomography (OCT) (Fig. 14.7). Electrode thickness varied between 80 and 120 μm depending on the exact cross-section, so 100 μm was subtracted from the measurement of the distance of the electrode to the internal limiting membrane. The thickness of the retina was defined as the distance from the inner surface of the retinal pigment epithelium to the internal limiting membrane.

Impedance was measured using Second Sight Medical Products (Inc.) proprietary software. Impedance measurements were taken at the beginning and end of each stimulating session.

Data suggests that distance from the retinal surface is a critical factor in determining both threshold and impedance. For a given electrode size, electrodes that are close to the retinal surface have lower thresholds and higher impedances. As shown in Fig. 14.8b, we see a positive correlation between threshold current and electrode distance from the retina. These psychophysical data are consistent with retinal electrophysiology data, suggesting that the distance of the electrodes from the retina is a significant concern [31, 38]. Stimulus current requirements are likely to be minimized when the array is in close position to the retina, minimizing power consumption by the stimulator and allowing for smaller electrodes to generate phosphenes within safe charge density limits.

On the whole, subject impedances tended to decrease postoperatively over time. However, impedances are also negatively correlated with the distance of the electrode from the retinal surface, as shown in Fig. 14.8c. These data are consistent with the notion that electrodes that are close to the surface of the retina have higher impedances (due to the adjacent retinal tissue) than electrodes that have lifted from the retina (where fluid with higher conductivity intervenes between the electrode and the retinal surface) [32, 65]. Indeed, consistent with this hypothesis, threshold is negatively correlated with the impedance (Fig. 14.8a).

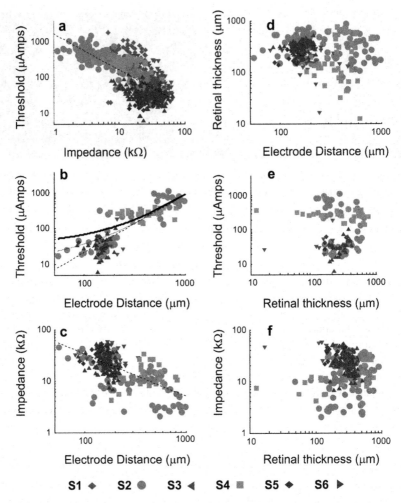

Fig. 14.8 Correlations between threshold, impedance, electrode distance and retinal thickness. Each subject is shown with a different symbol shape. *Straight lines* represent the best fitting linear regression on log-log axes. (**a**) Impedance versus threshold. (**b**) Electrode distance from the retinal surface vs. threshold. The *curved solid lines* show predicted thresholds for 260 (*lower, thin solid line*) and 560 μm (*upper, thick solid line*) electrodes based on the model of Palanker et al. [51]. (**c**) Electrode distance from the retinal surface vs. impedance. (**d**) Distance versus retinal thickness. (**e**) Retinal thickness versus threshold. (**f**) Retinal thickness versus impedance. Reprinted from [21], with permission

14.3.2 Pulse Train Integration and Temporal Sensitivity

It is important to understand, at the single electrode level, how the electrical signal is integrated over time to modulate visual sensitivity or suprathreshold brightness. The retina delivers information about the visual scene to higher visual centers through its time-varying spike signal [3, 23], so it would be potentially beneficial to encode the light signal using pulse sequences that trigger "naturalistic" patterns of activity in retinal cells. Indeed, it has been shown that using short electrical pulses results in phase-locked spikes in ganglion cells up to 250 Hz [25, 64].

Until recently, the general assumption has been that the rate of ganglion cell firing is simply monotonically related to the "intensity" of the stimulus. However, this idea of "simple rate coding" has recently come into question as it has been shown that the visual system is sensitive to spike timing on a much finer scale (<10 ms) [60] and spike train variability cannot simply be described by a Poisson distribution [9, 70]. Indeed, the data in this section show that the human visual system is not only highly sensitive to changes in overall pulse train frequency, but to the distribution of the pulses within a given window of time. Although we can only speculate as to the underlying physiological mechanism that is involved in integrating these pulse signals, it clearly shows that being able to control the precise timing of spikes may prove to be as important as controlling their absolute rate.

The most comprehensive study of temporal integration comes from patients implanted with the Second Sight Medical Products, Inc. epiretinal prosthesis [33]. We describe here a recently published model suggesting that visual sensitivity for electrical stimulation can be described by a relatively simple linear-nonlinear model that predicts the relationship between electrical stimulation and brightness for any temporally-varying stimulation pattern. This model can not only be used to determine the "optimal" pattern of stimulation given a variety of engineering constraints (such as stimulating at safe levels of charge density and minimizing overall charge, for example), but in addition, its biological plausibility provides insight into the neural pathways that underlie the perceptual effects of electrical stimulation. Data described in this section were collected on two subjects (S05 and S06) using the Second Sight Medical Products, Inc. retinal prosthesis.

Threshold values were collected as described in Sect. 14.3.1. Suprathreshold brightness-matching was carried out on single electrodes using a two-interval, forced-choice procedure. Each trial contained two intervals with each interval containing a pulse train of a different frequency. For example, interval 1 might contain a 15 Hz pulse train and interval 2 might contain a 45 Hz pulse train. Subjects were asked to report which interval contained the brighter stimulus. A one-up, one-down staircase method was used to adjust the amplitude of the higher frequency pulse train based on the observer's response. Using this method, we were able to obtain an isobrightness curve that represented the current amplitude needed to maintain the same subjective brightness across a wide range of frequencies.

Data were modeled using a linear-nonlinear model (Fig. 14.9) similar to models of auditory stimulation in cochlear implant users [66], retinal ganglion cell spiking behavior during temporal contrast adaptation [6, 16, 59], and human psychophysical temporal sensitivity in normal vision [74]. The stimulus was convolved with a temporal low-pass filter using a one-stage gamma function:

$$r_1(t) = f(t) * \delta(t, 1, \tau_1) \tag{14.1}$$

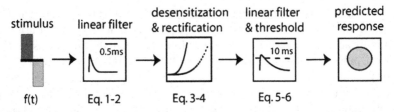

Fig. 14.9 Model schematic. The time varying stimulus is convolved with a linear filter. The result of this convolution is passed through a static nonlinearity and convolved with a secondary linear filter. We assumed that a stimulus was at visual threshold (or a given brightness level) when threshold reached a specific value. Reprinted from [33], with permission

where $f(t)$ is the electrical stimulation input pattern, t is time (ms), and δ is the impulse response function with time constant τ_1. The gamma function used to model this impulse response can be generally described as:

$$\delta(t,n,\tau_1) = \frac{e^{-\frac{t}{\tau_1}}}{\tau_1(n-1)!}\left(\frac{t}{\tau_1}\right)^{n-1},\tag{14.2}$$

where t=time, n=the number of identical, cascading stages, and τ_1 is the time constant of the filter (the one-stage gamma function in (14.1) is simply an exponential function).

We assumed that the system became less sensitive as a function of accumulated charge. This was computationally implemented by calculating the amount of accumulated cathodic charge at each point of time in the stimulus, c(t), and convolving this accumulation with a second one-stage gamma function having a time constant τ_2. The output of this convolution was scaled by a factor ε and then subtracted from r_1 (14.1),

$$r_2(t) = r_1(t) - \varepsilon(c(t) * \delta(t,1,\tau_2)).\tag{14.3}$$

r_2 was then passed through a power nonlinearity,

$$r_3(t) = \left(r_2(t)\right)^{\beta}\tag{14.4}$$

and convolved with a low-pass filter described as a three-stage gamma function with time constant τ_3,

$$r_4(t) = r_3 * \delta(t,3,\tau_3).\tag{14.5}$$

We assumed that the response reached threshold (or the point of equibrightness during suprathreshold experiments) when

$$\max t(r_4) >= \theta\tag{14.6}$$

where θ is a fixed constant.

Patients typically reported that phosphenes appeared white or yellow in color, and round or oval in shape. At suprathreshold, percepts were reported as brighter and the shape occasionally became more complex than a simple round or oval shape. The shapes were reported as being approximately 0.5–2 in. in diameter at arm's length, corresponding to roughly 1–3° of visual angle. Occasionally, a dark spot rather than a white or yellow percept was reported. In this case, the patient would use the relative contrast of the spot for detection (threshold) or "brightness comparison" (suprathreshold).

After optimizing the model using a subset of the full set of data, the best-fitting parameters values were averaged for τ_1, τ_2, τ_3, ε, and β across the electrodes used for optimization. These mean values were then used to predict threshold and suprathreshold data for novel electrodes.

Figure 14.10 shows subject thresholds (gray squares) and model predictions (solid line) for a single biphasic pulse presented on a novel electrode for both subjects. Figure 14.11 shows threshold data and predictions for pulse trains containing either 2 (**B**) or 15 (**C**) pulses, whose frequency was varied between 3 and 3,333 Hz. In summary, the model and parameter values generalized to successfully predict data on novel electrodes. The data shown in these two figures are from a subset of over ten experiments conducted over five electrodes from each of the two subjects.

The ability of the model to predict suprathreshold responses to novel pulse train waveforms not used to optimize model parameters was then examined (Fig. 14.12).

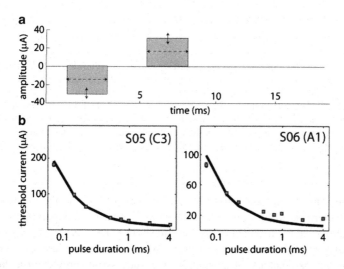

Fig. 14.10 Single pulse threshold. These data are from electrodes C3 and A1, from patient S05 and S06, respectively. Stimuli (**a**) were single, biphasic, charge-balanced square pulses, whose pulse width (*dashed arrow*) varied in duration from 0.075 to 4 ms. For each pulse width, the amplitude was varied (*solid arrow*) to determine perceptual threshold. In the data plots (**b**), the x-axis represents pulse width (plotted logarithmically) and the y-axis represents the current amplitude (µA) needed to reach threshold. The *solid black line* represents the prediction of the model. Reprinted from [33], with permission

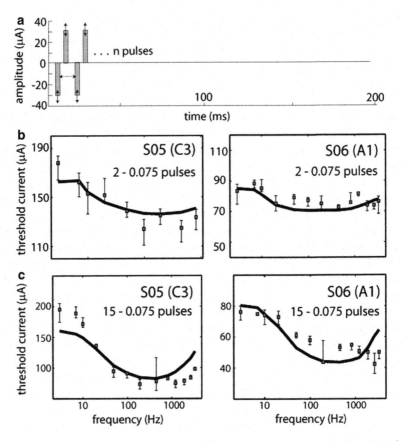

Fig. 14.11 Variable duration pulse train threshold. These data are from electrodes C3 and A1, from patient S05 and S06, respectively. Stimuli (**a**) were pulse trains whose frequency was varied between 3 and 3,333 Hz. Pulse trains contained either 2 (**b**) or (**c**) 15 pulses. The amplitude of all pulses within the train was varied simultaneously to determine threshold (see Methods for a full description of the threshold detection task). The x-axis represents pulse train frequency (Hz) (plotted logarithmically) and the y-axis represents the current amplitude (μA), per pulse, needed to reach threshold. The *black line* represents the prediction of the model. Reprinted from [33], with permission

Again, the same fixed values for τ_1, τ_2, τ_3, ε, and β based on the electrodes and stimulus patterns used for optimization were used, and the only parameter allowed to vary across each experiment was the threshold parameter θ. The novel waveforms consisted of repeated bursts of three pulses with a variable inter-burst delay. The model and parameter values generalized to successfully predict these data from a novel stimulation pattern on a novel electrode.

This model, like those describing the perception of light stimuli, presumably approximates the responses of neuronal populations. In the case of our threshold experiments, it is possible that firing within a relatively small number of retinal cells mediated detection. It has been previously shown that subjects with normal

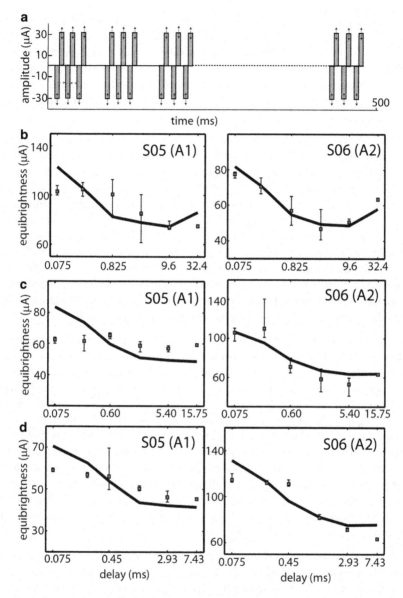

Fig. 14.12 Bursting pulse triplets, suprathreshold. These data are from electrodes A1 and A2, from patient S05 and S06, respectively. All pulse train stimuli (**a**) were either 15 (**b**), 30 (**c**), or 60 (**d**) pulse trains that were 500 ms in duration, consisting of bursts, or triplets, of groups of three pulses. Each burst consisted of 0.45 ms biphasic pulses with no inter-phase delay. The x-axis represents the inter-pulse delay between the set of three bursting pulses (plotted logarithmically), and the y-axis is current amplitude (μA), per pulse, needed to reach equibrightness. All stimuli were brightness matched to the maximally separated, or evenly distributed, pulse trains (32.4 ms delay for (**b**), for example). The *black line* represents the prediction of our model. Reprinted from [33], with permission

vision can reliably detect a single photon of light [30], suggesting that a very small increase over the baseline firing rate of ganglion cells is probably sufficient to mediate behavioral detection. Thus, thresholds in our subjects may have been mediated by a relatively small number of spikes: these spikes might, of course occur either in a single cell or across several cells. At suprathreshold our model presumably approximates the population response of a larger number of cells each producing one or multiple spikes.

τ_1 (14.1). The parameter τ_1 represents the time course of the first stage of current integration. Estimates of τ_1 in our model vary between 0.24 and 0.65 ms, with a mean of 0.42 ms, a value very similar to electrophysiology estimates of the integration of current by ganglion cells [25, 39, 64]. In contrast, long-latency spiking in ganglion cells, occurring >8–60 ms after the beginning of electrical stimulation [25, 27, 39], originate from bipolar cells since a cocktail of synaptic blockers completely suppresses this late-phase spiking in ganglion cells. The time constant associated with the inhibitory input from amacrine cells is on the order of 100–200 ms [25]. The similarity of τ_1 to time constants of current integration by ganglion cells suggests that direct stimulation of ganglion cells (rather than indirect stimulation via pre-synaptic input) may be primarily responsible for integration of stimulation current within the retina, particularly with pulse widths longer than 1 ms [27].

ε and τ_2 (14.3). The parameters ε and τ_2 represent desensitization as a consequence of accumulated charge, where ε represents the strength of desensitization and τ_2 represents the time constant over which charge was integrated. There are two possible sources for this change in sensitivity. One possibility is that injected charge directly results in a hyperpolarization of the membrane resting potentials within individual ganglion cells. Shifts in resting potentials, analogous to slow contrast adaptation effects, can be produced in ganglion cells by injection of hyperpolarizing current [6]. However, it is as likely that inhibition from presynaptic cells was involved in the desensitization we observed. Inhibitory presynaptic influences on spiking in response to electrical stimulation have been described by Fried et al. [25], particularly for longer pulses. It seems likely that the desensitization stage of our model simply approximates a series of complex adaptive processes, with time courses varying between milliseconds to tens of seconds [6, 16, 59].

β (14.4). β describes a power input-output nonlinearity. Power nonlinearities are frequently used in linear-nonlinear models describing spiking behavior in ganglion cells [6, 16]. A similar nonlinearity has been used in modeling human behavioral data of light stimuli [74]. One possibility is that as the intensity of stimulation increases, neurons with shallower input-output nonlinearities are recruited. Alternatively, this change in the power function may be driven by changes in the input-output nonlinearity within individual cells. It has been found in models of retinal spiking that the slope of the nonlinearity changes as a function of increased contrast [6, 59].

τ_3 (14.5). τ_3 determines the integration period of the final low pass filter. Thresholds decrease as a function of frequency for a fixed number of pulses, with an asymptote at around 100–200 Hz, with the effect being most noticeable for the pulse train containing 15 pulses. τ_3 may represent the slow temporal integration that occurs in cortex. Similar integration times have been found in simple cell recordings in cat striate cortex [55].

A successful retinal prosthesis will need to produce percepts consisting of regions of constant brightness across a range of brightness levels, while satisfying a complex set of engineering constraints: charge densities must remain relatively low, it is technically difficult to produce very high current amplitudes, and absolute charge must be minimized to maximize battery life. Models of the perceptual effects of electrical stimulation, such as that described here, will be critical in allowing electrical stimulation protocols to be selected that best satisfy these many constraints.

14.4   Suprathreshold Brightness

A visual prosthesis should produce regions of constant brightness across a range of brightness levels, and ideally these suprathreshold brightness levels should be consistent with the apparent brightness of objects as they appear to those with normal vision. To date, all three groups have examined how apparent brightness changes as a function of stimulation intensity.

14.4.1   Brightness Using the Retinal Implant AG System

Brightness as a function of stimulus intensity has been measured in patients implanted with the Retinal Implant AG device [78, 80]. These tests have tended to use a slightly more clinical methodology than the psychophysical measures reported for patients implanted with the Second Sight LLC implant. Among other tests (described below), patients were asked to rate the perception of brightness elicited by applying biphasic voltage impulses from 1 to 2.5 V presented on four electrodes in a square configuration (3 ms pulse duration, presented in a random order) using a scale from 5 (very strong) to 0 (none). When there were six steps between 1 and 2.5 V (corresponding to a charge increase of approximately 0.23 mC/cm^2 between each stimulus assuming a linear scale) the apparent brightness of the elicited spots varied from scale 0 to 5 in a linear manner.

This group has also carried out brightness matching experiments using pairs of pulses that differed by as much as 0.8 V (10 s interval between each pulse). A difference in brightness between two consecutive pulses was discerned if a difference in charge of at least 161 μC/cm^2 was applied. If equal charges were applied within both stimulation intervals, the second flash always was perceived as slightly dimmer irrespective of the stimulation level.

Subjective brightness-size interactions were observed at medium stimulation levels and at certain frequencies. The subjective size of the phosphene elicted by four electrodes increased from 1 to 5 mm (at arm's length) if the voltage was increased from 1.5 to 2.5 V [79].

14.4.2 Brightness Using the Intelligent Medical Implant System

Thresholds for each patient were obtained from a Weibull fit and resulted in an average threshold of 25.3 ± 7 nC, which is considerably lower than the result obtained in a previous study involving acute stimulation with the same patients [57]. In one subject, thresholds were measured over a period of 4 months. In this case, thresholds were determined by stimulating at charge levels between 0 and 122 nC [58], using a two-alternative forced-choice procedure. In total, 23 runs were conducted over the course of the 4 months. Although the threshold varied from 8.0 to 35.9 nC between subjects, the data suggest that the thresholds were stable over the entire 4 month period. Visual percepts depend on amplitude levels and electrode location [57].

14.4.3 Brightness Using the SSMP A16 System

In these measurements, two subjects implanted with the A16 epiretinal prosthesis were asked to rate the brightness of a test pulse in comparison to the brightness of a standard of fixed current amplitude. Stimulation for test and reference pulses always consisted of a single biphasic, cathodic-first, charge-balanced square wave pulse, with a pulse duration of 0.975, and a 0.975 ms inter-pulse interval. The reference pulse was fixed at a current amplitude chosen to be roughly 2.5 times the threshold amplitude for a single pulse on that electrode.

We used a classic brightness matching procedure based on that of Stevens [68]. Before beginning each testing session, subjects were repeatedly stimulated with the reference pulse and were told, "*This reference pulse has brightness of 10 and we will present it to you before we begin each trial. Your task is to compare the brightness of the test pulse in each trial to the brightness of this reference pulse. If the test pulse seems to be twice as bright as the reference pulse then give it a rating of 20. If the test pulse seems to be half as bright as the reference pulse, then give it a rating of 5.*"

Once the subject reported feeling confident of having a clear idea of the brightness of the reference pulse, we began the experiment. All subject ratings were provided verbally. On each trial, subjects were first presented with the reference pulse and were reminded that this pulse should be considered as having a brightness of 10. This reference pulse was quickly followed by the test pulse. Subjects were then asked to verbally rate the apparent brightness of the test pulse, as compared to the reference pulse (Fig. 14.13).

The test pulse was always presented on the same electrode as the reference pulse, and had a current amplitude that varied pseudo-randomly from trial to trial using the method of constant stimuli. Subjects were not told which test pulse current value had been presented on each trial, and no feedback was provided. Each test current amplitude was presented four times, and the mean and the standard error of brightness ratings for each stimulation amplitude was calculated across these four repetitions.

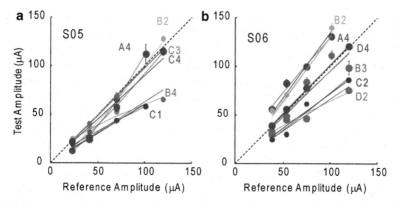

Fig. 14.13 Brightness matching data for both subjects. For each subject, equibrightness measurements were conducted at with pulse amplitude on the reference electrode fixed at five different amplitude levels. The x-axis represents the amplitude of the pulse on the reference electrode. The y-axis represents the PSE on each of six test electrodes and the reference electrode brightness matched to itself. 520 μm electrodes are represented by large symbols, 260 μm electrodes by small symbols. The *dashed line* represents equal amplitude on test and reference electrode. The *different labels* represent measurements for a specific electrode (e.g., C1). Reprinted from [28], with permission

Brightness matching judgments were also carried out, subjects made brightness judgments (which interval contains the brighter stimulus) between a pulse train presented on a reference electrode and a pulse train presented on a test electrode using a two-interval forced choice procedure. Intensity of the test electrode was adjusted through a staircase procedure and data were fit with a cumulative normal distribution to find the point of subjective equality. Subjects could reliable differentiate between pulse pairs separated by less than 20 μA in the discrimination paradigm.

Both brightness rating and brightness discrimination judgments could be well fit by a classic Stevens function, $B = aC^b$, where B is the brightness rating made by the subject, C is stimulus current amplitude, and a and b are free parameters. Data could still be fit when b was fixed to be the median of the best-fitting values of b across all four electrodes for that subject, suggesting that it may be possible to normalize brightness across an entire array of electrodes by measuring a single parameter for each electrode.

14.5   Spatial Vision

The data described above show that it is possible to control the perceptual brightness of a stimulus presented on a single electrode through either the timing or amplitude of stimulation. The data described below examine how multiple electrodes interact to create a patterned image, and the first studies examining whether these implants might offer the potential to provide functional vision.

When evaluating spatial vision or functional vision tasks it is important to recognize that subjects may develop "strategies" to perform these tasks, especially when given previous training with feedback within tasks that have a constrained set of response alternatives. This is a particular concern for two or four alternative forced choice tasks with training and/or feedback. These are of course perfectly valid psychophysical techniques, and these tests are based on standard clinical tests of visual acuity, but it needs to be remembered that in these cases the subject is performing a constrained discrimination task, not an identification task. For example, these tasks may not be particularly revealing about whether subjects are doing these tasks on the basis of percepts that would be meaningful outside the laboratory environment. To take an extreme case, if stimuli are "jumbled" in space, then each of two alternatives might produce a pseudo-random percept, but these percepts would be perceptually distinct. In this case the subjects could perform the task perfectly with training but the implant would be useless for functional vision.

14.5.1 Spatial Vision with the Retinal Implant AG System

As described above, a set of computerized, standardized tests for patients with visual prostheses was developed to quantify the functional abilities of patients implanted with the Retinal Implant AG device [78, 80].

Electrical stimulation of rows, columns and blocks of four electrodes allowed some patients to clearly distinguish horizontal from vertical lines under four-alternative forced choice conditions. Under optimal conditions, dot alignment and direction of dot movement could also be differentiated when three neighboring electrodes were switched on simultaneously or sequentially at 1 s intervals.

This study also reports evidence examining letter-reading and stripe pattern recognition using the Retinal Implant AG system. Using the direct stimulation (DS) 4×4 array, electrode stimulation configurations were used to represent letters. These images were perceived as 5 cm in diameter when presented at a 60 cm distance [80]. Patient 1 correctly determined the orientation of a letter "U" (20/24 times) when using a four alternative forced-choice task (4 AltFC). Patient 2 correctly discriminated between the letters C, O, I, L, and Z using this same 4 AltFC paradigm. Additionally, when using the light-sensitive chip, this same subject was able to differentiate the letters L, I, T, and Z when presented on a screen 62 cm away. Both Patient 1 and Patient 3 could determine the direction of lines or stripe patterns using the light sensitive chip (11/14 and 11/12, respectively). To date it is not clear why some subjects could perform some tasks, but not others, and to what extent performance on these tasks was mediated by practice with feedback with individual stimuli.

14.5.2   Spatial Vision with the Intelligent Medical Implant System

The Intelligent Retinal Implant (IRI) was implanted in four patients with bare light perception (BLP) or less. These patients were then tested on at least 20 separate occasions over a period of approximately 12 months. Across these session both thresholds and pattern recognition was evaluated.

During stimulation sessions the patients were able to distinguish between different points in space when spatially-segregated electrodes were activated. This point-to-point discrimination task was successful both horizontally and vertically. When presenting multi-electrode stimulation, patients were able to recognize simple patterns such as horizontal bars [56] in a forced choice procedure. Simple patterns (vertical/horizontal bar, a cross) administered via activation of appropriate electrodes were also distinguishable by patients in forced choice procedures.

14.5.3   Spatial Vision with the SSMP A16 System

Assuming that each electrode on a 2-dimensional electrode array can produce individual, punctate phosphenes, visual resolution is simply limited by the pitch or spacing of the electrodes. Using the Second Sight A16 system, visual acuity performance was evaluated in a single blind human subject (S06) to determine whether his spatial visual resolution could approach the level expected from the spacing between the electrodes in the A16 electrode array [14].

The first experiment tested whether or not an oriented contour could be generated using the retinal prosthesis in direct stimulation mode (see Sect. 14.2.3 for a more in-depth description of the direct stimulation and camera modes). In each trial, a single row of 4 electrodes was used to stimulate a row and then a column (with a 1 second delay between the two stimuli), and the subject was instructed to draw on a board the pattern they perceived. The predicted percept would be a right-angle cross of the two lines with a 90° angle of intersection. A head-mounted camera system was used to record the movement of the marker on the board at arm's length from the subject and the digital data output was analyzed offline. Over 14 trials, the subject drew 2 lines with an average angle of 87.4° (1.8° standard error).

In a second experiment, S06 was asked to report the orientation of a high-contrast, square-wave grating presented on a screen. The orientation of these gratings was either horizontal, vertical, diagonal right orientation, or diagonal left orientation. Thus, the chance performance on this task was 25%. These data were collected in camera mode. In each session, high-contrast gratings of different spatial scales (2.77-2.00 logMAR; Snellen equivalent 20/11,777-20/2000) were randomly interleaved. The probability of detection was calculated for each spatial frequency and the data was fit with a logistic psychometric function. The subject performed significantly above chance for all trials down to 2.21 logMAR. At the critical sampling frequency, each black and white bar falls directly on one row of electrodes. The resolution was, therefore, directly limited by the spacing of the electrodes.

Taken together, these data suggest that the visual resolution of a blind patient implanted with a SSMP A16 retinal prosthesis is limited only by the spacing of the electrodes in the array.

14.6   Models to Guide Electrical Stimulation Protocols

Achieving useful percepts via electrical stimulation requires satisfying a variety of safety and engineering constraints. Useful percepts will require stimulation at frequencies higher than subjects' perception of visible flicker (frequencies above the "critical flicker frequency"). Second, safety concerns dictate relatively stringent charge density limits, since high charge densities have the potential to compromise the integrity of electrode material [12, 13] and cause damage to stimulated neural cells [48, 49, 67]. Third, the maximum current amplitude that can be produced may in some cases be limited by the compliance voltage of the stimulator. A final set of constraints include limits in the amount of power available to the implant given the need for a long battery life, and power limits inherent in transmitting power inductively, resulting in a need to minimize overall charge.

The models described in this chapter provides an example of how the optimal stimulation pattern needed to produce a percept of a given brightness level can be determined given a set of constraints. A particular example is given in Fig. 14.14, which shows example predictions of threshold current amplitude (graph a), charge density (graph b), and overall charge (graph c) for a 500 ms pulse train presented on an electrode of typical sensitivity across a range of pulse widths and frequencies. The dashed lines represent examples of safety and engineering constraints that might restrict the potential set of stimulation patterns. In the example shown here, a current amplitude limit of 200 µA, and a charge density limit of 0.35 mC/cm². Given these example constraints, our model predicts that the most charge efficient stimulation pattern, for the conditions and prosthetic device tested here, is a 50 Hz pulse train consisting of 0.089 ms pulses. Similarly, for a given compliance voltage, the most efficient operation (in terms of energy delivered to the electrodes vs. energy dissipated in the current regulator) is when the voltage drop across the electrodes is near this compliance voltage. These engineering constraints may result in the most efficient pulse being at the highest current that can be supplied, making it advantageous to manipulate brightness using either frequency or pulse width. These models can also be used to calculate the most energy efficient pulse width (chronaxie) [17, 26]. Depending on the assumed constraints of the stimulation protocol, models such as these can estimate the best stimulation protocol.

Of course this ability to evaluate engineering and safety trade-offs across different pulse patterns need not be restricted to the simple stimulation patterns used in this example. Our hope is that this model (or similar models) can be generalized

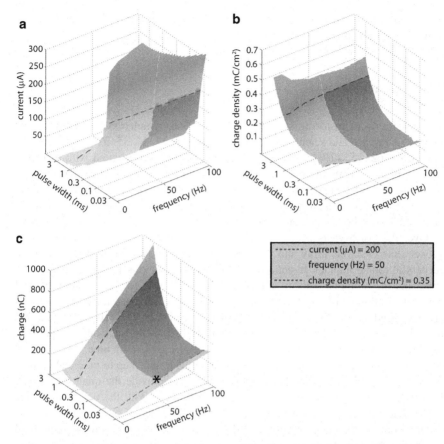

Fig. 14.14 Efficiency predictions for a 500 ms pulse train. In each panel the x-axis represents pulse width on a logarithmic axis, and the y-axis represents frequency. *Red dashed lines* represent a current amplitude limit of 200 μA, *yellow dashed lines* represent the constraint that stimulation must occur above the critical flicker frequency of 50 Hz, and *blue dashed lines* represent the constraint of a charge density limit of 0.35 mC/cm². Light shading represents pulse widths and frequencies that fall outside these constraints. The z-axis represents current (**a**), charge density (**b**), and overall charge across the entire pulse train (**c**). Given these example constraints, our model predicts that the most charge efficient stimulation pattern is a 50 Hz pulse train consisting of 0.089 ms pulses, as shown by an *asterisk* in (**c**). (Please see online version for full-color representation). Reprinted from [33], with permission

to describe percepts over a wide range of brightness levels, across multiple electrodes. It is also to be hoped that models such of these will generalize to other devices, though it is of course quite likely that the models needed to explain subretinal stimulation will differ substantially from those developed to explain epiretinal stimulation. While models such as these may always be a crude approximation of perceptual effects, without them developing stimulation protocol procedures will remain an ad hoc procedure of trial and error.

14.7 Conclusions

The possibility of restoring sight through electrical stimulation has captured the interest of laymen and scientists for many years [11, 22, 24, 35, 37, 42, 44, 45, 50, 53, 54, 61, 62, 75, 77]. As we make progress, the goals associated with restoring sight in blind patients (as well as our appreciation of the difficulties that must be overcome to reach that goal) have become more sophisticated. In more recent studies, the effort has not been simply to create phosphenes, but to create images that are predictable over both space and time.

Over the last 5 years it has become apparent that maintaining close proximity between the electrode array and the retinal surface will be critical in developing a successful retinal implant. In addition to affecting threshold, separation between the array and the retina is likely to compromise the ability to produce small localized percepts. As thinner electrode array structures are developed and improved methods are developed for attaching the array to the to the retina [29, 71] it should be possible to maintain electrodes that remain flush with retinal tissue for indefinite periods, resulting in impedance and threshold values that are stable over time. A successful prosthesis will require arrays which are stable on the retina, map to predictable locations in space, and are of high enough resolution to provide the quality of visual information needed to perform useful real world tasks. With the use of electrode arrays that meet these criteria, it is likely that the influence of other factors such as progression of retinal degeneration and subject age will become more apparent both within threshold measurements, and within more complex measures of perception.

Over the last 5 years, work by a number of groups including ours has demonstrated that simple visual percepts generated by direct retinal electrical stimulation on a single electrode can be modeled relatively simply. This is true both for amplitude coding [28] and for manipulations of pulse timing within an electrode (frequency encoding).

Of course a wide variety of challenges remain, even in understanding the effects of stimulating a single electrode. For example, apparent brightness is not the only perceptual quality that needs to be considered. It is possible that different temporal patterns stimulate slightly different subpopulations of neurons, resulting in distinct percepts. Moreover, the experiments described here only considered pulse trains or stimulation periods of relatively short duration (a maximum of a few seconds). Longer periods of continuous stimulation (minutes or hours) may result in long-term adaptation, sensitization, and/or retinal rewiring. It is quite likely that frequent electrical stimulation over a time scale of weeks and months may result in changes in retinal connectivity and responsivity [46].

More importantly, it is of critical importance to better understand how neighboring electrodes interact in the spatiotemporal domain. The models described in Sects. 14.3

and 14.4 simply predict sensitivity at the single electrode level, the extension of models such as ours to the spatial domain is an obvious next step.

While preliminary results with "natural tasks" show promise, it is still not entirely clear what kind of spatial resolution is mediated by current prosthetic devices. One concern is that, to date, most (though not all) data have come from constrained two or four alternative forced choice tasks with training and/or feedback so the subject is performing a constrained discrimination task, not an identification task. A subject may be able to discriminate "horizontal" from "vertical" without the horizontal line appearing as a horizontal line, and the vertical line appearing as a vertical line – all that is necessary is for the two stimuli to be perceptually distinct. A second concern is the extent of variability across subjects – to date no group has reported successful performance across a wide range of tasks within all (or even a majority) of implanted subjects. A third concern is that there is still some doubt as to whether all electrodes in these arrays map neatly to the expected perceptual location in space. As described in this chapter, progress over the last 5 years has been rapid, and progress over the next five is likely to bring us still closer to a useful prosthetic array. While it is unlikely that we will be able to build devices that resemble "natural" vision in the next 5 years, it is possible that, even if with some "jumbling" of the sensory input (as is found in cochlear implants for hearing) the brain will learn to understand the new sensory representation (analogous to learning to interpret modern art sketches for people with normal vision). However, we will probably not know the full capacity of the human visual system to adapt to make use of retinal implants until these devices are in more common use.

References

1. Acland GM, Aguirre GD, Bennett J, et al. (2005), *Long-term restoration of rod and cone vision by single dose rAAV-mediated gene transfer to the retina in a canine model of childhood blindness.* Mol Ther, **12**(6): p. 1072–82.
2. Acland GM, Aguirre GD, Ray J, et al. (2001), *Gene therapy restores vision in a canine model of childhood blindness.* Nat Genet, **28**(1): p. 92–5.
3. Adrian ED, Matthews R (1928), *The action of light on the eye: Part III. The interaction of retinal neurones.* J Physiol, **65**(3): p. 273–98.
4. Aguirre GK, Komaromy AM, Cideciyan AV, et al. (2007), *Canine and human visual cortex intact and responsive despite early retinal blindness from RPE65 mutation.* PLoS Med, **4**(6): p. e230.
5. Alitto HJ, Usrey WM (2008), *Origin and dynamics of extraclassical suppression in the lateral geniculate nucleus of the macaque monkey.* Neuron, **57**(1): p. 135–46.
6. Baccus SA, Meister M (2002), *Fast and slow contrast adaptation in retinal circuitry.* Neuron, **36**(5): p. 909–19.
7. Bainbridge JW, Smith AJ, Barker SS, et al. (2008), *Effect of gene therapy on visual function in Leber's congenital amaurosis.* N Engl J Med, **358**(21): p. 2231–9.
8. Batten ML, Imanishi Y, Tu DC, et al. (2005), *Pharmacological and rAAV gene therapy rescue of visual functions in a blind mouse model of Leber congenital amaurosis. [see comment].* PLoS Med, **2**(11): p. e333.

9. Berry MJ, Warland DK, Meister M (1997), *The structure and precision of retinal spike trains.* Proc Natl Acad Sci USA, **94**(10): p. 5411–6.

10. Bi A, Cui J, Ma YP, et al. (2006), *Ectopic expression of a microbial-type rhodopsin restores visual responses in mice with photoreceptor degeneration.* Neuron, **50**(1): p. 23–33.

11. Brindley GS, Lewin WS (1968), *The sensations produced by electrical stimulation of the visual cortex.* J Physiol, **196**(2): p. 479–93.

12. Brummer SB, Robblee LS, Hambrecht FT (1983), *Criteria for selecting electrodes for electrical stimulation: theoretical and practical considerations.* Ann NY Acad Sci, **405**: p. 159–71.

13. Brummer SB, Turner MJ (1975), *Electrical stimulation of the nervous system: the principle of safe charge injection with noble metal electrodes.* Bioelectrochem Bioenerg, **2**: p. 13.

14. Caspi A, Dorn JD, McClure KH, Humayun MS, Greenberg RJ, McMahon MJ (2009), *Feasibility study of a retinal prosthesis: spatial vision with a 16-electrode implant.* Arch Ophthalmol, **127**(4): p. 398–401.

15. Chader GJ (2002), *Animal models in research on retinal degenerations: past progress and future hope.* Vision Res, **42**(4): p. 393–9.

16. Chander D, Chichilnisky EJ (2001), *Adaptation to temporal contrast in primate and salamander retina.* J Neurosci, **21**(24): p. 9904–16.

17. Coates S, Thwaites B (2000), *The strength-duration curve and its importance in pacing efficiency: a study of 325 pacing leads in 229 patients.* Pacing Clin Electrophysiol, **23**(8): p. 1273–7.

18. Congdon N, O'Colmain B, Klaver CC, et al. (2004), *Causes and prevalence of visual impairment among adults in the United States.* Arch Ophthalmol, **122**(4): p. 477–85.

19. Daiger SP, Bowne SJ, Sullivan LS (2007), *Perspective on genes and mutations causing retinitis pigmentosa.* Arch Ophthalmol, **125**(2): p. 151–8.

20. Dan Y, Atick J, Reid R (1996), *Efficient coding of natural scenes in the lateral geniculate nucleus: experimental test of a computational theory.* J Neurosci, **16**: p. 3351–62.

21. de Balthasar C, Patel S, Roy A, et al. (2008), *Factors affecting perceptual thresholds in epiretinal prostheses.* Invest Ophthalmol Vis Sci, **49**(6): p. 2303–14.

22. Dobelle WH (1974), *Introduction to sensory prostheses for the blind and deaf.* Trans Am Soc Artif Intern Organs, **20B**: p. 761.

23. Field GD, Chichilnisky EJ (2007), *Information processing in the primate retina: circuitry and coding.* Annu Rev Neurosci, **30**: p. 1–30.

24. Foerster O (1929), *Beitrage zur pathophysiologie der sehbahn und der spehsphare.* J Psychol Neurol (Lpz), **39**: p. 435.

25. Fried SI, Hsueh HA, Werblin FS (2006), *A method for generating precise temporal patterns of retinal spiking using prosthetic stimulation.* J Neurophysiol, **95**(2): p. 970–8.

26. Geddes LA (2004), *Accuracy limitations of chronaxie values.* IEEE Trans Biomed Eng, **51**(1): p. 176–81.

27. Greenberg RJ (1998), *Analysis of electrical stimulation of the vertebrate retina – work towards a retinal prosthesis.* Thesis, The Johns Hopkins University.

28. Greenwald SH, Horsager A, Humayun MS, Greenberg RJ, McMahon MJ, Fine I (2009), *Brightness as a function of current amplitude in human retinal electrical stimulation.* Invest Ophthalmol Vis Sci, **50**(11): p. 5017–25.

29. Guven D, Weiland J, Maghribi M, et al. (2006), *Implantation of an inactive epiretinal poly (di methyl) siloxane electrode arrays in dogs.* Exp Eye Res, **82**: p. 81–9.

30. Hecht S, Shlaer S, Pirenne MH (1942), *Energy at the threshold of vision.* Science, **93**(2425): p. 585–7.

31. Hesse L, Schanze T, Wilms M, Eger M (2000), *Implantation of retina stimulation electrodes and recording of electrical stimulation responses in the visual cortex of the cat.* Graefes Arch Clin Exp Ophthalmol, **238**: p. 840.

32. Heynen H, van Norren D (1985), *Origin of the electroretinogram in the intact macaque eye – II. Current source-density analysis.* Vision Res, **25**(5): p. 709.

33. Horsager A, Greenwald SH, Weiland JD, et al. (2009), *Predicting visual sensitivity in retinal prosthesis patients.* Invest Ophthalmol Vis Sci, **50**(4): p. 1483–91.

34. Hubel DH, Wiesel TN (1962), *Receptive fields, binocular interaction and functional architecture in the cat's visual cortex.* J Physiol, **160**: p. 106–54.
35. Humayun MS, de Juan E Jr, Dagnelie G, et al. (1996), *Visual perception elicited by electrical stimulation of retina in blind humans.* Arch Ophthalmol, **114**(1): p. 40–6.
36. Humayun MS, de Juan E Jr, Weiland JD, et al. (1999), *Pattern electrical stimulation of the human retina.* Vision Res, **39**(15): p. 2569–76.
37. Humayun MS, Weiland JD, Fujii GY, et al. (2003), *Visual perception in a blind subject with a chronic microelectronic retinal prosthesis.* Vision Res, **43**(24): p. 2573–81.
38. Jensen RJ, Rizzo JF III, Ziv OR, et al. (2003), *Thresholds for Activation of rabbit retinal ganglion cells with an ultrafine, extracellular microelectrode.* Invest Ophthalmol Vis Sci, **44**(8): p. 3533–43.
39. Jensen RJ, Ziv OR, Rizzo JF (2005), *Responses of rabbit retinal ganglion cells to electrical stimulation with an epiretinal electrode.* J Neural Eng, **2**: p. S16–21.
40. Jones BW, Watt CB, Frederick JM, et al. (2003), *Retinal remodeling triggered by photoreceptor degenerations.* J Comp Neurol, **464**(1): p. 1–16.
41. Lagali PS, Balya D, Awatramani GB, et al. (2008), *Light-activated channels targeted to ON bipolar cells restore visual function in retinal degeneration.* Nat Neurosci, **11**(6): p. 667–75.
42. LeRoy C (1755), *Ou l'on rend compte de quelques tentatives que l'on a faites pour guerir plusieurs maladies par l'electricite.* Hist Acad Roy Sciences (Paris), **60**: p. 87–95.
43. Lin B, Koizumi A, Tanaka N, et al. (2008), *Restoration of visual function in retinal degeneration mice by ectopic expression of melanopsin.* Proc Natl Acad Sci USA, **105**(41): p. 16009–14.
44. Lowenstein K, Borchardt M (1918), *Symptomatologie und elektrische Reizung bei einer Schußverletzung des Hinterhauptlappens.* Dtsch Z Nervenheilk, **58**: p. 264–92.
45. Mahadevappa M, Weiland JD, Yanai D, et al. (2005), *Perceptual thresholds and electrode impedance in three retinal prosthesis subjects.* IEEE Trans Neural Syst Rehabil Eng, **13**(2): p. 201–6.
46. Marc RE, Jones BW, Watt CB, Strettoi E (2003), *Neural remodeling in retinal degeneration.* Prog Retin Eye Res, **22**(5): p. 607–55.
47. Marg E (1991), *Magnetostimulation of vision: direct noninvasive stimulation of the retina and the visual brain.* Optom Vis Sci, **68**(6): p. 427–40.
48. McCreery DB, Agnew WF (1990), *Mechanisms of stimulation-induced neural damage and their relation to guidelines for safe stimulation*, in *Neural Prostheses Fundamental Studies,* Agnew WF, McCreery DB, Editors. Prentice Hall: New Jersey, p. 297.
49. McCreery DB, Agnew WF, Yuen TG, Bullara LA (1988), *Comparison of neural damage induced by electrical stimulation with faradaic and capacitor electrodes.* Ann Biomed Eng, **16**(5): p. 463.
50. Murphey DK, Maunsell JH (2007), *Behavioral detection of electrical microstimulation in different cortical visual areas.* Curr Biol, **17**(10): p. 862–7.
51. Palanker D, Vankov A, Huie P, Baccus S (2005), *Design of a high-resolution optoelectronic retinal prosthesis.* J Neural Eng, **2**: p. S105–20.
52. Pawlyk BS, Smith AJ, Buch PK, et al. (2005), *Gene replacement therapy rescues photoreceptor degeneration in a murine model of Leber congenital amaurosis lacking RPGRIP.* Invest Ophthalmol Vis Sci, **46**(9): p. 3039–45.
53. Penfield W, Rasmussen T (1952), *The Cerebral Cortex of Man.* New York: Macmillan. p. 135.
54. Pezaris JS, Reid RC (2007), *Demonstration of artificial visual percepts generated through thalamic microstimulation.* Proc Natl Acad Sci USA, **104**(18): p. 7670–5.
55. Reid RC, Victor JD, Shapley RM (1992), *Broadband temporal stimuli decrease the integration time of neurons in cat striate cortex.* Vis Neurosci, **9**(1): p. 39–45.
56. Richard G, et al. (2007), *Chronic epiretinal chip implant in blind patients with retinitis pigmentosa: long-term clinical results.* Invest Ophthalmol Vis Sci, **48**: ARVO E-Abstract 989.
57. Richard G, et al. (2008), *Visual perception after long-term implantation of a retinal implant.* Invest Ophthalmol Vis Sci, **49**: ARVO E-Abstract 245.

58. Richard G, et al. (2009), *Long-term stability of stimulation thresholds obtained from a human patient with a prototype of an epiretinal retina prosthesis*. Invest Ophthalmol Vis Sci, **50**: ARVO E-Abstract 634.

59. Rieke F (2001), *Temporal contrast adaptation in salamander bipolar cells*. J Neurosci, **21**(23): p. 9445–54.

60. Rieke F, Warland D, de Ruyter van Steveninck RR, Bialek W (1997), *Spikes: Exploring the Neural Code*. Cambridge, MA: MIT Press.

61. Rizzo JF III, Wyatt J, Loewenstein J, et al. (2003), *Perceptual efficacy of electrical stimulation of human retina with a microelectrode array during short-term surgical trials*. Invest Ophthalmol Vis Sci, **44**(12): p. 5362–9.

62. Schmidt EM, Bak MJ, Hambrecht FT, et al. (1996), *Feasibility of a visual prosthesis for the blind based on intracortical microstimulation of the visual cortex*. Brain, **119**(Pt 2): p. 507.

63. Schnapf JL, Kraft TW, Baylor DA (1987), *Spectral sensitivity of human cone photoreceptors*. Nature, **325**(6103): p. 439–41.

64. Sekirnjak C, Hottowy P, Sher A, et al. (2006), *Electrical stimulation of mammalian retinal ganglion cells with multi-electrode arrays*. J Neurophysiol, **95**(6): p. 3311–27.

65. Shah S, Hines A, Zhou D, et al. (2007), *Electrical properties of retinal-electrode interface*. J Neural Eng, **4**(1): p. S24–9.

66. Shannon RV (1989), *A model of threshold for pulsatile electrical stimulation of cochlear implants*. Hear Res, **40**(3): p. 197–204.

67. Shannon RV (1992), *A model of safe levels for electrical stimulation*. IEEE Trans Biomed Eng, **39**(4): p. 424–6.

68. Stevens SS (1960), *Psychophysics of sensory function*. Am Sci, **48**: p. 226–52.

69. Thylefors B, Negrel AD, Pararajasegaram R, Dadzie KY (1995), *Available data on blindness (update 1994)*. Ophthalmic Epidemiol, **2**(1): p. 5–39.

70. Uzzell VJ, Chichilnisky EJ (2004), *Precision of spike trains in primate retinal ganglion cells*. J Neurophysiol, **92**(2): p. 780–9.

71. Walter P, Szurman P, Vobig M, et al. (1999), *Successful long-term implantation of electrically inactive epiretinal microelectrode arrays in rabbits*. Retina, **19**(6): p. 546–52.

72. Wang H, Peca J, Matsuzaki M, et al. (2007), *High-speed mapping of synaptic connectivity using photostimulation in Channelrhodopsin-2 transgenic mice*. Proc Natl Acad Sci USA, **104**(19): p. 8143–8.

73. Wang X, Wei Y, Vaingankar V, et al. (2007), *Feedforward excitation and inhibition evoke dual modes of firing in the cat's visual thalamus during naturalistic viewing*. Neuron, **55**(3): p. 465–78.

74. Watson AB (1986), *Temporal sensitivity*, in *Handbook of Perception and Human Performance*, Boff K, Kaufman L, Thomas J, Editors. Wiley: New York.

75. Weiland JD, Humayun MS, Dagnelie G, et al. (1999), *Understanding the origin of visual percepts elicited by electrical stimulation of the human retina*. Graefes Arch Clin Exp Ophthalmol, **237**(12): p. 1007–13.

76. Yanai D, Lakhanpal RR, Weiland JD, et al. (2003), *The value of preoperative tests in the selection of blind patients for a permanent microelectronic implant*. Trans Am Ophthalmol Soc, **101**: p. 223–8; discussion 228–30.

77. Yanai D, Weiland JD, Mahadevappa M, et al. (2007), *Visual performance using a retinal prosthesis in three subjects with retinitis pigmentosa*. Am J Ophthalmol, **143**(5): p. 820–7.

78. Zrenner E (2007), *Restoring neuroretinal function: new potentials*. Doc Ophthalmol, **115**: p. 56–9.

79. Zrenner E, et al. (2007), *Psychometric analysis of visual sensations mediated by subretinal microelectrode arrays implanted into blind retinitis pigmentosa patients*. Invest Ophthalmol Vis Sci, **48**: ARVO E-Abstract 645.

80. Zrenner E, et al. (2006), *Subretinal chronic multi-electrode arrays implanted in blind patients*. Invest Ophthalmol Vis Sci, **47**: ARVO E-Abstract 551.

81. Zrenner E, et al. (2009), *Blind retinitis pigmentosa patients can read letters and recognize the direction of fine stripe patterns with subretinal electronic implants*. Invest Ophthalmol Vis Sci, **50**: ARVO E-Abstract 456.

Chapter 15
Findings from Chronic Optic Nerve and Cortical Stimulation

Edward M. Schmidt

Abstract This chapter reviews the experiments that have produced visual sensations in humans through electrical stimulation of the central nervous system. Initially, surface stimulation of the visual cortex, provided insight into how electrical stimulation of V1 could possibly provide a visual prosthesis for the blind. Intracortical microstimulation was then investigated that would allow lower power stimulation and increased density of microelectrodes. The stimulation of the optic nerve has also been investigated as a possible site for a visual prosthesis.

The next section is dedicated to what is known and what needs to be done for the development of a visual prosthesis.

The following section examines current research efforts directed towards the development of a visual prosthesis. They include optic nerve stimulation, cortical surface stimulation and intracortical stimulation of visual cortex. The CORTIVIS Program is a comprehensive development of an intracortical visual prosthesis. The lateral geniculate nucleus is also being studied as a site for a visual prosthesis.

The final section of this chapter deals with the developments that are needed for a functional visual prosthesis. They include microelectrode arrays, stimulation hardware, and low power image sensing and processing circuitry that can control the stimulators.

Abbreviations

2D	Two dimensional
3D	Three dimensional
EIC	EIC laboratories
HMRI	Huntington Medical Research Institute
ICMS	Intracortical microstimulation
IIT	Illinois Institute of Technology
LGN	Lateral geniculate nucleus

E.M. Schmidt (✉)
National Institutes of Health (retired)
e-mail: emschmidt@atlanticbb.net

G. Dagnelie (ed.), *Visual Prosthetics: Physiology, Bioengineering, Rehabilitation,*
DOI 10.1007/978-1-4419-0754-7_15, © Springer Science+Business Media, LLC 2011

MIPS   Multimode digital image sensor
MIT   Massachusetts Institute of Technology
NIH   National Institutes of Health
NY   New York
UC   University of Chicago

15.1 Background

Visual sensations produced by stimulation of the visual cortex in human patients were well known to German neurosurgeons, Kraus [34] and Foerster [27], as early as 1924. A number of reports has been published over the years describing the effects of electrical stimulation of the visual cortex in lightly anesthetized surgical patients [38, 39]. When their visual cortex was stimulated, patients usually report small spots of light called phosphenes.

Shaw [45] obtained a patent for a "Method and Means for Aiding the Blind". In his system, a photoelectric tube controlled the intensity and/or frequency of an electrical stimulus that was applied directly by internal electrodes, or indirectly by external electrodes to the visual areas of the brain. Although this appears to be one of the first concepts of a visual prosthesis, actual implementation of the system has not been found.

Button and Putnam [11] demonstrated, in blind subjects, visual responses to intracortical stimulation controlled by a photoelectric cell. This allowed the subjects to identify a light source by orientation of the cell. Of the three subjects, one was able to follow a flashlight carried by an attendant 15 ft away.

15.2 Cortical Surface Stimulation

The first chronic experiment to determine the effects of stimulating the visual cortex was carried out by Brindley and Lewin [9]. They implanted an array of 80 electrodes on the medial surface of the occipital pole in a 52 year-old woman who had been totally blind for 6 months. The electrodes were platinum squares 0.8 mm on a side. They were connected to 80 radio receivers mounted to the skull, beneath the pericranium. Alternate receivers were tuned to 6.0 or 9.5 MHz. Pressing a transmitter coil on the scalp above a receiver and applying the proper frequency provided stimulation currents to the associated electrode. With the technology available at the time, 80 receivers covered half of the cranium.

When electrodes that produced phosphenes within 10° of the fovea were stimulated, the patient reported a very small spot of light, or phosphene, and described it as "the size of a grain of sago at arm's length" or "like a star in the sky". Phosphenes further from the fovea were sometimes elongated, "like a grain of rice at arm's length". The most peripheral phosphenes were round like a cloud. There were three electrodes that produced a pair of phosphenes about a degree apart and two electrodes that produced a row of three phosphenes each about a degree apart from the next. When multiple phosphenes occurred, stimulus amplitude could not be adjusted to produce single phosphenes. For 13 electrodes, weak stimulation produced a single phosphene but higher-level stimulation produced a second phosphene in a different part of the visual field.

Other significant findings from this patient were:

1. Phosphenes always flickered regardless of stimulation parameters.
2. Phosphenes moved with eye movement.
3. Phosphenes could usually be resolved that were produced by electrodes spaced 2.4 mm apart.
4. Phosphenes usually ceased immediately at the end of stimulation, but after strong stimulation they could persist for up to 2 min.
5. Stimulation of multiple electrodes could produce simple patterns.

By improving the experimental prototype, Brindley and Lewin [9] believed that at least 200 electrodes per hemisphere could be implanted and would permit blind patients to read and navigate.

Dobelle and Mladejovsky [22] were able to conduct a series of acute experiments involving volunteers undergoing neurosurgical procedures for removal of tumors or other lesions to verify the results of Brindley and investigate the possibility of producing a visual prosthesis. Dobelle's data are based on 16 experiments in 15 volunteers. They were able to confirm most of Brindley's results from a single volunteer. A summary of the results obtained from Dobelle's experiments were:

1. Phosphene chromatic effects or flicker may or may not occur.
2. Phosphenes moved with eye movement.
3. Two-point discrimination was about 3 mm.
4. Phosphenes appear immediately when stimulation is begun and end immediately upon cessation of stimulation.
5. Phosphenes fade after 10–15 s of continuous stimulation.
6. Multiple phosphenes are co-planar.
7. Thresholds ranged between 1 and 5 mA, with 3 mA being typical.
8. Electrodes of 1, 3, and 9 mm^2 size had similar thresholds and percepts.
9. Brightness modulation can be achieved by changing pulse amplitude.

From these studies, it was apparent that to provide a blind person with a stable image, either the subject had to learn to use head movements instead of eye movements, or the camera used by the visual prosthesis had to move with eye movement. Also, long stimulation trains had to be interrupted to compensate for phosphene fading.

Dobelle's group chronically implanted four volunteers in the 1970s with a subdural 64-electrode array placed on the medial surface of the visual cortex of the right occipital lobe. The wires were terminated in a 72-pin micro-miniature connector encapsulated in a transcutaneous pyrolytic carbon pedestal, attached to the cranium by platinum bone screws.

Of these four volunteers, two had useful results for the future of artificial vision. One of them, blind for 10 years and implanted in 1975 at age 33, could perceive 46 useful phosphenes out of 60. Using six phosphenes with a layout similar to that of a Braille cell, he could read cortical Braille at approximately five words per min but he could only read tactile Braille at one word per min [23]. He could identify the orientation of white strips of tape on a blackboard by manipulating a video camera mounted on a joystick. His phosphene map stayed constant and his thresholds only had small changes over 10 years.

Another volunteer, blind for 7 years and implanted in 1978 at age 41 with an identical 64-electrode array could perceive 21 useful phosphenes. Over the last 25 years, his phosphene map and thresholds have stayed constant. In the late 1990s this volunteer benefited from the miniaturization of electronic components and advances in computer technology. He was the first blind volunteer to wear a miniature video camera mounted on his eyeglasses and a sub-notebook computer, a stimulator, and batteries in a waist pack [24]. Using an edge detection algorithm, the images from the video camera were processed by the computer, which selected the electrodes that produced phosphenes on or near the high-contrast areas of the images. The stimulator in turn generated the proper stimuli for the selected electrodes.

Compactness and portability of the system allowed the subject to detect and negotiate objects, follow a child walking slowly and close to him in a hallway, follow a strip of black tape on the floor, enter a room, grab a ski cap hung on the opposite wall, turn around, walk towards a mannequin and put the cap on its head. Accompanied by staff in the NY City subway system, an environment he was familiar with, he could get inside a subway car. He found it easier to differentiate the space between two cars and an open car door with his visual prosthesis than with his cane.

The results of the research done on these two volunteers, particularly the last one, were quite promising. If they could achieve all this using a single array with a limited number of phosphenes, the logical conclusion was that with two arrays, blind patients would have more phosphenes, creating images with higher resolution, therefore giving them more independence and mobility.

15.3   Intracortical Microstimulation

In cat motorsensory cortex, Stoney et al. [46] showed that thresholds for facilitation of spinal motorneuron pools by intracortical microstimulation (ICMS) could be as low as 2 µA, which is 1/100 of the threshold for producing similar effects with surface stimulation (Asanuma et al. [2]). These results led Dobelle & Mladejovsky to try ICMS in patients where the cortex was going to be surgically removed. This was not successful, possibly due to pathological involvement of the cortex in question [22].

In 1980 Bartlett and Doty [4] investigated the ability of primates to detect ICMS of the visual cortex. They advanced microelectrodes through the visual cortex and recorded the primate's threshold for detection of the stimulus. They found thresholds significantly lower than surface stimulation, with some thresholds as low as

2 μA (0.2 ms at 50 Hz). It was not apparent if the primates were responding to phosphenes similar to those produced by surface stimulation in humans. If the primates were seeing phosphenes then it appeared that it might be possible to produce an intracortical visual prosthesis requiring much less power than using surface stimulation. This question could only be answered in human subjects.

Dr. Hambrecht, who was Director of the Neural Prosthesis Program at the National Institutes of Health (NIH), assembled a team of scientists to determine if ICMS was suitable for use in a human visual prosthesis. Protocols were approved at the NIH and at the University of Western Ontario to test patients who were undergoing surgery for excision of epileptic foci in the visual cortex. Three patients were studied in Canada for 1h each [3] by first briefly stimulating the exposed cortex with a surface electrode and then inserting pairs of electrodes into the region where the patient reported phosphenes. As the electrodes were advanced through the cortex, the threshold for phosphene production dropped from as high as 5 mA at the surface to about 20 μA at 2–3 mm from the surface. Near threshold, the phosphenes were usually blue, yellow or red. The phosphenes did not flicker. With interleaved stimulation of two microelectrodes that were 0.7–1 mm apart, the patient reported "two blobs fusing." When the tip separation was 0.3 mm, the percept was a singular round shape.

The next step in developing a visual prosthesis was to chronically implant a blind human volunteer with an array of intracortical electrodes. Hambrecht [29] provided an excellent review of the next study and Schmidt et al. [44] provided the details of the human experiment. This study was limited to a 4-month investigation as set out in the approved protocol.

Thirty-eight microelectrodes were implanted in the visual cortex. They consisted of 12 single microelectrodes and 18 pairs. The spacing between pairs of microelectrodes was 250, 500 or 750 μm. Two of the microelectrode leads were broken at the time of implantation and only two of the remaining 36 microelectrodes failed to produce phosphenes. Due to the untimely breakage of a number of microelectrode wires, planned pattern recognition studies could not be conducted.

The phosphenes produced by ICMS were similar to those reported in the Canadian study [3]. A summary of the results obtained with ICMS were:

1. Phosphenes never flickered.
2. Phosphenes moved with eye movement and a group of phosphenes maintained their relative positions with eye movement.
3. Stimulation of microelectrodes, with tips separated by 0.5 mm, produced separate phosphenes.
4. Phosphenes appeared immediately after the beginning of stimulation and except for rare occasions, disappeared at the termination of stimulation.
5. When stimulation continued beyond a second, phosphenes usually disappeared.
6. By interrupting a long stimulation pulse train with brief pauses, the duration of phosphene perceptions could be increased.
7. Multiple phosphenes were co-planar.

8. Threshold currents were as low as 1.9 μA, while most of the microelectrodes had thresholds below 25 μA.
9. Thresholds for cathodic-first stimulation with a biphasic pulses was always less than anodic-first.
10. Varying the stimulus frequency, pulse width or amplitude modulated phosphene brightness.
11. Near threshold stimulation, the phosphenes were often reported to have colors of red, blue or yellow, but never green. When the stimulation levels were increased, the phosphenes generally became white, grayish or yellowish.
12. Brighter phosphenes could obscure dimmer phosphenes. The subject had to adjust current levels so that all phosphenes in a group could be seen at the same time.
13. Stimulation of some microelectrodes produced a second closely spaced phosphene at a higher current than the first. When three microelectrodes that produced two phosphenes were simultaneously stimulated, producing six phosphenes, they appeared in almost a vertical row and the subject identified them as a letter "I".

As specified in the approved protocol, at the end of the 4-month implantation period the electrode lead wires were removed, along with several of the microelectrodes, for examination. The volunteer never experienced residual side effects from the implant. Two years after the study was completed, she suddenly died. An autopsy revealed that she had a ruptured berry aneurysm located in the hemisphere opposite from where the microelectrodes had been implanted. The conclusion, as determined by an investigative panel, was that the experimental visual implant was not responsible for her death.

Although a limited amount of information was obtained from this volunteer, an intracortical visual prosthesis looks promising.

15.4   Optic Nerve Stimulation

The next feasible site for a visual prosthesis, after the retina, is the optic nerve. Prior to consideration of this site, considerable work was done on cats in the development of a spiral nerve cuff electrode [51]. Selective recruitment of different muscles innervated by the sciatic nerve could be accomplished by electrode selection within the cuff [28]. With this background, Veraart decided that a visual prosthesis based on optic nerve stimulation might be feasible.

The first optic nerve implant volunteer was a 59-year-old female that was totally blind due to retinitis pigmentosa [52]. The self-sizing cuff electrode contained four contacts. Electrical stimulation of the electrodes never produced any sensation other than vision. With each stimulation, using a given set of parameters, the patient reported multiple phosphenes in a cluster of 2–5, or arranged in rows, or clumps of 6–30. Stimulus currents as low as 30 μA were capable of eliciting phosphenes.

Although changing the stimulation parameters generated a large number of different phosphenes, the actual parameters were not reported with the retinotopic stimulation map

Pattern recognition studies were undertaken when the subject used a head-mounted video camera [53]. Either 4 or 24 phosphene locations were used in the tests. When the video image intersected one or more phosphene locations, a sequential pattern of stimulation was produced. Forty-five simple patterns were presented and after learning, the subject reached a recognition score of 63% with a processing time of 60 s. This study shows that information can be delivered through optic nerve stimulation, but at an extremely slow rate.

One of the last reported studies with this subject involved object localization, discrimination, and grasping [25]. After training, the subject reached 100% success rate in performing all three tasks. Localization was achieved in 20 s while discrimination required 40 s. Grasping required no more than 6 s. This study provides data that indicates optic nerve stimulation might be useful in daily life if more phosphenes can be generated simultaneously rather that sequentially.

A second blind volunteer was implanted, but only the details of the surgery have been reported [8]. Unfortunately, do to the retirement of Dr. Veraart, the optic nerve visual prosthesis program in Belgium may not continue. The implanted patients will however be followed by Dr. Delbeke.

For an optic nerve visual prosthesis to be useful, many more stimulations sites are required than can be obtained with cuff electrodes. The Utah slanted electrode array (USEA) [7] could theoretically provide up to 100 independent phosphenes. This assumes that the electrode array can be implanted in the optic nerve and each electrode produces an independent phosphene. For peripheral nerve, a pneumatically actuated impact insertion tool was developed [42]. How such a device can be used for optic nerve implantation remains to be determined. When one solves the electrode array implantation problem, the next question is how many electrodes are required to provide a useful optic nerve visual prosthesis.

15.5 What Is Known and What Needs to Be Done

From the results of optic nerve and visual cortex stimulation, the only type of visual prosthesis that we can consider for a blind patient, at this time, is a scoreboard type of display. Implanted subjects have shown that meaningful information can be obtained with 20–109 phosphenes, if they remain stable over time. The image would be a 2D set of phosphenes.

Biologically safe stimulation parameters for the different types of electrodes used in a visual prosthesis have to be determined.

Gray scale rendition can be obtained by varying stimulation parameters, but this depends on knowing the threshold current of each electrode. A stimulus-brightness curve for each electrode would have to be obtained in order to properly scale stimulation intensity for a given scene. If a very bright phosphene is near a dim phosphene, the dim phosphene may not be observed. A stimulus-brightness curve can be generated from a patient's verbal response, or using a reaction time task to the presentation of different stimulus intensities. Brighter phosphenes are observed sooner than dimmer phosphenes. To collect this data is a formidable task, but may prove useful with the initial implant patients. Most likely it will be an indispensable aspect of exploring the potential and limitations of all visual prostheses.

Colored phosphenes occur near threshold currents with no guarantee as to what color will be produced. Threshold currents will be different for the electrodes in the array and must be determined for each electrode. At this time it is doubtful if a meaningful color visual prosthesis can be developed.

With long trains of stimulation, phosphenes will fade. If brief pauses are provided in the stimulus train, the intensity can be maintained for a longer period. A suggested mode of operation for a long stimulus train is to stimulate for 320 ms and then pause for 32 ms before repeating the stimulation pulses. This strategy will have to be verified in patients.

When using arrays with a large number of electrodes, we are not sure where the phosphenes will appear on the subject's visual map. A rapid phosphene mapping technique is required so that the correct transformation of the camera scene to appropriate electrodes can be made.

Phosphenes move with eye movement so that if a patient looks at a phosphene it will move away. In order to center the object in the field of vision of the prosthesis, the patient will have to make compensating head movements if the camera that is sensing the scene, is mounted on an eyeglass frame. One approach to this problem is to develop a miniature camera that tracks eye movement and moves in the same manner. Another approach is to implant a miniature camera in the eye.

After an image is obtained from a video camera, the data needs to be processed before stimulus signals are applied to the appropriate electrodes. This can be as simple as threshold detection and any signal above threshold initiates a constant level of stimulation. Edge detection algorithms can be employed to minimize the number of electrodes that are stimulated. When reading black on white text, converting the positive image to a negative one would produce white letters on a black background, again reducing the number of electrodes stimulated.

To summarize what is needed for a visual prosthesis implant:

- Electrochemically safe electrode arrays
- Biocompatibility of implants
- Means of efficiently obtaining electrode threshold current
- Threshold current stability
- Means of efficiently mapping phosphene location
- Determine phosphene brightness versus current
- Determine length of stimulation and duration of pauses to stabilize brightness
- Develop a camera coupled to eye movement
- Image processing

15.6   Current Research Efforts

The research on a visual prosthesis using sites other than the retina can be divided into a number of sub categories. They are optic nerve stimulation, lateral geniculate, surface stimulation of visual cortex, intracortical stimulation of visual cortex, stimulation hardware, microelectrode arrays, miniature cameras and animal models. The ongoing research in each of these areas will be listed separately.

15.6.1   Optic Nerve Stimulation

Ren and co-workers at the Shanghai Jiao-Tong University in Shanghai, China has established a program called C-Sight to investigate implantation of penetrating microelectrodes in the optic nerve for a visual prosthesis [41]. They are investigating an image acquisition and processing system, a data telemetry system, a neural stimulator, and an implantable micro-camera system for an optic nerve visual prosthesis.

Another approach that is being actively studied in Germany is the use of regeneration microelectrode arrays. These electrodes consist of a wafer that has a number of holes into which nerve fibers can regenerate. The holes contain electrical contacts that enable single or a few nerve fibers to be stimulated. The optic nerve is cut and sutured to either side of the perforated microelectrode array. In rats, recovery of visual evoked potentials occurred in 2–8 weeks [30]. If regeneration through the perforated microelectrode arrays can be successful in primates and chronic stimulation of fibers can be shown to produce phosphenes then one could consider implanting this type of microelectrode in humans. One of the disadvantages of this type of microelectrode is that the optic nerve has to be cut and success of the implantation cannot be known for weeks or months. This might discourage some volunteers.

A group at Osaka University, Japan is investigating a different approach by stimulating the fibers in the optic nerve head inside the eye [26]. The advantages of this approach over the optic nerve cuff are that the exposure of fibers across the rim of the optic nerve head allows stimulation of small groups of fibers, and the intraocular surgical procedure is less invasive.

15.6.2   Cortical Surface Stimulation

After Dr. Dobelle died, his family donated the project, his patent and the technology to SUNY (State University of New York) at Stony Brook, in May 2006. Members from the staff at SUNY and Avery Biomedical Devices have teamed up to completely redesign the system used with the 16 patients implanted in Portugal.

The redesign of the electrode array and electronics package will be completed before seeking FDA approval to implant patients in the USA.

Chowdhury and colleagues, in Australia, has been investigating a cat model for evaluating prototype cortical surface electrode arrays for a visual prosthesis [16, 17]. At present, this group does not have any plans for implanting human subjects.

15.6.3   Intracortical Stimulation of Visual Cortex

Because the National Institute of Health (NIH) is funded year to year by Congress, long-term patient care cannot be guaranteed. Thus the NIH administration decided not to continue the Visual Prosthesis Program for that reason. The scientists in the program were given the task of finding an appropriate University hospital that had access to the engineering expertise needed to carry out the Visual Prosthesis Program. Troyk and co-workers, at the Illinois Institute of Technology (IIT), formed a consortium consisting of IIT, University of Chicago (UC) and their Medical Center, EIC, and Huntington Medical Research Institute (HMRI). The NIH technology was transferred to IIT. The role of IIT is to develop implantable microelectrode arrays that contain RF powered and controlled stimulator packages and establish safe stimulation parameters for the microelectrodes [49]. EIC provides the electrochemistry expertise to properly develop iridium oxide stimulating electrodes [50]. The University of Chicago conducts the primate psychophysics experiments [6] and the Medical Center implants the primates in preparation for human implants. HMRI conducts safety experiments and histological evaluations of all implants. The Wilmer Eye Institute at Johns Hopkins University has been added to the consortium for evaluating human implants. A human implant is envisioned within 2 years

The University of Utah has conducted a number of studies aimed at implanting microelectrodes in the visual cortex. Normann and co-workers have developed a micro-machined electrode array consisting of 100 microelectrodes [37]. His group has conducted a number of studies that could lead to a human implant in the near future. They have looked at the histological effects of implanting these electrodes [35], the results of acute implantation in human neocortex [31] and the thermal impact of active arrays implanted in the brain [33]. They envision a intracortical visual prosthesis system employing 625 microelectrodes. The system receives video information from a micro-camera mounted in eyeglasses, processes the images with a computer and transmits the information over a telemetry system to stimulators on the electrode arrays.

15.6.4   CORTIVIS Program

A consortium of European Research Institutions has formed under the coordination of Dr. Fernandez in Alicante, Spain, called CORTIVIS [18]. The aim of the consortium is to develop a visual prosthesis based on intracortical stimulation of the visual cortex. They have developed an image processing system that mimics the human retina. The signals from the retina module are converted into neuromorphic pulse-coded signals through a circuit that emulates the function of the retinal ganglion cells [1]. They are currently performing in vitro experiments (for biocompatibility, in vivo animal experiments (acute and chronic)) and working towards human implants. Initially they will use the Utah Electrode Array [37] while developing a 3D probe array.

15.6.5 Lateral Geniculate Stimulation

Pezaris and Reid at Harvard Medical School [40] have demonstrated in primates that microstimulation in the lateral geniculate nucleus (LGN), which is the relay between the retina and the visual cortex, produced localized visual percepts. To assess the effects of microstimulation of LGN in a primate, an eye movement task was used with visual targets presented on a computer screen or through microstimulation. Saccades made to electrical targets were comparable to saccades made to optical targets. They estimate that 200–300 stimulation sites are available in the LGN. This would be adequate for reading with a visual prosthesis. However, developing the required electrode arrays and implanting them in the LGN is a formidable task.

15.7 Microelectrode Arrays and Stimulation Hardware

The University of Michigan has a long history in the development of multi-site silicon stimulating probes [54, 55]. Their resent development is a 64-site wireless microstimulator (Interstim-2B) [36]. Up to 32 chips can be connected in parallel to drive 2,048 stimulation sites. This should be more than adequate for any currently planned visual prosthesis.

PolySTIM Neurotechnologies Laboratory in Montreal, Canada, Has developed a power efficient stimulator for an intracortical visual prosthesis [19].

Delbeke et al. [20] have developed a microsystem based stimulator for an optic nerve prosthesis.

15.7.1 Miniature Cameras

A group at Shanghai Jiao Tong University, Shanghai, China have developed a micro-camera that can be implanted in eye and powered by a solar array positioned in front of the iris [12]. Since phosphenes move with eye movement, an eye-mounted camera should help to stabilize the perceived image. The camera provides a 32×32 element image, which with their simulation studies allowed a subject to recognize simple scenes. Through simulations, they also found that a 12×12 array of pixels was sufficient to recognize Chinese characters [13]. With a 10×10 array, the recognition level dropped to slightly under 50%.

The retinal visual prosthesis group at the University of Southern California, USA, is also developing a camera implantable in the eye.

PolySTIM Neurotechnologies Laboratory has developed a CMOS multimode digital image pixel sensor (MIPS) for a visual prosthesis [43]. Three selectable operation modes are combined in the proposed MIPS: a high dynamic range logarithmic mode, a linear integration mode, and a novel differential mode between two consecutive images. This last mode allows 3D information for a cortical stimulator.

15.7.2 Animal Models

The major groups that are investigating the entire realm of aspects leading to human implants of a visual prosthesis are employing animal models at some stage of their work. Other groups are just looking at animal models and how they might apply to a visual prosthesis.

The group at IIT/UC [6] have chronically implanted arrays of microelectrodes in non-human primates to evaluate intracortical stimulation. One of the major findings was that the stimulation package originally developed under an NIH contract, as described on the IIT web site [32], could not be connected to the intended number of microelectrodes at surgery. Small electrode-stimulator modules had to be developed that used telemetry to transmit power and stimulation in formation. At MIT, Tehovnik and colleges [47, 48] have used moveable microelectrodes to map the generation of saccadic eye movements and study how these data might be applicable to a visual prosthesis. DeYoe [21] and Bartlett [5] at the University of Rochester used moveable microelectrodes to study stimulation parameters and laminar distribution of phosphene production in non-human primates. These studies will aid in the development of a human visual prosthesis.

15.7.3 Image Processing and Phosphene Mapping

Part of the CORTIVIS project is the development of a bio-inspired visual processing front-end that would be placed between the photosensor array and the stimulator for an intracortical visual prosthesis [18]. The images are processed by a set of separate spatial and temporal filters that mimic the functions of the photoreceptors, amacrine and bipolar cells in order to enhance specific features of the captured visual image.

The C-Sight Visual Prosthesis Group in China has been studying tactile phosphene mapping in sighted subjects using a head mounted display for the simulated phosphenes and a 19 in. touch screen to record the subject's tactile position [14, 15].

PolySTIM Neurotechnologies Laboratory has surveyed image processing strategies that can be used with a visual prosthesis [10].

15.8 Conclusion

With the wide range of research that is currently underway to develop a visual prosthesis it is possible that we will see several groups implanting humans in the next 5 years.

References

1. Ahnelt P, Ammermuller J, Pelayo F, et al. (2002), Neuroscientific basis for the design and development of a bioinspired visual processing front-end. EMBEC Abstract, p. 1692–3.
2. Asanuma H, Stoney SD Jr, Abzug C (1968), Relationship between afferent input and motor outflow in cat motorsensory cortex. J Neurophysiol, 31(5): p. 670–81.
3. Bak M, Girvin JP, Hambrecht FT, et al. (1990), Visual sensations produced by intracortical microstimulation of the human occipital cortex. Med Biol Eng Comput, 28: p. 257–9.
4. Bartlett JR, Doty RW (1980), An exploration of the ability of macaques to detect microstimulation of striate cortex. Acta Neurobiol Exp (Wars), 40(4): p. 713–27.
5. Bartlett JR, DeYoe EA, Doty RW, et al. (2005), Psychophysics of electrical stimulation of striate cortex in macaques. J Neurophysiol, 94: p. 3430–42.
6. Bradley DC, Troyk PR, Berg JA, et al. (2005), Visuotopic mapping through a multichannel stimulating implant in primate V1. Neurophysiol, 93(3): p. 1659–70.
7. Branner A, Stein RB, Normann RA (2001), Selective stimulation of cat sciatic nerve using an array of varying-length microelectrodes. J Neurophysiol, 85(4): p. 1585–94.
8. Brelen ME, DePotter P, Gersdorff M, et al. (2006), Intraorbital implantation of a stimulating electrode for an optic nerve visual prosthesis. Neurosurg, 104(4): p. 593–7.
9. Brindley GS, Lewin WS (1968) The sensations produced by electrical stimulation of the visual cortex. J Physiol (London), 196: p. 479–93.
10. Buffoni LX, Coulombe J, Sawan M (2005), Image processing strategies dedicated to visual cortical stimulators: a survey. Artif Organs, 29(8): p. 658–64.
11. Button J, Putnam T (1962), Visual response to cortical stimulation in the blind. J Iowa Med Soc, 52: p. 17–21.
12. Chai X, Li L, Wu K, et al. (2008), C-Sight visual prosthesis for the blind. IEEE BMES, 27(5): p. 20–8.
13. Chai X, Yu W, Wang J, et al. (2006), Recognition of pixelized Chinese characters using simulated prosthetic vision. Artif Organ, 31: p. 175.
14. Chai X, Zhang L, Li W, et al. (2007), Tactile based phosphene positioning system for visual prosthesis. Invest Opthalmol Vis Sci, 48(5): p. 662, E-Abstract.
15. Chai X, Zhang L, Li W, et al. (2008), Study of tactile perception based on phosphene positioning using simulated prosthetic vision. Artif Organs, 32(2): p. 110–5.
16. Chowdhury V, Morley JW, Coroneo MT (2004), Surface stimulation of the brain with a prototype array for a visual cortex prosthesis. J Clin Neurosci, 11(7): p. 750–5.
17. Chowdhury V, Morley JW, Coroneo MT (2004), An in-vivo paradigm for the evaluation of stimulating electrodes for use with a visual prosthesis. ANZ J Surg, 74: p. 372–8.
18. CORTIVIS http://cortivis.umh.es/.
19. Coulombe J, Carniguian S, Sawan M (2005), A power efficient electronic implant for a visual cortical prosthesis. Artif Organs, 29(3): p. 233–8.
20. Delbeke J, Wanet-Defalque MC, Gerard B, et al. (2002), The microsystems based visual prosthesis for optic nerve stimulation. Artif Organs, 26(3): p. 232–4.
21. DeYoe EA, Lewine JD, Doty RW (2005), Laminar variations in threshold for detection of electrical excitation of striate cortex by macaques. J Neurophysiol, 94: p. 3443–50.
22. Dobelle WH, Mladejovsky MG (1974), Phosphenes produced by electrical stimulation of the visual cortex, and their application to the development of a prosthesis for the blind. J Physiol (London), 243: p. 553–76.

23. Dobelle WH, et al. (1976) Braille reading by a blind volunteer by visual cortex stimulation. Nature (London), 259: p. 111–2.
24. Dobelle WH (2000), Artificial vision for the blind by connecting a television camera to the visual cortex. ASAIO J, 46: p. 3–9.
25. Duret F, Brelen M, Lambert V, et al. (2006), Object localization, discrimination, and grasping with the optic nerve visual prosthesis. Restor Neurol Neurosci, 24: p. 31–40.
26. Fang X, Sakaguchi H, Fujikado T, et al. (2006), Electrophysiological and histological studies of chronically implanted intrapapillary microelectrodes in rabbit eyes. Graefe's Arch Clin Exp Ophthalmol, 244(3): p. 364–75.
27. Foerster O (1929), Beitrage zur Pathophysiologie der Sehbahn und der Sehsphare. J Psychol Neurol Lpz, 39: p. 463–85.
28. Grill WM Jr, Mortimer JT (1996), Quantification of recruitment properties of multiple contact cuff electrodes. IEEE Trans Rehabil Eng, 4(2): p. 49–62.
29. Hambrecht FT (1995), Visual prostheses based on direct interfaces with the visual system. Brailliere's Clin Neurol, 4(1): p. 147–65.
30. Heiduschka P, Fischer D, Thanos S (2005), Recovery of visual evoked potentials after regeneration of cut retinal ganglion cell axons within the ascending visual pathway in adult rats. Restor Neurol Neurosci, 23: p. 303–12.
31. House PA, MacDonald JD, Tresco PA, Normann RA (2006), Acute microelectrode array implantation into human neocortes: preliminary technique and histological considerations. Neurosurg Focus, 20(5): p. E4.
32. http://neural.iit.edu/technology.htm
33. Kim S, Tathireddy P, Norman RA, Solzbacher F (2007), Thermal impact of an active microelectrode array implanted in the brain. IEEE Trans Neural Syst Rehabil Eng, 15(4): p. 493–501.
34. Krause F (1924), Die Sehbahnen in chirurgischer Beziehung und die faradische Reizung des Sehzentrums. Klin Wochenschr, 3: p. 1260–5.
35. Maynard EM, Fernandez E, Normann RA (2000), A technique to prevent dural adhesions to chronically implanted microelectrode arrays. J Neurosci Methods, 97(2): p. 93–101.
36. Najafi K, Ghovanloo M (2004), A multichannel monolithic wireless microstimulator. Conf Proc IEEE Eng Med Biol Soc, 6: p. 4197–200.
37. Normann RA, Maynard EM, Rousche PJ, Warren DJ (1999), A neural interface for a cortical vision prosthesis. Vision Res, 39(15): p. 2577–87.
38. Penfield W, Rasmussen T (1952), The Cerebral Cortex of Man. New York: Macmillan, p. 135–47, 165–6.
39. Penfield W, Jasper H (1954), Epilepsy and functional anatomy of the human brain. London: Churchill, p. 116–6, 404–40.
40. Pezaris JS, Reid RC (2007), Demonstration of artificial percepts generated through thalamic stimulation. Proc Natl Acad Sci USA, 104(18): p. 7670–5.
41. Ren Q, Zhang L, Shao F, et al. (2007), Development of C-Sight visual prosthesis based on optic nerve stimulation with penetrating electrode array. Invest Ophthalmol Vis Sci, 48: p. 661, E-Abstract.
42. Rousche PJ, Norman RA (1992), A method for pneumatically inserting an array of penetrating electrodes into cortical tissue. Ann Biomed Eng, 20(4): p. 413–22.
43. Sawan M, Trepanier A, Trepanier J-L, et al. (2006), A new CMOS multimode digital pixel sensor dedicated to an implantable visual cortical stimulator. Anal Integ Cir Sig Proc, 49(2): p. 925–1030.
44. Schmidt EM, Bak MJ, Hambrecht, FT, Kufta CV, et al. (1996), Feasibility of a visual prosthesis for the blind based on intracortical microstimulation of the visual cortex. Brain, 119: p. 507–22.
45. Shaw JD (1955), Method and mean for aiding the blind. United States Patent Number 2,721,316.
46. Stoney SD Jr, Thompson WD, Asanuma H (1968), Excitation of pyramidal tract cells by intracortical microstimulation: effective extent of stimulating current. J Neurophysiol, 31(5): p. 659–69.

47. Tehovnik EJ, Slocum WM, Carvey CE, et al. (2005), Phosphene induction and the generation of saccadic eye movements by striate cortex. J Neurophysiol, 93: 1–19.

48. Tehovnik EJ, Slocum WM (2007), Phosphene induction by microstimulation of macaque V1. Brain Res Rev, 53(2): p. 337–43.

49. Troyk PR, Bak M, Berg J, Bradley D, et al. (2003), A model for intracortical visual prosthesis research. Artif Organs, 27(11): p. 1005–15.

50. Troyk PR, Detiefsen DE, Cogan SF, et al. (2004), "Safe" charge-injection waveforms for iridium oxide (AIROF) microelectrodes. Conf Proc IEEE Eng Med Bio Soc, 6: p. 4141–4.

51. Veraart C, Grill WM, Mortimer, JT (1993), Selective control of muscle activation with a multipolar nerve cuff electrode. IEEE Trans Biomedical Eng, 40(7): p. 649–53.

52. Veraart C, Raftopoulos C, Mortimer JT, et al. (1998), Visual sensations produced by optic nerve stimulation using an implanted self-sizing spiral cuff electrode. Brain Res, 813: p. 181–6.

53. Veraart C, Wanet-Defalque MC, Gerard B, et al. (2003), Pattern recognition with the optic nerve visual prosthesis. Artif Organs, 27: p. 996–1004.

54. Yao Y, Gulari M, Hetke J, Wise K (2004), A low-profile neural stimulating array with on-chip current generation. Conf Proc IEEE Eng Med Soc, 3: p. 1994–7.

55. Yao Y, Gulari MN, Ghimire S, et al. (2005), A low-profile three-dimensional silicon/parylene stimulating electrode array for neural prosthesis applications. Conf Proc IEEE Eng Med Soc, 2: p. 1293–6.

Part IV
Towards Prosthetic Vision: Simulation, Assessment, Rehabilitation

Chapter 16
Simulations of Prosthetic Vision

Michael P. Barry and Gislin Dagnelie

Abstract Simulations of prosthetic vision can provide requirements and specifications for prosthesis designs and stimulus conditions; these requirements are expected to differ according to the visual task. Studies reviewed here include examinations of visual acuity, reading, face and object recognition, hand–eye coordination, way finding, visual tracking, and simple design feasibility. Based on these studies, visual acuity with prosthetic vision seems to depend most on the resolution of perceived phosphenes. Given usable visual acuity, all visual tasks that have been evaluated in simulations with variable dot counts demonstrate some significant dependence on the number simulated phosphenes provided. Some tasks also have more unique dependencies: Facial recognition seems quite sensitive to the number of gray levels and the relative size of dots and spacing. Wayfinding is most dependent on the angle of view captured by the camera. In many of the simulation studies practice was found to be an important factor for successful task performance. As visual prosthesis development becomes less limited by technological barriers, findings from simulation studies may become increasingly important for the design of implants and rehabilitation programs.

Abbreviations

´	Symbol for minutes of arc
DBS	Deep brain stimulation
HMD	Head-mounted display
LGN	Lateral geniculate nucleus of the thalamus
logMAR	Logarithm of the minimum angle of resolution
MPDA	Multi-photodiode array

M.P. Barry (✉)
Lions Vision Research and Rehabilitation Center, Wilmer Eye Institute, Johns Hopkins
University School of Medicine, 550 N. Broadway, 6th Floor, Baltimore, MD 21205, USA
e-mail: mbarry11@jhu.edu

G. Dagnelie (ed.), *Visual Prosthetics: Physiology, Bioengineering, Rehabilitation*, 319
DOI 10.1007/978-1-4419-0754-7_16, © Springer Science+Business Media, LLC 2011

16.1   Introduction

Although not perfect models of visual prostheses, simulations of prosthetic vision provide insight into how these prostheses can theoretically function. While tests of actual prostheses suffer the burdens of device construction and implantation and all the associated costs and approvals, simulations are relatively simple to implement and require less regulatory oversight. As such, before any sizable clinical trials were possible, simulations of visual prostheses were used to investigate the usability of prosthetic vision. Now, as clinical trials move forward, prosthetic vision simulations still provide insight on how different elements of the technology interact and affect performance. Simulations of theoretical device designs also help guide developers in building next-generation prostheses.

The first studies utilizing simulations of prosthetic vision were published in 1992 by a group at the University of Utah [3–5]. Each of these initial studies simulated simple square grids of dots by covering a small screen (1.7° of visual field across) with a film containing chemically etched holes. Using this simulation scheme, Cha et al. evaluated normally sighted subjects with tests of visual acuity [3], reading [5], and wayfinding [4]. As the availability of technology progressed over time, simulations of prosthetic vision evolved to software-based implementations of visual prostheses; however the basic categories of tests have persisted, with a few additions: face and object recognition [14, 15, 21, 29, 30, 33, 34], hand–eye coordination [14, 15, 21, 29], visual tracking [20, 31], and purely computational tests [25]; most of these tasks can be implemented and explored in virtual [1, 7, 13, 29, 32] as well as real [4, 13, 15, 29] environments. These simulation studies, taken together, provide a wide range of knowledge on what may be possible with actual prostheses, what resolution and other device properties may be required for specific tasks, and in which directions prosthetic development should proceed.

In this chapter we will summarize studies that have been performed in these different categories. We will open with some general remarks about the ways simulations are implemented and some of the basic parameters that can be varied. For the sake of consistency, the words "array" and "phosphene" will be used only to refer to actual prostheses and their associated percepts, while "grid" and "dot" or "simulated phosphene" will be used to refer to simulations of visual prostheses.

16.2   Simulation Techniques and Basic Parameters

In a typical prosthetic vision simulation, a sighted individual is presented visual stimuli that approximate what the visual prosthesis wearer is expected to perceive. Typically, these are images in which the original resolution has been reduced to represent the stimulating array that is to be implanted in a blind subject, with individual dots or squares of light representing the phosphenes elicited at each point of stimulation.

Figure 16.1 shows the implementation of a prosthetic simulation commonly used in our laboratory, where pixelized images are presented in a video headset,

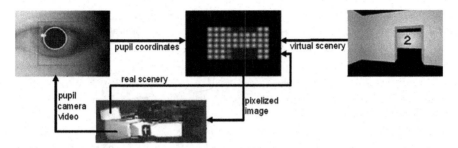

Fig. 16.1 Schematic arrangement in a typical prosthetic vision simulation. The filtering engine (*top center unit*) converts an incoming video stream from either a real or a virtual scene into pixelized imagery. The head-mounted display (HMD) in this arrangement is used to present the imagery to the subject and monitor the subject's gaze through a built-in video camera observing the pupil. A scene camera mounted on the HMD can be used to provide live video for filtering. The pupil-tracking software (*top left unit*) provides the filtering engine with near-real time gaze information, allowing the imagery to be stabilized on the subject's retina, simulating a fixed position of the stimulating array

with either a scene camera on the headset or rendering software under control of a gaming engine (HalfLife; Valve Software, Bellevue, WA). Other configurations may involve a monitor display, a hand-held or glasses-mounted camera, or other image capture and display methods. Central in all simulations is a processor that transforms the incoming video stream into an outgoing stream that fulfills the properties of prosthetic vision as they are envisioned by the experimenter.

16.2.1   Gaze Tracking and Image Stabilization

An important aspect of prosthetic vision with an external (head-worn, hand-held, or stand mounted) camera is the loss of the effects of eye movements to which every sighted person is accustomed. As illustrated in Fig. 16.2 (left panel) an eye movement executed by a sighted person makes the image of the object being observed shift across the retina, and hence across the projection areas in the visual cortex. The visual system deals with this by signaling to the visual cortex that an eye movement is being made, so the shift of the image is perceived as a stable rather than a shifting world. This situation changes dramatically (Fig. 16.2, central panel) if the image from a stationary external camera is presented to the visual system in the form of electrical impulses from a set of electrodes attached to the retina or higher visual centers: An eye movement executed by the prosthesis wearer will still signal the visual cortex that an image shift should happen, but since the camera and electrodes are stationary no such shift occurs, and the resulting percept is a disconcerting jump of the scene.

Retinal implants that perform image capture directly inside the eye will not have this problem, as the image will shift according to eye movements. Of the current

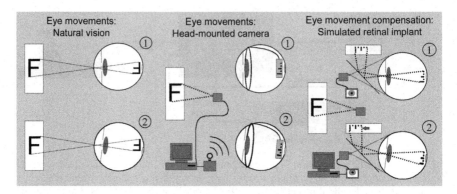

Fig. 16.2 The effect of eye movements on stimulation of the visual system in natural vision (*left panel*) and in prosthetic vision with an external camera, both without (*center panel*) and with (*right panel*) compensation through gaze tracking

implants, only the multi-photodiode array (MPDA) of Retina Implant AG provides this capability. For all devices with an external camera the situation can be remedied by tracking the prosthesis wearer's eye position and presenting a corresponding shift of the image to the implant. This would be done most easily by using a wide-angle camera and instantly panning the section to be presented to the prosthesis wearer in accordance with the current direction of gaze. Such accurate and instantaneous gaze tracking is not currently used, however.

Accurate prosthetic vision simulations should therefore have the ability to mimic gaze stabilization. In the diagram of Fig. 16.1 this is implemented through a pupil-tracking video camera built into the HMD, eye-tracking software (Arrington Research, Scottsdale, AZ), and a resulting offset of the filtered imagery according to the updated gaze position; typically this is done at 30 or 60 frames per second, but more rapid systems are now available.

16.2.2 Filter Engine Parameters

In order to present imagery that closely resembles what a prosthesis wearer is expected to perceive, the filtering engine needs to transform the incoming video frames according to a number of important aspects. Roughly, these can be categorized into four groups: raster spatial properties, dot spatial and temporal properties, and dynamic background noise.

16.2.2.1 Raster Spatial Properties

Typically, the experimenter will have a specific implant configuration in mind and will sub-sample the incoming image to match that configuration. For a retinal implant, the electrode arrangement will most likely be rectangular and regular, although hexagonal and/or radially expanding configurations could in principle be used, in order to conform more closely to the native properties of retinal processing. In all cases the incoming image is reduced in resolution by grouping the intensity and color values within the aperture of each prospective dot position. As an example, a typical 320×240 pixel camera image can be down-sampled to simulate a 10×6 implant by dividing it into 60 rectangular subfields of 32×40 pixels each, and averaging the pixel values within each rectangle to yield a single value that will be represented by the simulated phosphene. Color information will typically be discarded, since only grey scale values are thought to be meaningfully conveyed.

There are several instances where a regular grid of simulated phosphenes is not an adequate representation of what the implant recipient is expected to see. Most importantly, this is the case for implants beyond the retina. Stimulation of the optic nerve, LGN, or primary visual cortex should still provide a predictable phosphene array, depending on the accuracy of electrode placement, and these irregularities can be built into the simulated phosphene map.

Even for a retinal implant there may be distortions of the regular grid. In the normal retinal anatomy the centermost fovea does not contain any secondary neurons, so many neurons at $1°$–$2°$ eccentricity in the retina will correspond to locations much closer to fixation in the visual field, and stimulating those neurons will cause an apparent contraction of the image: phosphenes will be denser immediately around the point of fixation, and correspondingly sparser in a ring at $2°$–$4°$ eccentricity. In addition, the retinal rewiring process described in Chap. 3 will cause inner retinal neurons to migrate from their original positions, and may thus convey random scatter to the perceived phosphene positions. The magnitude of both effects can be estimated, but to our knowledge have not been taken into account in simulations of a retinal prosthesis. On the other hand the crude resolution of most current prostheses, with electrode separations of approximately $2°$, reduces the need for such refinements.

In addition to the overall arrangement of dots in the raster, several parameters can specify raster properties:

- *Dot number*: This quantity corresponds to the number of electrodes in the implant.
- *Dot density*: This quantity determines the center-to-center distance between dots, and is typically chosen to correspond to the inter-electrode distance of the implant. For rectangular grids it is common for density to be equal in the two perpendicular directions. Note that density is the inverse of center-to-center distance.
- *Dot spacing*: When viewing the dot grid one can envisage each dot as being situate at the center of a "unit cell," and the dot may or may not fill the entire cell. For round dots in a rectangular (rather than square) grating, dot spacing will be different in the two orthogonal directions. The space between dots and the background intensity light filling that space will be further discussed under *Sect. 16.2.2.2.*

- *Grid size*: This quantity has a direct relationship to the previous two; it is common for one of the three parameters to be kept constant, and study the trade-off between the remaining two.
- *Dot drop-out*: A subset of electrodes in an implant may prove non-functional after implantation or lose functionality over time; this loss of function can be caused by either the implant itself or by degeneration of the tissue substrate; to model this, a subset of dots may be omitted; typically this subset is chosen at random, and not altered while testing a given subject over multiple sessions, to investigate whether adaptation may occur to this localized absence of image information.

Effects of several grid parameter changes are shown in Fig. 16.3.

16.2.2.2   Dot Spatial Properties

Phosphenes elicited by localized electrical stimulation in blind individuals have generally been described as small round dots, varying in size from a pea to a quarter at arms length, and either sharp or fuzzy in appearance; some subjects have described rings or dark dots on a lighter background, depending on the stimulus conditions. This illustrates a basic problem when rendering images in even the simplest prosthetic vision simulation. The square pixelization commonly employed to hide a person's identity in the media (see Fig. 16.4, left panel) lend themselves to rapid image rendering and have been used extensively by one research group [16, 17, 24, 26–28], but may not be an optimal representation of what is described by patients undergoing stimulation. Other groups have spent considerable effort on

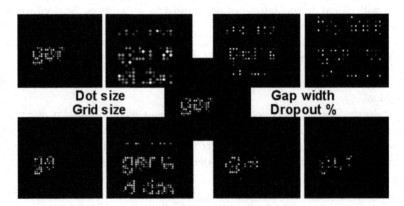

Fig. 16.3 Illustration of the effects of grid and dot parameters on the display of a text fragment with pillbox-shaped dots. All changes are relative to the "standard condition" in the center of the figure. In the *top right panel* grid size is changed without increasing dot size or number, whereas in the *bottom left panel* the dot number is changed, and in the *top left panel* dot size is changed while keeping the gaps separating the dots equal

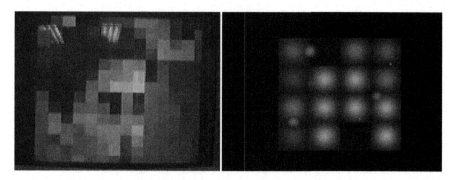

Fig. 16.4 Examples of pixelization used in prosthetic vision simulations. In both examples a rectangular raster was used. The *left panel* shows a 14×14 cell grid with square pixelization of a face (courtesy Dr. Wentai Liu), while the *right panel* shows a 4×4 grid with Gaussian dot profile as seen by a subject in our laboratory inspecting the scoop of a spoon

the creation of model phosphenes with precisely specified spatial properties. Generally, the following parameters are specified:

- *Shape*: Although most phosphenes seen by patients are not perfectly round, the most common shapes used in simulations have been bright circles on a dark background, as shown by the examples in Figs. 16.3 and 16.4 (right panel).
- *Profile/size*: There is a variety of ways in which the light representing the intensity in the scene can be distributed across the unit cell. The most common profiles chosen are rectangular and Gaussian; the extent to which the light in one cell merges with that in neighboring cells depends on the radius of the pillbox (0.495 in the example in Fig. 16.5; hence there is no overlap in the right half of the figure) or the value of σ (four values shown). If a Gaussian profile is chosen there is always some overlap, making the use of Gaussians much more computationally intensive in a real-time simulation. The increased speed of general purpose processors and the use of dedicated hardware have led to more frequent use of Gaussian profiles in recent simulations, since they correspond more closely to the reports of prosthesis wearers [22].
- *Intensity/contrast*: Most simulations use bright dots and modulate the peak intensity of the dots to represent local brightness in the scene, on a black background. Yet it is unlikely that prosthesis wearers will experience such high contrast percepts: Patients blind from outer retinal degenerations describe their world as grey rather than black. For this reason some simulation studies have explored the dependence of subject performance on contrast. In some cases this was done by only changing dot brightness but leaving the black background; this reduces brightness rather than contrast and has very little effect as the subject adapts to the lower light level. Increasing the background intensity, with or without a reduction in dot intensity, is an appropriate way to reduce contrast.

Some studies (e.g., Chap. 17) have modulated the radius of pillbox dots rather than their intensity, but a systematic comparison of the two methods across

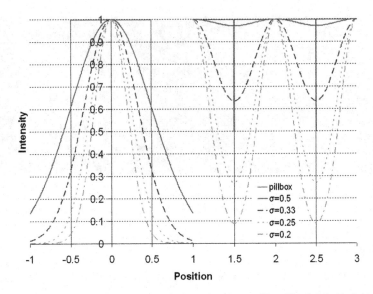

Fig. 16.5 Illustration of pillbox and several Gaussian dot profiles. The *left half* of the figure shows the profiles in isolation, while the *right half* shows the effect of interactions with neighbors in a square grid. Four different σ values have been chosen to illustrate the effect of neighbor interactions. Inter-dot distances and peak intensity values have been normalized. Note that the radius of the pillbox equals 0.5 in this example, and that a modest increase in σ (0.33–0.5) causes dramatic blurring, due to long-range Gaussian overlap

multiple tasks has not been performed; in principle a similar intensity modulation could be used with Gaussian dots, by modulating σ rather than peak intensity. To our knowledge this has not been attempted.

- *Grey scale/size resolution*: Natural vision is capable of resolving subtle differences in shading, even in the absence of color, but this is unlikely to be the case in prosthetic vision. For this reason a number of simulation studies have examined subjects' visual performance under conditions of reduced grey scale resolution. Typical resolutions used range from 2 to 16.
- *Homogeneity*: Most simulation studies use identically shaped dots, but it is inevitable that phosphenes perceived by prosthesis wearers will vary in intensity, size, shape, and other aspects. Until more information from prosthesis wearers is obtained it may be premature to build such inhomogeneities into simulations, but it may be an important future extension.

16.2.2.3 Temporal Properties

The rewiring taking place in the degenerating retina (see Chap. 3) and possibly other stages of visual processing are likely to include the loss of rapid signal processing and the creation of feedback loops. The effect of such changes will be that temporal properties of prosthetic vision will be much slower than in natural vision, as crudely represented in Fig. 16.6 by the "ghosting" of the visual percept of a maze.

Fig. 16.6 Tracing simulation shown in non-pixelized format to facilitate visualizing the smearing of the maze and the circle indicating stylus contact when both stylus and head-worn camera are moved

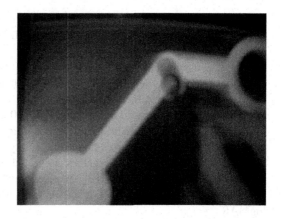

16.2.2.4 Dynamic Background Noise

The occurrence of photopsias (see Chap. 5) in patients with retinal degenerations is expected to persist when these individuals receive a visual prosthesis. For this reason some simulations have examined the effects of the presence of randomly distributed and slowly decaying dots of light on visual task performance during simulations. Some of these spontaneous dots can be seen in the right panel of Fig. 16.4.

16.2.2.5 Input Filtering/Windowing, Image Enhancement

There is a wide range of other image manipulations that can be studied through simulations in the quest to improve prosthetic visual performance. Among the obvious examples are spatial and temporal filtering of the input image to adjust the image properties to those of prosthetic vision. Yet the opposite approach, pre-emphasis filters such as edge detectors may improve a prosthesis wearer's ability to detect obstacles or recognize objects by their outline. Some simulation studies (see Chap. 17) have dealt with the question whether a weighted input window for each unit cell in the simulation grid, c.q., in the prosthetic array, may be beneficial for image understanding. Such studies have been few, and deserve further attention.

16.3 Optotype Resolution and Reading

16.3.1 Visual Acuity

Cha et al.'s first experiments centered on measuring visual acuity using Tumbling E stimuli [3]. They provided their subjects with pixelized vision spanning up to 1.7° of visual field, using square grids varying in dot counts from 100 to 1,024.

Cha et al. examined decreases in dot count from the 1,024-dot condition both by holding the dot density intact, resulting in smaller grids, and by holding the grid size constant, resulting lower dot densities.

Cha et al. found that visual acuity did not fall from that measured with the full 1,024-dot grid (approximately 0.11 logMAR) when dot count decreased as long as dot density was maintained. Visual acuity fell to approximately 0.70 logMAR, however, when they reduced the dot count to 100 and reduced dot density accordingly. They concluded that the size of the dot grid, and therefore the span of the prosthetic vision, did not matter, and visual acuity was closely tied to the density of the dots provided. These investigators suggested that an array of 625–1,024 electrodes, corresponding to about 210–350 electrodes/deg^2, could feasibly be placed in visual cortex to provide prosthetic vision with acuities as good as to 0.11–0.18 logMAR.

Between 2001 and 2003, Hayes et al., in a cooperative effort between the University of Southern California and Johns Hopkins University, examined visual acuity with prosthesis designs suitable for intraocular placement [21]. They created software implementations for 4×4, 6×10, and 16×16 dot grids, spanning 7.3°×7.3°, 11.3°×19.3°, and 19.3°×19.3°, respectively. These measurements translate to each grid having respective densities of 0.30, 0.28, and 0.69 dots/deg^2. These densities are much lower than those investigated by Cha et al. [3], but do reflect electrode densities possible for retinal prostheses with current technology.

Subjects in this study performed with visual acuities of approximately 1.96 logMAR with the 4×4 grid, 1.82 logMAR with the 6×10 grid, and 1.32 logMAR with the 16×16 grid. Like in the Cha et al. study [3], the subjects used head motions to scan stimuli and achieve a better visual acuity than that possible with a comparable static image. Unlike the Cha et al. study [3], however, these subjects seemed to benefit from larger grid spans across the visual field, particularly noticeable when comparing the 4×4 and 6×10 grids, which had nearly identical dot densities. This difference may be explained by the much sparser grids used in this study, and by the larger differences in grid size. Hayes et al. did confirm the importance of dot density, though, as the grid with the highest dot density offered the best visual acuity in this experiment.

Cai et al., from Tsinghua University in Beijing, studied the effects of introducing irregularity in the presentation of simulated phosphenes, as could be expected of the percepts in actual prosthesis-wearers [2]. They used a grid with a density of approximately 0.86 dots/deg^2, which offered visual acuities of about 1.36 logMAR. Using a scheme that assigned dot values without taking grid deformations into account, the irregularities induced a worsening of visual acuity by as much as 0.47 logMAR. Taking the deformations into account before down-sampling the original image, however, mitigated this acuity loss to only 0.22 logMAR. The authors therefore suggest that such a method of adapting down-sampling could help improve prosthesis-wearers' performance.

Chen et al., of the Universities of New South Wales and Newcastle in Australia, have conducted the most detailed visual acuity research in relation to prosthetic vision simulations to date [6–10]. Unlike previous groups, Chen et al. have investigated visual acuity using both grids with rectangular layouts and hexagonal layouts. Both grid layouts incorporated 100 dots in a square or approximately square grid. Dot densities remained at about 0.56 dots/deg^2 for hexagonal grids and rectangular grids that were shrunken to maintain density, and 0.51 dots/deg^2 for unaltered rectangular grids.

In their first simulation study, published in 2004, Chen et al. examined visual acuity with hexagonal and unaltered rectangular grids by using Landolt rings [10]. Both pillbox and Gaussian input filters and dot profiles were implemented; all of these had circular apertures. For more discussion on these filters, see Chap. 17 and [18, 19]. The investigators also varied the effective size of the simulated phosphenes. They found that the hexagonal layout seemed to provide an advantage over the rectangular layout, but not with statistical significance. For both filtering schemes, optimal acuities were reached at about 1.55 logMAR, at a sigma value of about 33% of the simulated phosphene separation for Gaussian filtering, and a kernel radius of about 50% of the dot separation for mean filtering. Visual acuities dropped to about 1.7 logMAR at non-optimal aperture sizes. These acuity measurements are reasonably consistent with the Hayes et al. and Cai et al studies, when considering grid densities.

In a follow-up paper, Chen et al. took a closer look at grid configurations and densities [7]. They tested subjects on Landolt rings with the same grids as in their previous study, as well as the shrunken rectangular grid mentioned above. Simulated phosphenes were depicted as round dots with Gaussian profiles, and images were processed via mean filtering used in their first study. Similar to their previous results, measured acuities ranged from 1.55 logMAR to 1.7 logMAR; however, in this study, the authors were able to demonstrate a significant benefit (about 0.5 logMAR) of hexagonal over non-shrunken rectangular grids for filter aperture sizes of 50 and 70% of dot spacing. The authors claimed to see some benefit the hexagonal grids over the rectangular grids shrunken to maintain density, but did not have enough subjects to make this difference significant.

In a later paper [6], Chen et al. examined how subjects approached the simulated handicap of prosthetic vision. They first found that about 80% of subjects could improve their performance with prosthetic vision after practice. Their subjects improved by about 1.1% each session, which lasted an average of 33 min, up to about 15–20 sessions.

As most subjects, during the course of testing with prosthetic vision, develop and utilize scanning to increase visual acuity, Chen et al. also investigated the types of scanning developed and their benefit over no scanning [8]. They found that fast (10°/s and greater), circular scanning motions helped subjects achieve acuities up to two times better than acuities inherently bestowed by the densities of the dot grids. This preference for circular scanning, however, may have been biased by testing only with Landolt rings, and no other stimuli. Horizontal and vertical scanning, as one of their subjects chose, may be just as or more beneficial in detecting non-circular targets.

Finally, related to scanning techniques, Chen et al. also investigated the use of head movements while using prosthetic vision [9]. They first had subjects perform the Landolt ring test with the simulator setup, but no actual simulation (i.e. simply performing the task with a video camera image). They found that head movements were minimal in these control tests, but subjects did adopt head motion when presented with simulated prosthetic vision. The speed of head motions seemed to increase by an average of 4°/s from the largest rings presented to the smallest (corresponding to 2.0–1.3 logMAR). The authors suggest that increased speed of head movements may provide greater access to high-frequency information than slower motions.

Based on these simulation studies, the density of phosphenes perceived by prosthesis wearers should most directly affect the visual acuity they can achieve. At comparable density levels, greater spans of the phosphenes across the visual field should also provide some benefit [21]. Apart from the design of the devices themselves, practice [6], techniques of scanning [8, 9], and appropriately compensating for aberrations in perceived configurations should also affect the visual acuities achieved [2]. These conclusions must be taken with skepticism, however, as simulations do not necessarily reflect percepts elicited by real prostheses. Even simple assumptions, e.g., that phosphene resolution improves with increased electrode density, deserve serious scrutiny: More densely packed electrodes would be required for higher-resolution percepts, yet there is no guarantee that electrodes placed so close together will create separate and distinct phosphenes. Moreover, these studies did not incorporate any thorough measures to restrict or monitor eye movements, and thus may have overestimated the resolution of prostheses with external image capture [11].

16.3.2 Reading

Cha et al.'s study of reading with simulated prosthetic vision [5] used the same setup as their visual acuity study [3]. Subjects read text that was either scrolled as they read, so no scanning was necessary, or used head or eye movements to scan a page of text. Without any complications of scanning, the subjects could read text, where letters had an optimal size of approximately 0.4° of visual angle, at rates of 200 words/min with grids containing 625 or 1,024 dots. Reading rates dropped significantly with lower dot counts, but interestingly, this did not depend on whether grid size or density was varied. Thus, unlike in visual acuity, the authors found that dot count is the primary factor in determining reading speed.

This was confirmed by tests using head scanning to read a page of text; reductions in dot count had the same effect for both reducing dot density and reducing grid size. Tests with head and eye scanning did additionally show, however, that the requirement of scanning significantly slows subjects' reading speeds. With head scanning, reading rates dropped to 120 words/min for grids containing 625 or 1,024 dots. Scanning text with eye movements proved to be more difficult, with reading rates of only 55 words/min, but the exact reason for this seems to be unclear.

Dagnelie et al., of Johns Hopkins University, followed Cha et al.'s study with a publication on reading with prosthetic vision in 2001 [14]. Like Cha et al., they found that reading speed did increase with increasing numbers of dots in the grids. Dagnelie et al. tested other parameters, as well, including the effects of dot dropout, high versus low contrast, number of gray levels, dot size and spacing, and font size. The authors found that even with 30% dot dropout, reading rates of 50 words/min were possible with grids containing 625 dots. Increasing dropout, reducing contrast, or altering character perception to less than 2 cycles/character width (i.e., 4 dots/char width) all generally slowed reading rates. The number of gray levels, however, did not appear to have a significant effect. Although the free-viewing reading rates published in this study are markedly lower than those reported by Cha et al. [5], the results are comparable when taking into account that Dagnelie et al. did not perform any tests with less than 30% dropout.

Sommerhalder et al., from the Geneva University Hospitals, Switzerland, in 2003 conducted reading tests [26] with a different approach than the previous two groups. Unlike Cha et al. and Dagnelie et al., Sommerhalder et al. projected a stabilized image onto the retina and denied subjects the option of scanning with eye movements throughout their experiments. Their reading tasks also differed, in that subjects were only required to read one word at a time, and subjects could not change their perspectives of the word by scanning. Under these conditions, the authors concluded that grids of at least 300 dots located over the fovea would be necessary for reading accuracies greater than 90%. For eccentricities of 10° or more, grids of greater size and/or resolution become increasingly necessary to maintain high reading accuracy. For example, they found that a grid located at 20° eccentricity and spanning 20°×7° required about 875 dots to achieve 63% reading accuracy. Sommerhalder et al. found that problems at high eccentricities relate to a "crowding effect," in which interference among stimuli reduces perception below the otherwise expected visual acuity. The authors suggest that training with eccentric reading can improve performance, through suppression of this "crowding effect" and/or reduction of reflexive eye movements.

Sommerhalder et al. expanded their study in 2004 to incorporate full-page reading [27], more akin to the experiments already conducted by Cha et al. and Dagnelie et al. For these experiments, the authors continued to enforce a gaze-stabilized view, and chose to model a optoelectronic retinal prosthesis, with which subjects could use eye movements to change the image presented by the array. When they stabilized the image of the grid, containing 572 dots spanning 10°×7° of visual field, over the fovea, they measured reading performance similar to that found by Cha et al. when eye-tracking was employed, about 65 words/min and nearly perfect reading accuracy. One of their three subjects was able to improve to a reading rate of 122 words/min, but neither of the other two subjects showed significant improvement with practice.

In the second phase of their 2004 study, Sommerhalder et al. stabilized the image of the grid at a visual eccentricity of 15° below the center of vision. At first, reading rates dropped to about 3 words/min, with reading accuracies of 85% for one subject and 13% for the other two. After approximately 2 months of practice,

55–68 sessions per subject, reading accuracies plateaued at an average of 94%. At this time, reading rates were still improving, and the final measured rates averaged 23 words/min. During these experiments, the authors also measured reading comprehension, and found that useful reading ability most likely requires at least 85% reading accuracy. The authors thus concluded that, in agreement with previous studies, the prosthetic generation of 600 discrete percepts should be sufficient to allow effective reading. An array of the same size placed as far out as 15° eccentricity would also be usable, but would require significantly more instruction and practice.

Kelley et al., of the Johns Hopkins University group, presented a study on reading with gaze-locked vision in 2004 [23]. They provided subjects with grids of 10×10 or 25×25 dots to view segments of text, which the subjects scrolled by controlling a mouse. The authors varied dot size, font size, contrast, and the application of image stabilization. The subjects in this study maintained reading accuracies of 95% or more for nearly all forms of free-view reading; only when characters spanned as few as 4.5 dots, accuracy fell to 70–80%. In gaze-locked trials, accuracy still remained above 90% for most conditions, and fell to about 60% for characters spanning 4.5 dots. Reduction of character resolution caused the greatest drop in reading speed, about 60–80%, followed by stabilization, reducing speed by 50–75%, and contrast reduction lowered speeds by 15–25%. Dot size did not seem to have a significant effect. The authors also point out that practice significantly improved performance, particularly for trials using stabilization or low contrast, as demonstrated in Fig. 16.7. This study points out the need for sufficient character resolution to have successful reading with simulated prosthetic vision; more importantly, it demonstrates that adequate training will allow reading under gaze-locked conditions, reducing concerns expressed by many that performance of complex tasks would be impossible with external image acquisition, unless eye movement feedback is built into the system.

Dagnelie et al. performed a another study on reading with prosthetic vision in 2006 [12]. Similar to their 2001 study [14], the authors tested the effects a wide variety of parameters on paragraph reading controlled by mouse scrolling, with no restrictions on viewing. These parameters included dot size, spacing and count, dropout, number of gray levels, contrast level, and text size. The authors found significant effects of each of these parameters, and determined that reading with 90% accuracy was possible provided that characters were at least three dots wide and dropout levels did not surpass 50%. Reading speed dropped below 20 words/min whenever accuracy fell below 90% or, at low contrast, the presented grid was smaller than the width of two characters. The authors suggest that an array containing 256 electrodes, with 30% dropout, should be sufficient to allow accurate paragraph reading, with a maximum reading rate of about 30 words/min.

According to these studies, reading with prosthetic vision should certainly be possible. Reading accuracy appears to depend strongly on the resolution provided to the visual system, requiring at least three distinct phosphenes across for each character [12]. Provided useful reading accuracy (i.e. at least 85% accuracy [27]), reading rates seem to depend most upon the number of distinct phosphenes a

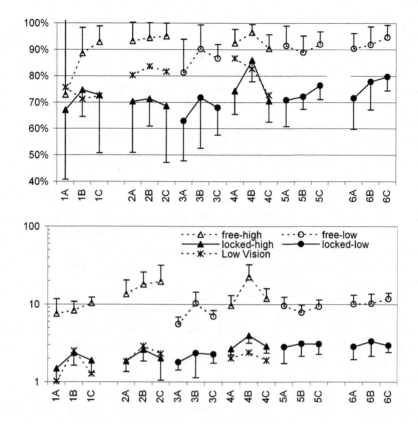

Fig. 16.7 Reading accuracies (*top*) and speeds (*bottom*) for two stabilization (free-locked) and two contrast (low-high) conditions [23]. Trials were presented in six sets (1–6), each composed of three consecutive blocks of 16 trials (A–C). Each set took between one and three 1-h sessions to complete. Error bars denote the between-subject standard deviation among five subjects. The "low vision" points represent the performance of the one subject with severely reduced visual acuity and contrast sensitivity. Notice the effect of practice in on gaze-locked performance

prosthesis-wearer can perceive [5]. Without any dropout, arrays producing about 600 distinct percepts should be enough to allow read rates of about 50–70 words/min.

16.4   Face and Object Recognition

Dagnelie et al. published the first study involving a recognition task in 2001 [14]. In their experiment, subjects were asked to identify pixelized faces 12° wide among four options. The number of dots in the grid simulation, percentage of dot dropout, and the number of gray levels strongly affected subjects' ability to recognize these faces. Instances where grid parameters reduced the sampling frequency below 8 cycles per face width, grid size fell below 256 dots and 7 deg^2, dropout exceeded 50%, or when fewer than four gray levels were used, caused face recognition abilities to fall to chance. Recognition accuracy of 80% or greater was consistently possible only in 98% contrast conditions, specifically with a grid size of 625 dots spanning $11° \times 11°$, or a grid of 256 dots about 70' wide each.

Thompson et al., of the same Johns Hopkins University group, with collaborators from the University of Southern California, published a very similar study to the one above in 2003 [30]. In their 2001 experiment [14], however, they used a dot size of 23' of visual field as a baseline parameter, whereas the base dot size was increased to 31.5' in their 2003 experiment. Among other factors, this increased the basic grid size from $7° \times 7°$ to $9.6° \times 9.6°$. While no parameter combination generated an average accuracy of 80% or more when contrast was set to 12.5%, several parameter conditions with 99% contrast did allow facial recognition with 80% or more accuracy. Based on their results, it appeared that a dot density around 1 dots/deg with 4.5-arcmin dot spacing was optimal among their parameter combinations. Increasing the grid size from 256 dots to 625 and 1,024 dots also consistently improved recognition accuracy.

Hayes et al. [21] also reported on object recognition using prosthetic vision simulations. Using grids with 4×4, 6×10, or 16×16 dots, with various levels of contrast and dynamic noise, subjects were asked to visually describe and, if possible, identify common objects without touching them. Contrast and noise did not appear to have significant effects, but grid size did. The 4×4 and 6×10 grids had the same dot size and spacing, but the 6×10 grid provided a significant advantage over the 4×4 grid. This would suggest that grid size, measured by dot count or visual span, is important for recognition tasks. The 16×16 grid was significantly more useful for object recognition than the 6×10 grid, but this could also be an effect of increased dot density.

Dagnelie et al. published a study on visual discrimination of white target squares on a black background ("modified checkerboard") in 2006 [15]. While this study did not ask subjects to recognize specific features of these targets, it did evaluate the subjects' abilities to discern and count these targets when gaze-locking was employed. For most of the subjects, the time to count all the targets on a modified checkerboard did not change with the number of targets, as a result of their counting strategies, but the addition of gaze-locking did significantly increase counting time, particularly before practice. Srivastava et al., of the Illinois Institute of Technology in cooperation with Dagnelie et al., published similar experiments with this counting task in 2009 by simulating a cortical prosthesis [29]. In these experiments, gaze-locking was enforced consistently and levels of dropout varied so that 325–650 dots were used. Search times were comparable with the 2006 study [15], after practice, and levels of dropout did not seem to affect the performance of most subjects.

Zhao et al., of Shanghai Jiao Tong and Peking Universities in China, reported results of testing subjects with object and scene recognition tasks in 2008 [34] and 2010 [33]. Subjects could freely view $4.5° \times 4.5°$ grids of either square or circular dots with various dot densities and two different methods of image processing: binary (black–white) output through contrast enhancement, and edge detection.

The authors set their threshold of recognition at 60% accuracy. They found that for common objects, subjects passed this threshold between grid resolutions of 16×16 and 24×24 dots, and for scenes, subjects reached this threshold around a grid resolution of 48×48 dots. When using a slightly lower grid resolution, the authors observed that simple binary image processing was more helpful than use of edge detection. At higher resolutions, however, edge detection seemed to be more beneficial.

Recognition tasks, in general, appear to be sensitive to the number dots contained within a grid [21, 33, 34]. Particularly for faces, high image quality of near-100% contrast, at least four gray levels, and at least 256 dots appears to be required for 80% or greater recognition accuracy [14]. This is not surprising, since the coarse traits of faces resemble each other, and successful discrimination is based on finer traits and shading.

16.5   Visually Guided Behavior

16.5.1   Hand–Eye Coordination

Along with their early visual acuity and facial recognition tasks. Dagnelie and colleagues conducted experiments of hand–eye coordination using both virtual reality and live video input [14]. In their virtual reality experiment, four subjects viewed a room with a table and chairs through pixelized vision with less than 20% contrast. Their task was to pick up objects off the floor and place them on the table, releasing them only when they rested on the surface. Out of 12 total attempts to transfer objects in this virtual scene, subjects were successful six times. In the live video experiment, with about 90% contrast and a 250 ms delay, subjects were able to transfer objects with the assistance of tactile feedback. Technological limitation prevented systematic studies, but these experiments did serve as an example for how prosthetic vision could be used for coordination tasks.

The same group later expanded their study of hand–eye coordination with simulated prosthetic vision, reported in Hayes et al., 2003 [21]. In this study, subjects were asked to perform two tasks. The first asked them to pour ten pieces of candy from one cup to another, without touching the second. Some subjects were able to do this successfully in the hardest condition, using a grid of only 4×4 dots. Only one subject required a grid of 16×16 dots. The authors concluded that, on average, a 6×10 grid of dots would be sufficient for a simple hand–eye coordination task.

The second task asked of these subjects was to cut along the outside of a black square outline on a white piece of paper. Times to completion and errors both fell with increasing grid size, where satisfactory performance was only achieved with a 16×16 grid. Hayes et al. reasoned that this task, unlike the first, requires constant reevaluation to acquire the position of the scissors relative to the border. The authors thus concluded that for more complex tasks the 6×10 grid would be insufficient, and larger and/or denser grids would be required for acceptable performance.

Dagnelie et al. and Srivastava et al. continued their studies with modified checkerboards, respectively in 2006 [15] and 2009 [29], by asking subjects to cover these targets with black checkers. Once a subject had correctly covered a target, the target would no longer be visible through the simulation grid. Dagnelie et al. found that, using a 6×10 grid, subjects could successfully learn how to perform this task with as little as 0.85% error, even in gaze-locked conditions.

Srivastava et al. utilized gaze-locking throughout their experiment (see Chap. 18 and [29]), and varied dot counts between 325 and 650 dots. Similar to the results of the counting task, practice seemed to substantially reduce any effects of dropout on time or error. After practice, these subjects were able to complete the task without any errors through the simulated cortical prosthesis.

As suggested by the Hayes et al. study [21], hand–eye coordination tasks seem to benefit from increases in grid size. It is unclear, however, whether this benefit is derived from a greater dot count, the increased visual span of the grid, or both. As seen with Dagnelie et al. [15] and Srivastava et al., gaze-locking and dropout in simulations do seem to mandate practice if normal performance is desired, but do not strongly hinder performance after the initial learning period.

16.5.2 Wayfinding

Cha et al. [4] provided subjects with simulated prosthetic vision similar to that in their visual acuity and reading experiments. The authors varied dot density, dot count, overall grid size (up to $1.7° \times 1.7°$), and the visual angle captured by the camera and projected onto this simulated prosthesis. They asked subjects to navigate through a maze with white walls, floor, and ceiling and black obstacles. The capture angle was found to be critical for this task: performance increased with this angle so long as individual stimuli did not become too small; performance declined once the capture angle was expanded past 18 times the angular subtense of the grid. At the optimal viewing angle, performance correlated well with the number of dots, almost regardless of dot density. The authors concluded that a cortical prosthesis with 25×25 or 32×32 electrodes, perceptually spanning 1.7° and incorporating 30° of a camera's view, could be used effectively for high-contrast obstacle avoidance and wayfinding in a familiar environment.

Dagnelie et al. published a report in 2007 of a similar pair of experiments on wayfinding [13]. In the first experiment, subjects used 4×4, 6×10, and 16×16 grids, respectively spanning $11° \times 11°$, $16° \times 27°$, and $27° \times 27°$. The camera's viewing angle was fixed at 37°. The authors found that, with increasing dot count, the subjects' wayfinding performance improved. For experienced subjects, the 6×10 grid was sufficient for this task. In their second experiment, subjects used the 6×10 grid to navigate through a virtual environment and were presented additional parameters of dynamic noise and dot dropout. The authors observed that noise did not have a significant effect, and dropout of 30% led to a slight decrease in performance. These findings match well with those of Cha et al. [4], particularly as the only differences between the 4×4 and 6×10 grids were size and dot count, and not density.

In 2008, Boyle et al., of Queensland University of Technology in Australia, conducted experiments where subjects were asked to judge the quality of pixelized images [1]. No viewing constraints were used, and no actual wayfinding was performed in this study. Subjects, instead, viewed an original image and then judged which among a set of binary, 25×25 dot versions of that image they would find most useful for navigation. The authors applied various algorithms to highlight important or salient stimuli in an image, and investigated various approaches to zooming the image in on important features. The authors found that subjects did not favor any kind of feature-based processing when no zoom was applied. In the second phase of their study, however, subjects appeared to prefer a zooming method that trimmed away unimportant pixels based on a saliency map generated by a program from the University of Southern California. The final image was still 25×25 dots, but contained information from a smaller portion of the original image. The authors suggest that such saliency detection and zoom could be used for actual prosthetic devices.

Wang et al. in our laboratory published experiments on wayfinding in virtual environments later in 2008 [32]. Subjects viewed a virtual environment through a stabilized image of 6×10 dots spanning $16.2° \times 27°$. The authors investigated factors affecting time to traverse the virtual environment, including contrast, background noise, and dot dropout. Of these, only dot dropout had significant effect on performance. When dropout was set to 30%, completion time increased by 40%. This accents the importance of the number of dots available to subjects for wayfinding, as well as the importance of array integrity in visual prosthesis implantees.

Srivastava et al., in the same publication as their target counting and checker placing tasks [29], had subjects navigate similar virtual environments as those of Wang et al. and Dagnelie et al., using a cortical rather than a retinal grid layout. Unlike with Wang et al. and Dagnelie et al., however, these subjects did suffer a significant deterioration in performance with dot dropout. This may be a result of the large visual angle subtended by the cortical grid, causing a random reduction from 650 to 325 dots (at 50% dropout) to leave a sparser set than what was available in retinal simulations.

Jointly, these studies suggest that, so long as an optimal viewing angle is obtained [4], wayfinding performance is primarily dependent upon the number of dots presented in a grid simulation [4, 13]. Based on Dagnelie et al.'s [13] and Wang et al.'s [32] studies, current 6×10 electrode arrays placed in the retina should be sufficient for basic wayfinding, but performance will deteriorate in cases of significant electrode dropout.

16.6 Visual Tracking

Hallum et al., of the Universities of New South Wales and Newcastle, Australia, examined the abilities of subjects to track a target using grids generated with three different aperture weighting schemes [20]. Each of these grids contained 23 dots in a hexagonal layout, spanning about $7.4° \times 5.4°$, on a screen spanning $16.7° \times 16.7°$. A target sized 36 arcmin^2 moved across the screen, generally in an S-shaped pattern, and subjects were asked to keep the grid centered over the target by moving a joystick. The first filtering scheme the authors employed was analogous to the setup of Cha et al. [3–5], where the stimulus was only sampled at the 23 locations corresponding to dot positions. The second scheme regionally averaged the underlying stimulus to determine dot characteristics. The third scheme employed a Gaussian sensitivity profile with $\sigma = 34.4'$. Subjects could freely view the screen binocularly. The authors found that, after practice, the use of a Gaussian profile permits more accurate fixation on, saccades to, and pursuit following the targets. The authors advocate use of this filtering technique to actual prostheses to improve performance in visual tasks, but they do not discuss the possibility that the unrestricted gaze and the choice of stimuli may have skewed their findings.

Wang et al. published a study with a similar experimental setup in 2008 [31]. Subjects viewed, monocularly, an $0.94°$ target moving across a screen, either with natural vision or through a 10×10 cell grid. The subjects were asked to follow the target with their eyes, monitored by an infrared eye tracker. When simulating prosthetic vision, the rendered grid would move with the subject's eye movements, corresponding to an array placed over the fovea, superior retina, or nasal retina. The authors found that, compared to using natural vision, subjects took about 65% longer to detect sudden target movements in simulated prosthetic vision. This is comparable to the 20% increase in reaction time over natural vision reported by Hallum et al. [20], when considering differences in viewing conditions between these two studies. The authors also found that horizontal eye movements were more unstable in simulated prosthetic vision, particularly for stabilized stimulus placement on the nasal retina. Simulating superior retinal placement offered more stability than nasal placement, but not as much as foveal placement. The authors conclude that, although initiation is slower and movement is less stable, pursuit eye movements should be possible with optoelectronic retinal prostheses and that, among peripheral retinal locations for a prosthesis, superior placement may be more beneficial than inferior placement.

16.7   Computational Simulations

Pezaris et al., of Harvard Medical School, are pursuing a visual prosthesis within the lateral geniculate nucleus (LGN) of the thalamus [25]. In 2009, these authors published results of a study they conducted to evaluate four different basic prosthesis designs with various values of electrode spacing. Unlike all other simulation studies mentioned in this chapter, however, Pezaris et al. used a purely computational approach to determine the likely benefit of each design's implementation.

The authors specifically investigated prosthesis designs where electrode tips are placed in a 3D grid throughout the LGN or along a 2D slice through LGN, and where activation is brought through two different forms of deep brain stimulation (DBS) electrodes. According to their analyses, the 3D grid would be the most useful for creating many phosphenes throughout the visual field. Among the possibilities for laying out a 2D grid in LGN, with much less of the visual field being stimulated, placement along a sagittal slice midway through the LGN would seem to be best. DBS electrodes that can contain as many as 60 microwires could create stimulation similar to the 2D plane. Traditional DBS electrodes, however, would only offer three to four large conductive cuffs on their shafts, which would provide stimulation more akin to solid curves or arcs. While the implementation of an LGN prosthesis through a traditional DBS electrode may be technically and surgically simpler, it would provide little useful stimulation. On the other hand, a full 3D grid, which would provide selective stimulation for the entire visual field, would be much more challenging. Future studies may shed more light on exactly how much benefit each design could provide.

16.8   Conclusion

Many of the designs found by these simulation studies to offer desirable performance in common visual tasks exceed current technological realities. For example, numerous reading studies concluded that grids containing about 600 dots are required for reading speeds of 50–70 words/min. Some studies, however, purposely used grid configurations that are currently in use by visual prostheses. These studies, limiting grid sizes to about 4×4 or 6×10 dots, found that even simple prosthesis designs can be used for modest performance on everyday visual tasks. While not as applicable to guiding performance expectations for current devices, simulations that represent more complex designs do guide developers in expanding this technology and provide specific goals for prosthesis structure and function.

References

1. Boyle JR, Maeder AJ, Boles WW (2008), *Region-of-interest processing for electronic visual prostheses.* J Electron Imaging, **17**(1): p. 013002.
2. Cai S, Fu L, Zhang H, et al. (2005), *Prosthetic visual acuity in irregular phosphene arrays under two down-sampling schemes: a simulation study.* Conf Proc IEEE Eng Med Biol Soc, **5**: p. 5223–6.
3. Cha K, Horch K, Normann RA (1992), *Simulation of a phosphene-based visual field: visual acuity in a pixelized system.* Ann Biomed Eng, **20**: p. 439–49.
4. Cha K, Horch KW, Normann RA (1992), *Mobility performance with a pixelized vision system.* Vision Res, **32**(7): p. 1367–72.
5. Cha K, Horch KW, Normann RA, Boman DK (1992), *Reading speed with a pixelized vision system.* J Opt Soc Am A, **9**(5): p. 673–7.
6. Chen SC, Hallum LE, Lovell NH, Suaning GJ (2005), *Learning prosthetic vision: a virtual–reality study.* IEEE Trans Neural Syst Rehabil Eng, **13**(3): p. 249–55.
7. Chen SC, Hallum LE, Lovell NH, Suaning GJ (2005), *Visual acuity measurement of prosthetic vision: a virtual–reality simulation study.* J Neural Eng, **2**(1): p. S135–45.
8. Chen SC, Hallum LE, Suaning GJ, Lovell NH (2006), *Psychophysics of prosthetic vision: I. Visual scanning and visual acuity.* Conf Proc IEEE Eng Med Biol Soc, **1**: p. 4400–3.

9. Chen SC, Hallum LE, Suaning GJ, Lovell NH (2007), *A quantitative analysis of head movement behaviour during visual acuity assessment under prosthetic vision simulation.* J Neural Eng, **4**(1): p. S108–23.

10. Chen SC, Lovell NH, Suaning GJ (2004), *Effect on prosthetic vision visual acuity by filtering schemes, filter cut-off frequency and phosphene matrix: a virtual reality simulation.* Conf Proc IEEE Eng Med Biol Soc, **6**: p. 4201–4.

11. Dagnelie G (2008), *Psychophysical evaluation for visual prosthesis.* Annu Rev Biomed Eng, **10**: p. 339–68.

12. Dagnelie G, Barnett D, Humayun MS, Thompson RW, Jr. (2006), *Paragraph text reading using a pixelized prosthetic vision simulator: parameter dependence and task learning in free-viewing conditions.* Invest Ophthalmol Vis Sci, **47**(3): p. 1241–50.

13. Dagnelie G, Keane P, Narla V, et al. (2007), *Real and virtual mobility performance in simulated prosthetic vision.* J Neural Eng, **4**(1): p. S92–101.

14. Dagnelie G, Thompson RW, Barnett D, Zhang W (2001), *Simulated prosthetic vision: Perceptual and performance measures.* In: *Vision Science and its Applications*, OSA Technical Digest. Washington, DC: Optical Society of America: p. 43–6.

15. Dagnelie G, Walter M, Yang L (2006), *Playing checkers: Detection and eye–hand coordination in simulated prosthetic vision.* J Mod Opt, **53**: p. 1325–42.

16. Fornos AP, Sommerhalder J, Rappaz B, et al. (2006), *Processes involved in oculomotor adaptation to eccentric reading.* Invest Ophthalmol Vis Sci, **47**(4): p. 1439–47.

17. Fornos AP, Sommerhalder J, Rappaz B, et al. (2005), *Simulation of artificial vision, III: do the spatial or temporal characteristics of stimulus pixelization really matter?* Invest Ophthalmol Vis Sci, **46**(10): p. 3906–12.

18. Hallum LE, Cloherty SL, Lovell NH (2008), *Image analysis for microelectronic retinal prosthesis.* IEEE Trans Biomed Eng, **55**(1): p. 344–6.

19. Hallum LE, Suaning GJ, Lovell NH (2004), *Contribution to the theory of prosthetic vision.* ASAIO J, **50**(4): p. 392–6.

20. Hallum LE, Suaning GJ, Taubman DS, Lovell NH (2005), *Simulated prosthetic visual fixation, saccade, and smooth pursuit.* Vision Res, **45**(6): p. 775–88.

21. Hayes JS, Yin VT, Piyathaisere D, et al. (2003), *Visually guided performance of simple tasks using simulated prosthetic vision.* Artif Organs, **27**(11): p. 1016–28.

22. Humayun MS, Dorn JD, Ahuja AK, et al. (2009), *Preliminary 6 month results from the Argus™ II epiretinal prosthesis feasibility study.* Conf Proc IEEE Eng Med Biol Soc, **1**: p. 4566–8.

23. Kelly AJ, Yang L, Dagnelie G (2004), *The effects of stabilization, font scaling and practice on reading in simulated prosthetic vision.* Invest Ophthalmol Vis Sci, **45**: p. ARVO E-abstr. #5436.

24. Perez Fornos A, Sommerhalder J, Pittard A, et al. (2008), *Simulation of artificial vision: IV. Visual information required to achieve simple pointing and manipulation tasks.* Vision Res, **48**(16): p. 1705–18.

25. Pezaris JS, Reid RC (2009), *Simulations of electrode placement for a thalamic visual prosthesis.* IEEE Trans Biomed Eng, **56**(1): p. 172–8.

26. Sommerhalder J, Oueghlani E, Bagnoud M, et al. (2003), *Simulation of artificial vision: I. Eccentric reading of isolated words, and perceptual learning.* Vision Res, **43**(3): p. 269–83.

27. Sommerhalder J, Rappaz B, de Haller R, et al. (2004), *Simulation of artificial vision: II. Eccentric reading of full-page text and the learning of this task.* Vision Res, **44**(14): p. 1693–706.

28. Sommerhalder JR, Fornos AP, Chanderli K, et al. (2006), *Minimum requirements for mobility inunpredictible environments.* Invest Ophthalmol Vis Sci, **47**: p. ARVO E-abstr. #3204.

29. Srivastava NR, Troyk PR, Dagnelie G (2009), *Detection, eye–hand coordination and virtual mobility performance in simulated vision for a cortical visual prosthesis device.* J Neural Eng, **6**(3): p. 035008.

30. Thompson RW, Jr., Barnett GD, Humayun MS, Dagnelie G (2003), *Facial recognition using simulated prosthetic pixelized vision.* Invest Ophthalmol Vis Sci, **44**(11): p. 5035–42.

31. Wang L, Yang L, Dagnelie G (2008), *Initiation and stability of pursuit eye movements in simulated retinal prosthesis at different implant locations.* Invest Ophthalmol Vis Sci, **49**(9): p. 3933–9.
32. Wang L, Yang L, Dagnelie G (2008), *Virtual wayfinding using simulated prosthetic vision in gaze-locked viewing.* Optom Vis Sci, **85**(11): p. E1057–63.
33. Zhao Y, Lu Y, Tian Y, et al. (2010), *Image processing based recognition of images with a limited number of pixels using simulated prosthetic vision.* Inf Sci, **180**: p. 2915–24.
34. Zhao Y, Tian Y, Liu H, et al. (2008), *Pixelized images recognition in simulated prosthetic vision.* IFMBE Proc, **19**: p. 492–6.

Chapter 17
Image Analysis, Information Theory and Prosthetic Vision

Luke E. Hallum and Nigel H. Lovell

Abstract Recent years have seen markedly improved clinical outcomes in cochlear implantees. This improvement is largely attributed to improvements in speech processing algorithms. In light of these improvements, researchers are prompted to ask, "Could image analysis improve clinical outcomes in retinal implantees?" We discuss our approach to image analysis, microelectronic retinal prostheses, and the perception of low-resolution images, which we believe can be used to help constrain the design of an implant. We hope that our approach, and developments thereof, will ultimately contribute to improved clinical outcomes in retinal implantees.

Abbreviation

APRL   Artificial preferred retinal locus

17.1 Introduction

It makes intuitive sense that the cochlear implant involves a speech processor. This processor is typically implemented in programmable microelectronics that the subject wears behind the ear; it lies between the device microphone and the electrode array implanted in the subject's cochlea. The processor analyzes incoming acoustic signals, deriving parameters that determine the current waveforms injected at each electrode. Recent years have seen markedly improved clinical outcomes in cochlear implantees. This improvement is largely attributed to improvements in speech processing algorithms [9]. In light of these improvements, researchers are prompted to ask, "Could

L.E. Hallum (✉)
Graduate School of Biomedical Engineering, University of New South Wales,
ANZAC Parade, Sydney 2052, Australia
and
Center for Neural Science, New York University,
New York, NY 10003, USA
e-mail: hallum@cns.nyu.edu

G. Dagnelie (ed.), *Visual Prosthetics: Physiology, Bioengineering, Rehabilitation*,
DOI 10.1007/978-1-4419-0754-7_17, © Springer Science+Business Media, LLC 2011

image analysis – analogous to the cochlear implant's speech processing – improve clinical outcomes in retinal implantees?" This chapter is concerned with image analysis and the perception of low-resolution images.

In this chapter we review two studies of ours [6, 8]. In the first study prosthetic vision was simulated via low-resolution images that we refer to as "phosphene images." Normally sighted subjects were required to track a moving target. The phosphene images were rendered using three image analysis schemes, and we examined the effects of each scheme on subjects' tracking performance. The second study was numerical. We used information theory to quantify the amount of information contained in the phosphene images, and how different image analysis schemes affect this information. This numerical approach ultimately served as a reasonably good model of subjects' tracking performance in [8]. Further, we used this numerical approach to suggest how image analysis schemes could be further improved.

The two above-mentioned studies help constrain the design of a retinal prosthesis. They directly address the topic of image analysis and its integration with a retinal implant. They effectively ask, "How should an implant process incoming images?" This contribution is taken up in the Sect. 17.5. There, we also discuss generalizing the approach to visual tasks beyond fixation, saccade, and pursuit, for example, visual acuity tasks and reading. We begin this chapter by situating image analysis, functionally speaking, with respect to the prosthetic device as a whole, and by describing the experimental framework within which low-resolution perception is investigated.

17.2   Situating Image Analysis

The major components of a retinal prosthesis, as it is envisioned and prototyped by a number of groups [4], are the camera, the image analyzer, the communication link and the implant, as discussed elsewhere in this volume. Physically, the image analyzer is external to the body and, functionally, lies between the acquisition of high-resolution images of the world ("scenes") and the generation of signals for transmission to the implanted component of the device (via the communication link). The camera, worn on spectacles by the subject, captures high-resolution spatiotemporal images of the real world. The image analyzer is implemented in programmable microelectronics, a digital signal processor, worn on the body. It processes data captured by the camera, effectively converting scenes to an array of numbers which determine the current waveforms at each electrode. These numbers are delivered to the implant by the communication link. As with the cochlear implant, this link ideally comprises radio-frequency signals transmitted transcutaneously; as opposed to a percutaneous link, a radio-frequency link poses a lesser risk of infection and allows the external unit to be physically uncoupled from the body. The implanted analog electronics decode the incoming radio-frequency signals and drive an array of electrodes. These electrodes are embedded in a flexible substrate that is affixed to the inner layer of the retina. Recent clinical trials involve arrays of 60 electrodes (Argus™ II, http://clinicaltrials.gov/ct2/show/NCT00407602).

Here we will step the reader through the processing stream that generated the phosphene image in Fig. 17.1. First, the high-resolution image (left panel) was filtered.

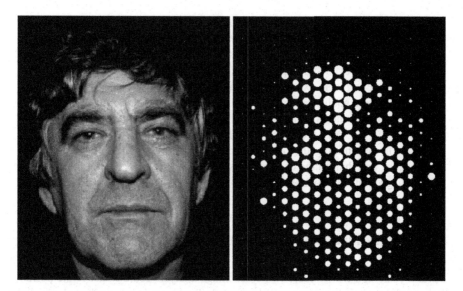

Fig. 17.1 A basic simulation of prosthetic vision. The high-resolution image (*left*) is filtered and subsampled. The samples are used to modulate the sizes of luminous phosphenes which, together, comprise the phosphene image (*right*). Image source: http://pics.psych.stir.ac.uk/

This filtering anti-aliases the resulting phosphene image, and may be used to emphasize salient features of the high-resolution image. Second, the filtered image was sampled at 400 locations, each pertaining to a phosphene. Each of these samples was then quantized to one of 16 levels, and used to modulate the size of its corresponding phosphene. Quantization makes for more accurate simulation since, as with the cochlear implant, the microelectronic retinal prosthesis is likely to elicit only a small number of different percepts at each location in the visual field.

17.3   The Experimental Framework

The right-hand panel in Fig. 17.1 shows a phosphene image of a face. This image depicts the sort of vision that the microelectronic retinal prosthesis aims to provide the otherwise profoundly blind implantee, that is, a relatively small number of discrete, luminous blobs. Phosphene images like this one, and also phosphene images that vary over time, may be used in psychophysical experiments and presented to normally seeing observers. These experiments draw on well established, psychophysical methods in vision research wherein perception is inferred through the measurement of subjects' behavior. For example, the reading speed of subjects presented low-resolution, phosphenized text could be measured. In this way, experimenters may use simulation data to predict clinical outcomes in actual implantees. This approach, which we call "visual modeling," is analogous to "acoustic modeling" of cochlear implants wherein colored noise bands, modulated by speech, or in recent cases by music, are played to normally hearing listeners. These listeners form a more readily accessible cohort than one comprising actual cochlear implantees. The behavior of these normally hearing listeners is then used to investigate improved signal processing strategies and electrode array designs.

Visual modeling is an experimental framework that allows us to test predictions concerning image analysis, microelectronic retinal prostheses, and the perception of low-resolution images (for discussion, see [7]). It has been used extensively in the field since the early work of Cha [2], who argued that "three parameters are important in determining the quality of a pixelized image: the number of pixels, their density, and their range of intensities." By contrast, we are using the approach to test image analysis schemes.

17.4 Tracking a Low-Resolution Target

Several years ago we wondered whether image analysis could improve the performance of the phosphene image observer [8]. To this end, we conducted a visual modeling experiment that compared the effects of three different image analysis schemes on subjects' performance. The task involved the fixation of, saccading to, and the pursuit of a small, high-contrast target. The first image analysis scheme that we investigated, referred to as scheme $Q0$, was trivial: images were not preprocessed but simply down-sampled. This scheme allowed for the comparison of Kichul Cha's results [2] with our results. The second scheme ($Q1$) blurred images using a uniform-intensity filter kernel, that is, the spatial equivalent of a boxcar filter. The spatial width of the kernel was identical to the spatial separation of phosphenes. This scheme allowed for the comparison of our results and the widespread approach to visual modeling which uses uniform-intensity kernels [5, 10]. The third scheme ($Q2$) involved pre-filtering with a Gaussian kernel. The standard deviation of the kernel was equal to one-third of the separation of phosphenes. We were interested in this kernel due to the nature of a Gaussian: it is dually compact in the spatial and Fourier domains, and is often used to model components in the early visual system of mammals. The standard deviation of this Gaussian, however, was chosen without quantitative reasoning. We hypothesized that scheme $Q2$ would afford subjects improved performance as compared to $Q1$.

For complete details as to the experiment the reader is referred to the original publication [7]. Here we summarize the details. A computer monitor was viewed from a fixed distance by 20 subjects each trained for 3 h. A hexagonal array of 23 phosphenes was freely viewed. Phosphenes were separated by approximately 1°, and therefore excited retinal loci separated by approximately 300 μm. Phosphenes were size-modulated (see Fig. 17.1), and of fixed intensity (white). The phosphene array was moveable via a joystick; subjects were required to track a moving target (a small, white square on a black background) using the central phosphene of the array. The target initially appeared in the center of the monitor (for fixation), and after a short, random interval it jumped (eliciting a visuomanual saccade in the subject) and then described a randomly generated, S-shaped course into the monitor's periphery (eliciting pursuit) at an average velocity of approximately 2°/s. Trials were counterbalanced so as to assess tracking for each of $Q0$, $Q1$, and $Q2$. Note that the setup, whilst good for examining the issues at hand, measured visuomanual behavior. Ultimately, prosthetic visual fixation, saccade, and smooth pursuit will likely involve head movements (of a head-mounted camera) and a stabilized retinal image. Thus we take several of the statistics of the measured behaviors as a model for what will ultimately involve head motion and stabilized retinal images.

Our data showed that, indeed, image analysis had a significant effect on the tracking performance of subjects. After practice, schemes $Q1$ and $Q2$ made for superior performance in all tasks (fixation, saccade, and pursuit) as compared to $Q0$. Scheme $Q2$ made for improved fixation accuracy of 35.8% (8.3 min of visual arc) as compared to scheme $Q1$ (as measured by mean deviation from the target), and for improved pursuit accuracy of 6.8% (3.3 min of visual arc). Schemes $Q1$ and $Q2$ made for comparable saccade accuracy. These results suggest that image analysis, when functionally integrated with a prosthetic device, can be programmed so as to result in improved visual outcomes in implantees. Furthermore, the results advocate a scheme of Gaussian kernels for fixation- and pursuit-related tasks.

The scanning data from the above-described experiment, that is, the way that subjects moved the phosphene array relative to the moving target, were also of interest. These data made us think about preferred retinal loci, and whether implantees would use some phosphenes in their visual field in preference to others. Hence, we coined the term "artificial preferred retinal locus" (APRL).[1] In the case of scheme $Q0$, which afforded subjects poor tracking accuracy, subjects adopted nystagmus-like scanning behaviors, that is, vigorous and wide-ranging scanning. This was apparently an attempt to effectively increase the spatial sampling rate of the array and in doing so render the phosphene image more informative as to the moving target's location. In this case, the associated APRL may be modeled as a bivariate function, centered on the array's central phosphene. That function is uniform in intensity, covering an area that encompasses most of the 23-phosphene array.

The scanning associated with scheme $Q1$ was similar to that of $Q2$. For those schemes, subjects adopted scanning that had dynamics approximately equal to the target's motion. In terms of an APRL, the bivariate function was centered on the central phosphene of the array, and was normal with a standard deviation equal to approximately 0.475°. Since tracking for $Q2$ was more accurate than tracking for $Q1$, the APRL associated with scheme $Q2$ was relatively narrow. See Fig. 17.2. We believe that scanning and modeling the APRL have important implications for developing a model of the phosphene observer, which we discuss further below.

Subsequently to the above-described experiment, we wondered, "Is scheme $Q2$ somehow optimal?" To address this question, we drew upon techniques in communication theory [6]. Specifically, we used the mutual-information function to

---

[1] For example, see the way in which sufferers of scotoma develop new preferred retinal loci in the vicinity of the fovea [11].

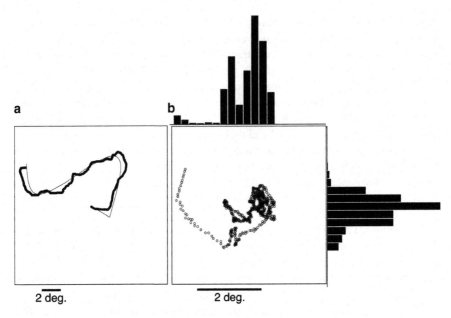

Fig. 17.2 Raw and analyzed scanning data from a single trial [8]. (**a**) Subjects were required to track a small, moving target with a phosphene array. The target's motion is shown by the *thin, solid line*. The target was initially stationary at the center of the monitor. The target then stepped (down and right by approximately 2°) before following an S-shaped curve to the monitor periphery. The *thick, dotted line* shows the subject's tracking. Specifically, the *thick, dotted line* shows the location of the center of the phosphene array. Each dot shows the position of the array center at successive points in time. (**b**) The scanning signal (target position minus tracking position) from the trial in (**a**). The average of this signal across trials and subjects is well modeled by a bivariate normal, indicated by the histograms for this trial

measure the amount of information contained in phosphene images, and how that information differed with different image analysis schemes. We used a numerical setup. We presented targets to the phosphene image in a way that was consistent with scanning behaviors, that is, the APRLs, found in [8]. These mutual-information measurements were then reconciled with the tracking performance. We found that, to an extent, the mutual-information function was a good model of tracking performance.

As a quick aside, the following paragraphs canvas the mutual-information function [1]. The mutual-information function is typically applied to communication channels, like the one shown in Fig. 17.3. There, a time-varying signal, $x(t)$, is input to a noisy channel, Q, and output in modified form as $y(t)$. The mutual-information function can be used to measure the information that $y(t)$ carries about $x(t)$. In other words, after receiving the signal $y(t)$, what is consequently known about the signal $x(t)$? Mutual information is typically used to assess the nature of the channel, Q.

The mutual-information function is written as follows:

$$I(p(x(t)); Q) = H(y(t)) - H(y(t) \mid x(t)).$$

Fig. 17.3 A simple communication channel. The input signal, $x(t)$, to the noisy channel, Q, results in the output signal, $y(t)$

The symbols I and H may be read as "information" and "uncertainty" respectively. Note that information is a function of the probability density that describes the input, $p(x(t))$. Information is also a function of Q, that is, the nature of the channel. Information is equal to the receiver's reduction in uncertainty regarding the channel input after having received the channel output (the term $y(t)|x(t)$ may be read as "the input after having received the output").

To illustrate the use of the mutual-information function, and its interpretation, consider the following example. A discrete random series, X, may take values a, b, c or d. All four values occur independently and with equal probability ($p = 0.25$). The series X generates an output series, Y. Let's suppose that inputs a and b generate output symbol e, and inputs c and d generate output symbol f. That is the nature of the channel, Q, which maps inputs to outputs. Using the mathematical definition of uncertainty, H, we can show that the information, $I(p(X); Q)$, contained in each output symbol is 1 bit. That is, $I(p(X); Q) = 1$ bit/symbol. One bit of information effectively affords the receiver one "yes/no" decision. In other words, 1 bit of information halves his/her uncertainty regarding the input, X. This makes intuitive sense: After receiving the value f, he would consequently know that either c or d were input to the channel. Prior to receiving f, he could do no better than guess that the input was a, b, c or d. Upon receiving f, his uncertainty was halved. For this example, the information carried by an output symbol can be increased if the specificity of the channel, Q, is increased. In other words, it is desirable if the output describes the input with less ambiguity. If inputs a and b generate the output e, whilst c generates f and d generates g, then we can arrive at an average information measure of 1.5 bits/symbol.

The simple communication channel of Fig. 17.3 is analogous to the above-mentioned process that converts high-resolution images of a small, moving target to phosphene images. In that process, the target varies in space, $x(s)$, where s is a vector denoting space. The target is input to the image analysis scheme, Q, which, here, is a spatial filter, not a temporal one. The phosphene image, $y(s)$, is analogous to an output symbol, and is rendered according to the output of the analysis scheme, Q. We developed this analogy in [6] by using the scanning model, that is, the APRL, as $p(x(s))$. See Fig. 17.4.

We found that image analysis scheme $Q2$ imparts more information to the phosphene image than $Q1$ [6]. Specifically, $Q2$ imparts approximately 5 bits/image as compared to $Q1$'s 1 bit/image. This improvement of 4 bits/image is the case when the target is presented to the phosphene image in a spatial pattern corresponding to the scanning behavior found in [8]. Those scanning behaviors were

High-resolution image   Image analysis scheme, *Q2*   Phosphene image

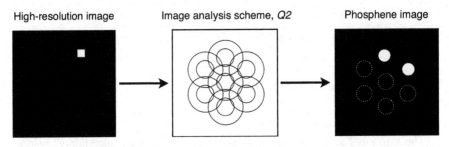

Fig. 17.4 The set-up of the numerical experiment involving a seven-phosphene image [6]. The high-resolution image (*left*) comprises a small, high-contrast target. The image is analyzed by the image analysis scheme (*middle*). The scheme comprises seven identical, Gaussian filter kernels; the *circles* indicate the first and second standard deviations of the kernels. Each kernel operates on the image and produces a response. Each response modulates the size of a phosphene comprising the phosphene image (*right*). In the example shown, the target activates two phosphenes since its location is "seen" by those two phosphenes, but no other phosphenes. For clarity, the phosphenes that are not activated are indicated with *dashed outlines*

modeled by a bivariate normal with standard deviation equal to 0.25 times the phosphene-to-phosphene spacing. This finding provides some theoretical reasoning for subjects' accuracy in [8]: $Q2$ afforded better fixation and pursuit accuracy because it provides 4 bits/symbol more information to the phosphene image observer.

Furthermore, our numerical experiment suggested that the image analysis scheme $Q2$ can be improved [6]. Specifically, the prediction is that, by using Gaussian kernels with standard deviation equal to 0.6 times the separation of phosphenes, performance will be further improved. This new scheme would make for approximately 8 bits of information in the phosphene image, assuming that subjects' scanning behaviors were unchanged. Indeed, the numerical experiment provides a prediction: that an image analysis scheme comprising Gaussian kernels with standard deviation 0.6 times the separation of phosphenes should afford superior tracking as compared to $Q2$. Experimental work that examines this prediction, and similar predictions, will be the subject of future work.

How are these measures of information to be interpreted? As discussed above, 1 bit of information affords the phosphene image observer a single "yes/no" decision. For localizing a target, a single bit of information effectively allows the observer to divide the visual field into two halves and ask, "Which half of the visual field contains the target?" Therefore, when considering scheme $Q1$, affording the phosphene image observer 1 bit of information per image is akin to informing the observer whether the target lies within the left half, or the right half, of the area covered by the phosphene array or not. This being the case, we would expect the observer to deviate, on average, from the target by approximately one-quarter of the diameter of the array. In contrast, scheme $Q2$, affording the phosphene image observer 5 bits of information per image is akin to informing the observer which 1/32 of the area covered by the phosphene array contains the target. In this latter case, we would expect the observer to deviate from the target by less than that amount; the mean deviation would be reduced by approximately a factor of 4 (rather than 16, since the field is two-dimensional).

17.5 Discussion

We have reviewed two studies of ours that address image analysis, its use in microelectronic retinal prostheses, and the perception of low-resolution images. These studies form the beginning of an approach that integrates theory and experiment and aims to better constrain the design of a prosthesis. Effectively, the approach seeks to answer the question, "How should high-resolution images be analyzed before rendering phosphene images?"

In our psychophysical experiments, subjects fixated and pursued a small, moving target that was rendered on an array of phosphenes [8]. There, we showed that image analysis, which converts the high-resolution image of the moving target to the phosphene image, can indeed be used to improve subjects' performance. During trials, subjects scanned the phosphene array over the target, using some phosphenes in preference to others. We modeled this scanning using a bivariate function, and we termed this model the "artificial preferred retinal locus" (APRL). The experiments in [6] were numerical. There, we used the APRL in conjunction with various image analysis schemes and measured the information contained in the phosphene image. We found that the scheme affording subjects the best tracking performance imparted the most information to the phosphene image. Further, we found an optimal scheme of Gaussian kernels for image analysis which we predict would afford further improvements in performance.

Our approach contributes to the existing literature on image analysis, its use in microelectronic retinal prosthesis, and the perception of low-resolution images. We have established an exchange between information theory and visual modeling, that is, the simulation of prosthetic vision using normal observers. This exchange allows for the design and implementation of image analysis schemes on the basis of quantitative reasoning, and for those schemes to be verified via psychophysical methods. For example, our numerical experiment predicts that the optimal Gaussian scheme for fixation and pursuit involves kernels with standard deviation equal to 0.6 times the phosphene-to-phosphene spacing [6]. This prediction may be tested in a visual modeling experiment, prior to a test in actual implantees who are capable of performing simple visual tasks. Alternative approaches to image analysis often simply cite "biological inspiration." For example, an edge-detection scheme is often thought to be justified because edges are known to be of particular salience to the visual system. These approaches may have merit, but the design of the image analysis scheme seems arbitrary, and usually is not tested using visual modeling.

It is important to consider the computational cost of an image analysis scheme. For our purposes, the operation of a kernel on an image involves A real multiplications and $A - 1$ real additions, where A is the area of the kernel in pixels. Therefore, the relative cost of image analysis scheme $Q2$ is proportional to the area of the $Q2$ kernel divided by that of the $Q1$ kernel. In many image processing applications, Gaussian kernels are restricted to a circular support with a radius of three standard deviations. Our circular averaging kernels ($Q1$) had a diameter equal to the separation of phosphenes. Therefore, the ratio of areas of these kernels is 3.24. That is, in our tracking study [8], the computational cost of using $Q2$ was 3.24 times that of $Q1$.

So far, our approach involves models of scanning that vary in space, but not time. However, scanning is likely to be better described by models that are spatiotemporal. A spatiotemporal model of scanning would describe not only which phosphenes were used in preference to others, but how the outputs of many phosphenes were used in combination over time. In other words, a spatiotemporal scanning model would describe how subjects tended to sweep the phosphene array across the high-resolution target. Our psychophysical data suggest that the temporal nature of scanning is important (see also [3]). For example, the image analysis scheme $Q0$ compelled subjects to use nystagmus-like scanning, rapidly moving the array back and forth across the underlying target. Rather than using the information contained in the phosphene image at a single instant, subjects integrated the phosphene array activity over short periods, and used that integrated information to guide behavior. Developing scanning models, that is, APRLs, to include second- and higher-order statistics is the subject of ongoing work.

Our approach concerns visual fixation and pursuit. In the future, we aim to extend the approach to include other tasks, such as reading. To do so, our psychophysical experiment [8] could be modified to involve the identification of commonly used words, as opposed to the tracking of a small, moving target. In this new experiment, subjects would employ scanning behaviors that were specific to reading. Then, images of these commonly used words could be used as stimuli in the numerical set-up of [6], and an image analysis scheme could be tailored to these images. Overall, it is likely that different image analysis schemes would apply to different visual tasks. For example, a Gaussian scheme, like $Q2$, may be suited to tracking a small, moving target, but some other scheme involving some other class of kernels may be better suited to reading, for example, oriented Gabor functions.

17.6 Conclusion

We have discussed our approach to image analysis, microelectronic retinal prostheses, and the perception of low-resolution images. We believe that this approach can be used to help constrain the design of an implant. The approach is analogous to the acoustic modeling of cochlear implants which involves normally hearing listeners. That approach has made important contributions to the improvement of clinical outcomes in cochlear implantees since 1990. We hope that our visual modeling approach, and developments thereof, will ultimately contribute to improved clinical outcomes in retinal implantees.

Acknowledgments We thank Shaun Cloherty for comments on an early draft of the manuscript.

References

1. Blahut RE (1987), *Principles and practice of information theory.* Addison-Wesley: Norwood, MA.
2. Cha K (1992), *Functional capabilities with a pixelized vision system: application to visual prosthesis.* PhD dissertation, University of Utah.
3. Chen SC, Hallum LE, Suaning GJ, Lovell NH (2007), *A quantitative analysis of head movement behaviour during visual acuity assessment under prosthetic vision simulation.* J Neural Eng 4: p. S108–S123.
4. Dowling J (2005), *Artificial human vision.* Expert Rev Med Devices 2: p. 73–85.
5. Fornos AP, Sommerhalder J, Rappaz B, *et al.* (2005), *Simulation of artificial vision, III: do the spatial or temporal characteristics of stimulus pixelization really matter?* Invest Ophthalmol Vis Sci 46: p. 3906–3912.
6. Hallum LE, Cloherty SL, Lovell NH (2008), *Image analysis for microelectronic retinal prosthesis.* IEEE Trans Biomed Eng 55: p. 344–346.
7. Hallum LE, Dagnelie G, Suaning GJ, Lovell NH (2007), *Simulating auditory and visual sensorineural prostheses: a comparative review.* J Neural Eng 4: p. S58–S71.
8. Hallum LE, Suaning GJ, Taubman DS, Lovell NH (2005), *Simulated prosthetic visual fixation, saccade, and smooth pursuit.* Vision Res 45: p. 775–788.
9. Rubinstein JT, Miller CA (1999), *How do cochlear prostheses work?* Curr Opin Neurobiol 9: p. 399–404.
10. Thompson, Jr., RW, Barnett GD, Humayun MS, Dagnelie G (2003), *Facial recognition using simulated prosthetic pixelized vision.* Invest Ophthalmol Vis Sci 44: p. 5035–5042.
11. Timberlake GT, Mainster MA, Peli E, *et al.* (1986), *Reading with a macular scotoma. I. Retinal location of scotoma and fixation area.* Invest Ophthalmol Vis Sci 27: p. 1137–1147.

Chapter 18
Simulations of Cortical Prosthetic Vision

Nishant R. Srivastava

Abstract Cortical stimulation for restoring vision presents researchers with many challenges and questions. The extent of the human visual cortex varies up to 50% from one individual to another, cortical folding and sulci limit the area of implantation, and surgical difficulties make it difficult to implant electrodes to produce phosphenes in the whole visual space. Researchers are faced with question such as: which electrodes to use – surface electrodes that are easy to implant or intracortical fine-metal electrodes that have lower current requirements and have five times better resolution? How many phosphenes will be enough to give limited, but useful vision? How will cortical physiology affect phosphene maps? Will percepts be distinct dots or complex in nature? What will be the long term response to stimulation? Will the brain adapt to seeing through dotted images? Some of these questions can be answered by conducting human psychophysical tests.

Abbreviations

fMRI	functional Magnetic resonance imaging
LGN	Lateral geniculate nucleus
V1	Striate cortex or primary visual cortex
V2	Prestriate cortex or secondary visual cortex
V3	Third visual complex

18.1 Introduction

In the 1990s, Cha et al. simulated arrays varying from 100 electrodes (10×10 arrays) to 1,024 electrodes (32×32 arrays), represented by small dots in a video display mounted on ski goggles to test the requirements of a cortical prosthesis

N.R. Srivastava (✉)
Department of Biomedical Engineering, Pritzker Institute of Biomedical Science and
Engineering, Illinois Institute of Technology, 3255 S. Dearborn, WH 314, Chicago, IL 60616, USA
e-mail: srivnis@gmail.com

G. Dagnelie (ed.), *Visual Prosthetics: Physiology, Bioengineering, Rehabilitation*,
DOI 10.1007/978-1-4419-0754-7_18, © Springer Science+Business Media, LLC 2011

device [4–6]. The images were captured by a head mounted video camera covered by a perforated mask. They concluded that, with 625 dots (25×25 array) in a visual field of 1.7°, a visual acuity of 20/30 and a reading speed of 100 words of paragraph text per minute could be achieved. They found that 625 or more dots, with a field view of about 30°, allowed normal walking speed through a maze with obstacles.

For these tests, they used identically-sized simulated phosphenes on regularly-spaced grids. Cortical stimulation experiments, however, have shown that, for a cortical prosthesis, a regular grid structure is not a true representation of the phosphene map. The maps will vary depending upon the area in the visual cortex used for implanting the electrodes and the type of electrodes used. For conducting proper simulation studies, simulated phosphene maps have to correspond to the map generated as per the targeted electrode location and the type of electrode used.

Some groups might target the medial surface for implantation, and a few groups might target the lateral surface or a combination of electrodes on the lateral and the medial surface. Every research group that is targeting the cortex for electrode implantation will have to generate an estimated percept map depending on their choices of electrodes and array location, and use this map to guide expectations for this device's performance. To generate this map, the representation of visual space on the cortex, cortical structure, and the corresponding biological responses have to be understood.

18.2   Representation of Visual Space on the Visual Cortex

One of the first published visual maps by Holmes shows representation of different visual fields on calcarine cortex, with a linear relationship of visual space to the cortex [12]. These maps were later modified by Horton and Hoyt [13]. The modified map shows the horizontal meridian running at the base of the calcarine fissure with iso-eccentricity contours from 2.5° to 40°. This map shows that the space follows a logarithmic representation till 40° of eccentricity, with the foveal area of visual field represented on a larger area of cortex. Recent experiments have supported these modified maps [17, 18]. This logarithmic representation of visual space on the visual cortex is known as cortical magnification.

In an fMRI study by DeYoe and colleagues, a consistent retinotopic organization was observed on responsive visual cortex both medially and ventrally [7]. The foveal representation was located posteriorly, near the pole, and greater eccentricities were represented anteriorly on the surface. Responses observed to visually expanding checkered rings extended from the collateral sulcus on the ventral surface, crossed the calcarine fissure and passed dorsally out onto the exposed lateral surface. The responses did demonstrate cortical magnification. The data also showed that as the eccentricity increased, there was an anterior progression of activation along with alteration of visual field meridian at transfer from one visual area to other. In another fMRI study by Levy and colleagues, it was observed that the visual cortex has a hierarchical organization that begins with the precise visual field maps in V1, V2 and V3, which degrades on the lateral and ventral regions [16]. The lateral and ventral regions contain coarse eccentricity maps with crude representations of the polar angle. It has been observed that visual space is represented in many visual clusters. Wandell and colleagues reported observing nine human visual field map representations [22]. These observed visual field maps preserved spatial structure.

Any electrode implantation will have to consider the logarithmic nature of these representations. Electrodes implanted near the occipital pole will cover a very small visual area corresponding to a few degrees of eccentricity, and phosphenes corresponding to electrodes on the lateral surface may be perceived more eccentrically, but will lose angular specificity. If the electrodes lie in an area with multiple cluster maps, then the position of the phosphenes in visual space will be almost random. Any psychophysical study designed to estimate the performance of a cortical prosthesis device will have to consider these factors.

18.3 Cortical Stimulation Studies

Brindley and Lewin placed eighty 0.64 mm^2-platinum electrodes between the medial surface of the occipital pole of the right cerebral hemisphere and the falx cerebri of a 52-year-old blind female patient [1–3]. They compared the observed phosphene maps with the Holmes map and found correlations between the two. This experiment was performed in 1968, when the revised map by Horton and Hoyt was not available. If the map published by Brindley and Lewin is compared with the map published by Horton and Hoyt, it is observed that the phosphene map does show cortical magnification. The phosphenes at the periphery of the Brindley and Lewin phosphene map were larger in size, which might be result of cortical magnification. A few irregularities were observed, such as an electrode placed close to a certain group of electrodes produced a phosphene away from the phosphenes corresponding to the group.

Dobelle and Mladejovsky published phosphene maps generated from surface electrodes placed on the right medial surface of a patient [8, 9]. These maps show discrepancies from the expected responses when compared to a logarithmic visual space map, both in terms of eccentricity of phosphenes observed, and polar angles expected from published visual maps.

Dobelle and colleagues published another set of results with 64 platinum disk surface electrodes implanted 3 mm apart on the medial surface of right occipital lobe [10]. The phosphenes followed the expected cortical magnification on visual space. A set of electrodes placed in a line close to calcarine fissure was expected to define the horizontal meridian; instead, it was almost perpendicular to it, close to the vertical meridian. It was hypothesized that the electrodes crossed into the V2 area hence showed the mirror image of the expected response from V1. This shows that the phosphenes might be produced by stimulating higher areas of extrastriate cortex. On the lateral surface, the boundaries of V1, V2 and V3 are not clearly defined.

Even with a very small area of V1 exposed on the lateral surface, though, studies have still shown the generation of phosphenes [14, 15, 19].

Schmidt and colleagues stimulated the lateral surface of visual cortex using intracortical electrodes [19]. The electrodes were placed approximately up to 22 mm away from the occipital pole on the lateral surface. Most of the phosphenes mapped were within 30° eccentricity, with few phosphenes up to 40° eccentricity. Few of the phosphenes were observed at the eccentricities and polar angles expected when compared to the visual maps discussed earlier, and a few phosphenes did not exhibit either the expected eccentricity or the polar angle.

Lee and colleagues stimulated the occipital cortex and the adjacent cortices using surface electrodes in 23 epilepsy patients [15]. The experiment shows that as electrodes are implanted away from the occipital pole, more anteriorly towards the frontal cortex, the response to stimulation varies from simple form phosphenes, to intermediate responses like triangles and diamond shapes, to complex responses like observations of color and evoking movement percepts. The initial 1–2 cm (approximate) from the occipital pole show simple responses, and 2–3 cm (approximate) show intermediate responses. If electrodes are placed on the lateral surface, they should be limited to a distance of 3 cm from the occipital pole.

Kaido and colleagues investigated retinotopic maps on the lateral surface of the occipital cortex in humans [14]. The researchers observed phosphenes of up to 80° eccentricity, stimulating up to 40 mm anterior to the occipital pole, on X-ray scale, which shows that the whole lateral surface of occipital cortex might generate phosphenes. Polar angles were preserved in a coarse manner. If electrodes are implanted too far anterior from the occipital pole, however, complex forms might be observed, as in stimulation studies by Lee and colleagues. [15].

18.4   Variability in Occipital Cortex

Stensaas and colleagues studied primary visual cortex of 52 hemispheres and found the average total area to be 2,134 mm^2 [21]. The average striate cortex exposed on all four surfaces was 689 mm^2, about 33% of striate cortex, and the other 67% of striate cortex (average 1,445 mm^2) was buried in fissures [21]. For inter-electrode distances of 3 mm, we might be able to place only 60–80 electrodes on V1 [1–3, 8–10, 21]. It was found that, on average, only 3% (55 mm^2) of primary human visual cortex extends to the occipital surface of the brain. For the average exposed V1 area of 55 mm^2, if intracortical electrodes are used with inter-electrode distances of 0.5 mm, it might be possible to place 200 electrodes [19]. The variability in the visual cortex within individuals will affect cortical implantation. If surgical methods are developed to implant electrodes over the complete exposed striate cortex, then the implanted electrode numbers might vary by 30%.

Dougherty and colleagues used data from fMRI and prepared 2-D flattened representations of the cortical manifold [11]. Using these flattened maps they calculated V1, V2 and V3 sizes. The left hemisphere of V1 was about 200 mm² larger than the right hemisphere of V1. Mean areas of V2 and V3 in left and right hemispheres were not significantly different. Dorsal V1, V2 and V3 regions were found to be larger than ventral V1, V2 and V3 regions. This will allow creating more phosphenes by implanting electrodes on the dorsal surface than on the ventral surface. They also found that the surface area of V2 representing eccentricities of 2°–12° was roughly 75% of that of V1, and that of V3 was only 56% the size of V1's corresponding area. They hypothesized that V2 either receives only a portion of the V1 output, or it has a more efficient representation of V1 output. They found that cortical magnification does not differ significantly between left and right hemispheres for V1, V2 and V3.

These results guide us to estimate the number of the electrodes which can be implanted on the targeted area and construct a phosphene map for it. These studies show that cortical surfaces can vary about 50% from one individual to other. Hence, the number of electrodes that can be implanted can vary up to 50%. This has a direct impact on the number of simulated phosphenes that should be used for psychophysical studies. Such psychophysical studies will have to consider this variation when generating dotted images, and for every placement, will have to consider a dropout rate of phosphene from 25 to 50%.

18.5   Phosphene Map Estimation

If electrodes are implanted on the medial surface of striate cortex, a phosphene map as shown in Fig. 18.1 can be expected [1–3, 8–10]. The phosphene size increases at higher eccentricities, and the distance between phosphenes increases,

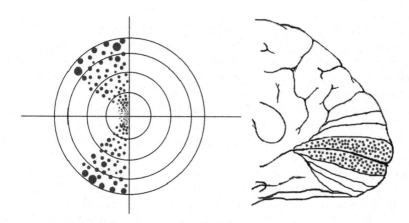

Fig. 18.1 V1 area on the medial surface shown by dots as the electrodes with the corresponding expected phosphenes in the visual space. Lateral blank visual space corresponds to the area buried in the calcarine fissure

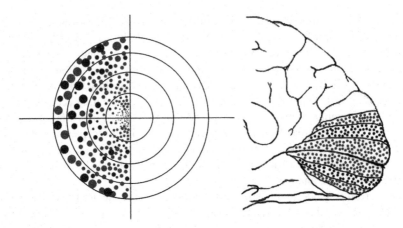

Fig. 18.2 The visual map expected with placement of electrodes on the medial wall of the V1, V2 and V3 areas. Note the additional phosphenes, mostly in the lateral visual field, as compared with Fig. 18.1; V2 and V3 phosphenes are largely intermixed

reflecting the distortion due to cortical magnification [1–3, 8–10]. The gap shown on the lateral visual field corresponds to the two-thirds of area V1 that lies within the calcarine fissure. This area is inaccessible with existing surgical techniques for both surface and the intracortical electrodes. It might be possible to generate phosphenes on the lateral visual field by stimulating V2 and V3, which have representations similar to that of V1. If electrodes are implanted on V2 and V3, along with V1, the mirror image correspondence of V1 to V2 and V2 to V3 might help to generate phosphenes on the lateral surface as shown in Fig. 18.2.

Schmidt and Kaido have shown the generation of phosphenes over a wide region of visual space, while stimulating the lateral surface, and concluded that this area can be used to create phosphenes for a cortical visual prosthesis [14, 19]. From the study of Lee and colleagues we observe that the limitation of area for electrode placement area on the lateral surface area is about 3 mm from the occipital pole to get a simple percept [15]. The fMRI study by DeYoe shows that for a 3 cm radius from the occipital pole, phosphenes might be observed throughout the central 25° of the visual field [7]. The expected phosphene map is represented in Fig. 18.3. This figure is derived from an fMRI study, but as observed with intracortical electrode experiments, few phosphenes might be observed in 40°–45° [19]. If it is assumed that 50% of phosphenes from this hypothesized map are between 25° and 45° eccentricity, this will give us a map in which 50% dots of phosphenes are dropped from the initial 25° of the eccentricity map in Fig. 18.3, and redistributed across 25°–45° eccentricity, giving us a map as shown in Fig. 18.4.

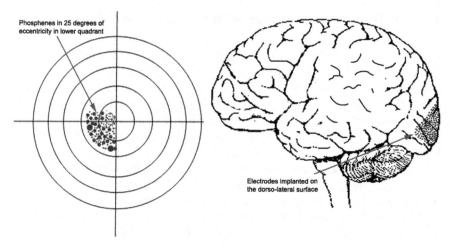

Fig. 18.3 Visual map showing phosphenes within 25° eccentricity expected with electrodes on the lateral surface area in a radius of 3 cm from the occipital pole. This map corresponds to fMRI studies

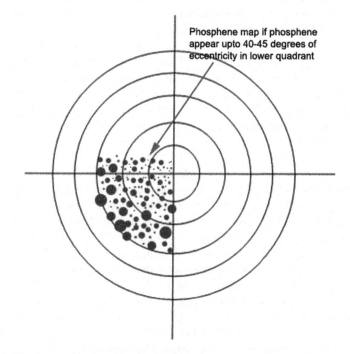

Fig. 18.4 Visual map with phosphenes in 40°–45° eccentricity. This map corresponds to observations by Schmidt in 1996, in which few phosphenes were observed at about 40°–45° [19]. If we get this scenario where the phosphenes are generated up to 40°–45° then we will have a larger field of view than we expect from fMRI studies and lower density of phosphenes

18.6 Psychophysical Studies with the Estimated Maps

In the field of visual prosthetics, very limited psychophysical studies have been conducted to judge the performance of a cortical prosthesis. After the studies by Cha and colleagues, many psychophysical studies were conducted targeting other areas of vision restoration or for judging different image processing schemes, but detailed psychophysical studies addressing cortical structure and biological limitations have been missing [4–6]. In 2007 psychophysical tests were conducted at John Hopkins University by Srivastava and colleagues for testing the expected response of a cortical prosthesis, being developed at the Illinois Institute of Technology, Chicago, using 650 intracortical electrodes targeting the dorso-lateral surface [20]. Tests were conducted on five volunteer subjects, three men and two women, using a dotted phosphene map similar to Fig. 18.3. Individual variations in the cortex and failure of some electrodes to elicit a phosphene were also incorporated in these studies.

Area and layout of the visual cortex may vary by up to 50% between individuals [7], hence the anatomical area for electrode implantation will similarly vary, hence in few patients the area available for electrodes might be less than the average. In addition, some electrodes, or stimulation sites, might fail during the surgical procedure itself, or during long-term implantation. To study the effect of fewer phosphenes than might be expected from the electrode count, dropout effects were included in our studies. Performance was judged under 0% dropout, 25% dropout and 50% dropout conditions. Three different tasks were selected to observe if, with these limited-field phosphene maps, persons can adapt to perform different tasks of eye-hand coordination and mobility. These tasks were judged to be representative of the basic tasks required to be done in daily day-to-day life. These experiments were conducted using eye-tracking to simulate the normal movement of phosphenes in visual space.

The purpose of the first two psychophysical experiments were to determine the subject's ability to perform detection by counting the white fields on a checkerboard and eye hand coordination by placing black checkers on the white fields of the checkerboard. The accuracy of counting and placing along with the time taken to complete the task were used to judge performance. Subjects were able to inspect the board by scanning the head-mounted camera in parallel with the board, or change their viewing angle by tilting their heads. The decision by the subjects for scanning the boards was intuitive for each individual. By giving the subject the ability to scan the board, spatial and temporal integration was achieved. The experimental setup is shown in Fig. 18.5. The leftmost image shows the phosphene map for the dorso-lateral surface with 650 electrodes implanted on a region of 3 cm radius area, with occipital pole as the center. The center image shows the checker board used, and the rightmost image shows the dotted image seen by the test subjects in a headset with eye tracking.

Counting time for counting all the white fields and the number of fields reported by the subject were recorded for each trial. The significant factors effecting the study were individual differences in subjects ($F = 25.29$, $p < 0.0001$), increase

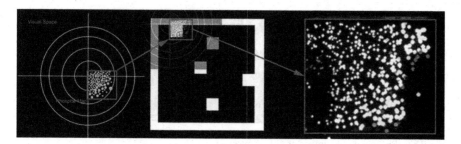

Fig. 18.5 Checkered board as observed through a phosphene-like map

in practice ($F=23.58$, $p<0.001$), dropout ($F=8.41$, $p=0.0040$) and increase in complexity ($F=50.75$, $p<0.001$). The dropout had a weak significance compared to the other factors showing that the effects of dropout could be overcome by practice. The same analysis was done for each individual subject, and practice was the significant factor in all subjects except the low vision subject.

Similar to counting time, placing time was also recorded. The significant factors effecting placing timing were variation due to individuals ($F=14.02$, $p<0.0001$), increase in practice ($F=30.12$, $p<0.001$), the dropout ($F=9.03$, $p=0.0040$) and the number of white fields on the board ($F=109.42$, $p<0.001$). Dropout had a weak significance compared to the other factors. For individual analysis, all the subjects show a significant reduction in task time with increasing practice. Three subjects showed an increase in the time with an increase in the dropout level, but with a modest significance level. This shows that the practice effect dominates over the dropout effect, and a decline in performance due to dropout can be overcome by practice.

In the third experiment, subjects' ability to recognize a pathway and orient oneself to follow it, without memorizing or recall, was judged by observing the performance of the subjects' maneuvering in virtual mazes. While for the counting and placing tasks, the subjects were observing the checkerboards with a camera, and could therefore adjust the level of detail by changing their viewing distance and angle, the full scene of the virtual mobility experiment, was imaged onto the phosphene map, and subjects were trained to use a game controller to change their vantage point, as if moving through virtual space. Thus, depending on the task, the acceptance angle of the camera or other image source may have to be adjusted by the subject to get an optimum view. This can also be achieved by selecting the part of the captured image to be presented in the phosphene map. This is shown in Fig. 18.6 where the virtual maze is shown on the left, which was made to fit on the central phosphene map by adjusting the image scale in the graphics processor. The resulting dotted image is shown in the right image.

The effect of learning on the performance was observed with increased practice of the experimental task. The time to complete each maze and the number of way-finding errors were recorded. The factors affecting the experiment were individual differences ($F=34.21$, $p<0001$), increase in practice ($F=90.04$, $p<0001$), and the dropout ($F=6.06$, $p=0.0189$). Dropout had a low significance, similar to the counting

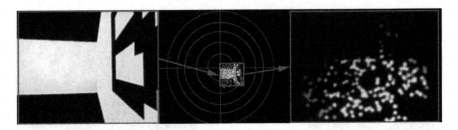

Fig. 18.6 Phosphene-like image of virtual maze

and placing experiments, and a separate analysis checking the effects within individuals yielded practice as the only significant effect. Other factors and interactions had no statistical significance.

The results from these experiments demonstrated that even with a limited number of phosphenes in one visual hemifield, with limited eccentricity in visual space, it is possible to attain a level of proficiency with which prosthesis wearers can perform simple tasks, albeit with practice.

A significant finding from these studies was that the degradation in the image, due to a lower number of phosphenes (higher dropout), could be largely overcome by an increase in practice and use by the tested subject. This result can be significant for individuals with a small area of visual cortex available for electrode implantation, who may still be candidates for prosthesis implantation. This result also is reassuring for researchers who worry about the failure of electrodes, or neuronal stimulation sites, during surgical implantation and over the lifetime of a prosthesis, and what effect this may have on performance. These results also may be of value to visual prosthesis researchers using other substrates, such as the retina, optic nerve, or lateral geniculate nucleus (LGN). As an indication for what they can expect from such devices. They may also provide a guide to vision scientists conducting simulation studies, in regard to how biological response and surgical difficulties might affect their results.

As human studies are conducted and more real maps are obtained, similar experiments might be repeated using more realistic maps, and potentially providing better predictions of performance. The experiments conducted by Cha and colleagues gave 625 electrodes for getting a limited sense of vision. In the experiments mentioned in this chapter by Srivastava and colleagues, it was observed that even with 325 electrodes, and an incorporated estimate of cortical distortion in the phosphene maps, the subjects learned and adapted to performing the basic tasks [20].

Another observation of this study was that producing a large number of phosphene in a limited visual field might not be helpful, since many of them might potentially overlap, and the effective result would not be a major improvement in recognition ability. It would be more helpful to create distinct phosphenes on larger visual area, which might require development of improved surgical techniques.

The relatively crude and limited information provided by the cortical visual prosthesis under development by various research groups, if successful, might help blind subjects obtain some assistance for conducting simple tasks in daily life.

References

1. Brindley GS (1970), *Sensations produced by electrical stimulation of the occipital poles of the cerebral hemispheres, and their use in constructing visual prostheses.* Ann R Coll Surg Engl, **47**(2): p. 106–8.
2. Brindley GS (1982), *Effects of electrical stimulation of the visual cortex.* Hum Neurobiol, **1**(4): p. 281–3.
3. Brindley GS, Lewin WS (1968), *The sensations produced by electrical stimulation of the visual cortex.* J Physiol, **196**(2): p. 479–93.
4. Cha K, Horch K, Normann RA (1992), *Simulation of a phosphene-based visual field: visual acuity in a pixelized vision system.* Ann Biomed Eng, **20**(4): p. 439–49.
5. Cha K, Horch KW, Normann RA (1992), *Mobility performance with a pixelized vision system.* Vision Res, **32**(7): p. 1367–72.
6. Cha K, Horch KW, Normann RA, Boman DK (1992), *Reading speed with a pixelized vision system.* J Opt Soc Am A, **9**(5): p. 673–7.
7. DeYoe EA, Carman GJ, Bandettini P, et al. (1996), *Mapping striate and extrastriate visual areas in human cerebral cortex.* Proc Natl Acad Sci USA, **93**(6): p. 2382–6.
8. Dobelle WH, Mladejovsky MG (1974), *Phosphenes produced by electrical stimulation of human occipital cortex, and their application to the development of a prosthesis for the blind.* J Physiol, **243**(2): p. 553–76.
9. Dobelle WH, Mladejovsky MG, Girvin JP (1974), *Artificial vision for the blind: electrical stimulation of visual cortex offers hope for a functional prosthesis.* Science, **183**(123): p. 440–4.
10. Dobelle WH, Turkel J, Henderson DC, Evans JR (1979), *Mapping the representation of the visual field by electrical stimulation of human visual cortex.* Am J Ophthalmol, **88**(4): p. 727–35.
11. Dougherty RF, Koch VM, Brewer AA, et al. (2003), *Visual field representations and locations of visual areas V1/2/3 in human visual cortex.* J Vis, **3**(10): p. 586–98.
12. Holmes G (1918), *Disturbances of vision by cerebral lesions.* Br J Ophthalmol, **2**(7): p. 353–84.
13. Horton JC, Hoyt WF (1991), *The representation of the visual field in human striate cortex. A revision of the classic Holmes map.* Arch Ophthalmol, **109**(6): p. 816–24.
14. Kaido T, Hoshida T, Taoka T, Sakaki T (2004), *Retinotopy with coordinates of lateral occipital cortex in humans.* J Neurosurg, **101**(1): p. 114–8.
15. Lee HW, Hong SB, Seo DW, et al. (2000), *Mapping of functional organization in human visual cortex: electrical cortical stimulation.* Neurology, **54**(4): p. 849–54.
16. Levy I, Hasson U, Avidan G, et al. (2001), *Center-periphery organization of human object areas.* Nat Neurosci, **4**(5): p. 533–9.
17. McFadzean R, Brosnahan D, Hadley D, Mutlukan E (1994), *Representation of the visual field in the occipital striate cortex.* Br J Ophthalmol, **78**(3): p. 185–90.
18. McFadzean RM, Hadley DM, Condon BC (2002), *The representation of the visual field in the occipital striate cortex.* Neuroophthalmology, **27**(1–3): p. 55–78.
19. Schmidt EM, Bak MJ, Hambrecht FT, et al. (1996), *Feasibility of a visual prosthesis for the blind based on intracortical microstimulation of the visual cortex.* Brain, **119 (Pt 2)**: p. 507–22.
20. Srivastava NR, Troyk PR, Dagnelie G (2009), *Detection, eye-hand coordination and virtual mobility performance in simulated vision for a cortical visual prosthesis device.* J Neural Eng, **6**(3): p. 035008.
21. Stensaas SS, Eddington DK, Dobelle WH (1974), *The topography and variability of the primary visual cortex in man.* J Neurosurg, **40**(6): p. 747–55.
22. Wandell BA, Brewer AA, Dougherty RF (2005), *Visual field map clusters in human cortex.* Philos Trans R Soc Lond B Biol Sci, **360**(1456): p. 693–707.

Chapter 19
Phosphene Mapping Techniques for Visual Prostheses

H. Christiaan Stronks and Gislin Dagnelie

Abstract Mapping of the visual world onto the visual system occurs in a highly ordered manner, yet with substantial interindividual variability. Since the retinal map of the scene at the photoreceptor level is fully determined by the optical projection of the eye, it is likely that a proximal map generated by a retinal prosthesis closely adheres to the same geometric projection. Once the nerve signals enter the optic nerve, this orderly map is redistributed, and while maps at more proximal levels still follow general rules, special mapping techniques in individual LGN or cortical prosthesis recipients will be required to allow reconstruction of spatial relationships in the outside world by means of a disorderly array of phosphenes.

This chapter provides an overview of mapping techniques that have been used in a number of laboratories; discuss the strengths and weaknesses of each; and suggest ways in which various techniques can be combined.

Abbreviations

HMD   Head mounted display
MDS   Multidimensional scaling
TMS   Transcranial magnetic stimulation

19.1 Importance of Mapping

Ever since researchers first started eliciting phosphenes in blind patients through electrical stimulation there has been a need to specify the location of the phosphenes in the visual field. Over 30 years ago phosphene mapping was defined by

H.C. Stronks (✉)
Lions Vision Research and Rehabilitation Center, Wilmer Eye Institute, Johns Hopkins University School of Medicine, 550 N. Broadway, 6th floor, Baltimore, MD 21205, USA
e-mail: hstronk1@jhmi.edu

G. Dagnelie (ed.), *Visual Prosthetics: Physiology, Bioengineering, Rehabilitation*,
DOI 10.1007/978-1-4419-0754-7_19, © Springer Science+Business Media, LLC 2011

Everitt and Rushton as "determining the position of each phosphene in the visual field" [13]. Phosphene mapping is an important step in determining the functionality of visual prostheses. Mapping phosphenes allows the characterization of how evoked phosphenes by stimulation of the different electrodes of a visual prosthesis cover the visual field. After the phosphene map is obtained, clinical fitting procedures can be applied to adjust the visual input stage (e.g., the video image) and visual processing strategies to provide the prosthesis wearer with a proper percept representative of the outside world.

In general, there are two different approaches to obtain a phosphene map; absolute and relative phosphene mapping. Absolute maps describe the position of phosphenes in absolute coordinates in the field of view, while relative maps provide information about the spatial relationships between phosphenes, in terms of distance and angle. Both methods have advantages and disadvantages, as discussed below.

Retinal prostheses likely yield predictable phosphene maps, since the representation of the outside world on the retina (i.e., the retinotopical organization) is determined by simple geometry and is constant across subjects [12]. Nevertheless, retinal neural organization has been shown to change during prolonged periods of visual impairment (e.g. [15, 25] and Chap. 3), so phosphene mapping might still prove important in patients with long-term vision loss. Since retinotopy is largely preserved in the optic nerve, implants in the optic nerve are likewise expected to yield predictable phosphene maps, although the accuracy and stability of such a map will depend on the ability to precisely position the electrodes, due to the high density and thin caliber of the nerve fibers.

Phosphene maps obtained in cortical prosthesis users are relatively unpredictable, since multiple maps of the visual field are represented in different cortical areas which may cause widely spaced electrodes to evoke phosphenes in different cortical areas, where they may or may not fall in the same area of the visual field [11]. Moreover, the presence of sulci and gyri in the cortex may lead to unexpectedly large distances between phosphene locations in the visual field. Finally, cortical organization differs from person to person and, more importantly, the functional organization in longer-term visually impaired individuals may be substantially different from normal-sighted people due to the plasticity of the visual cortex [22]. Therefore, phosphene mapping will be especially important in cortical prosthesis recipients.

This chapter deals with various phosphene mapping techniques. Comparable mapping techniques will be discussed together and will be presented in roughly chronological order, starting with the earliest report on mapping techniques from Brindley and Lewin. Wherever possible, comparable studies (e.g., cortical and retinal prostheses, simulation studies etc.) will be discussed together in the text. The chapter concludes with results from our laboratory, a short overview of phosphene mapping methods, and suggestions will be made which methods to use in different situations.

19.2 Early Absolute and Relative Mapping Procedures in Subjects with Cortical Prostheses: Pointing Techniques

Mapping of phosphenes already proved to be highly informative during the pioneering work of Brindley and Lewin. They acknowledged the importance of absolute and relative phosphene mapping in the 1960s. In one of their studies they chronically implanted a subject with no functional vision with an array of 80 subdural extra-cortical electrodes [3]. Absolute maps were obtained by letting the subject point towards the perceived phosphene with the left hand in a hemispherical bowl with a radius of 0.59 m, i.e., approximately at arm's length. The right hand grasped a small knob inside the bowl for tactile reference. Relative maps were obtained by sequentially stimulating two electrodes and asking the subject to describe the spatial relations between the two phosphenes, such as distance and compass angle.

The early experiments of Brindley and Lewin already showed that phosphene maps were not a simple reflection of the electrode array projected into the visual field of the subject. Rather, phosphene maps roughly corresponded to the classical cortical maps constructed by examination of gunshot victims of WWI (e.g. [17]), that showed the nonlinear projection of the visual field onto the cortical surface. Their experiments also showed that these phosphene maps were not very regular. Phosphenes lying in a straight line in the visual field could be evoked by electrodes lying in a triangular configuration on the cortical surface. It was also shown that activation of distant electrodes could result in phosphenes overlapping in the visual field. In addition, stimuli delivered well above threshold by a particular electrode could result in additional phosphenes being elicited in distant locations. These early experiments strongly indicated the need for proper phosphene mapping, since phosphene configuration may differ substantially from electrode organization.

Dobelle and Mladejovsky [10] performed similar experiments in acute sessions on normally sighted patients undergoing occipital lobe surgery. One patient received a sub-chronic implant for a period of 2½ days and most of the data were obtained from this subject. Phosphene mapping was performed by letting the subject point to where the phosphene was perceived. Maps were then created by drawing the phosphenes in a visual field map. Relative maps were obtained by asking the subject to describe how different phosphenes interrelated. Though the phosphene mapping techniques are not discussed in detail, the authors provide detailed analyses of the phosphene maps they obtained and include a critical discussion on the mapping techniques employed.

In conjunction with the classic cortical maps, they found that for a given inter-electrode distance, phosphene spacing varied depending on the area of the cortex being stimulated. Phosphenes close to the center of the field of vision (e.g., elicited by stimulation of electrodes near the occipital poles) were usually closer together than those in the periphery. Moreover, when the electrode array spanned a cortical fissure (*sulcus*), a gap between phosphenes in the visual field was observed, which could be explained by the fact that the electrodes did not penetrate deep enough to stimulate the cortical tissue within the sulcus [10].

The authors also recognized some of the most important advantages and disadvantages of the phosphene mapping techniques employed both by them and by Brindley and Lewin. While absolute mapping provides the scale of the map, relative mapping provides the detailed interrelationship of phosphenes [3], making both techniques complementary. Relative mapping is more time consuming and may therefore not be suitable for acute testing in the operating room. Nevertheless, relative mapping using (near-) simultaneous phosphene presentation may be preferable over absolute maps obtained by sequential activation of different electrodes, since phosphenes move with eye-position, making absolute localization in the visual field difficult. Another general disadvantage of (absolute) mapping by pointing is that phosphenes elicited by different electrodes may be too close together to be resolved properly due to inaccuracies in pointing, especially in blind subjects who have no visual feedback [10, 26].

The pointing method described in the 60s by Brindley and Lewin was applied in much the same way by Gothe et al. [16], who investigated cortically evoked phosphenes by means of transcranial magnetic stimulation (TMS). TMS is a method to affect cortical activity, and sometimes evoke phosphenes, by electromagnetic stimulation through the intact scalp and skull. Gothe et al. instructed sighted subjects and individuals with residual vision to use a laser pointer in a room with dimmed lights to indicate the position of phosphenes onto a semicircular screen that was placed 120 cm before the subject and extending 33° on each side. Subjects without residual vision were instructed to point in the direction of the percept.

They found that the number of cortical locations from which phosphenes could be evoked increased with the amount of residual vision. In normal-sighted subjects, stimulation of all areas of the occipital lobe yielded phosphenes, while in totally blind subjects only 20% proved responsive. These results were somewhat surprising, since Brindley and Lewin reported that activation of almost all of their 80 electrodes yielded perceivable phosphenes [3].

19.3 The Computer Era: Refining the Pointing Method of Phosphene Mapping

Following the pioneering studies of Brindley and Dobelle, relative phosphene mapping was improved during the 1970s when computers became available. Handmade maps could be digitized using the relative coordinates of each phosphene [9]. Everitt and Rushton [13] proposed a method to combine all the available data in a patient by digitizing and pooling relative maps and using an iterative "best fitting" procedure to obtain a reliable relative phosphene map. They were actually able to use these maps to present figures and letters by direct electrode stimulation which yielded patterns that were recognizable by the subject.

Dobelle and associates worked out a fully computerized protocol to overcome the problems of eye drift and inaccuracy of pointing [26]. Subjects were presented with pairs of simultaneously evoked phosphenes, minimizing the effect of eye-drift. One electrode was stimulated for 1 s, the other 3 s and the subject was asked to report the spatial relationship between the "short flash of light" and the "long flash of light". The subject entered the relative position of two phosphenes into a computer through two key presses on a touch-tone telephone pad, using "5" as a reference key. Thus, "1" encoded "above and left", and "2" "directly above" etc. By mapping different phosphene combinations according to their relative X and Y coordinates, the authors were able to construct relative phosphene maps.

This procedure was applied on a male patient blinded by a gunshot wound who was subsequently chronically implanted with a subdural 64-electrode array on the occipital cortex (knows as the striate cortex, V1, or area 17). The computerized procedure not only resulted in a relative phosphene map, it also enabled the authors to construct an accurate and detailed layout of the cortical surface under the electrode array [11]. With this map they could accurately predict where sulci were situated under the electrode array and even how deep a given sulcus was by determining the magnitude of the shift in phosphene location of adjacent electrodes. Boundaries of the striate and peristriate cortical areas could be identified by a reversal in phosphene direction when adjacent electrodes were stimulated (these areas contain reversed maps of the visual field). The calcarine fissure along the medial wall of the occipital lobe – known to separate cortical areas representing areas above and below the horizontal meridian – could accurately be identified by a sudden shift in adjacent electrodes evoking phosphenes in the upper and lower visual field. Furthermore, electrodes that could elicit phosphenes in different visual field locations, depending on the current level, were found to lie alongside sulci: the intervening portion of the visual field projected to the portion of cortex in the sulcus, and could therefore not be activated with the surface electrodes used at the time.

The methods for phosphene mapping essentially did not change much during the decades following Dobelle's work. Bak and colleagues [1] mapped size and absolute position of phosphenes by instructing intracortically stimulated (sighted) subjects during acute recordings to fix their gaze and point with their finger on a white screen with calibrated markers where the phosphene was perceived. Later, for absolute phosphene mapping in a chronically implanted subject, they used a dart board with 12 sectors and five annular zones for tactile feedback. The subject was asked to place a dart at the location of the phosphene while keeping her gaze fixed. For relative mapping they used the computerized method from Dobelle and associates and improved the resolution by deploying a joystick that could detect 16, instead of eight, relative angles [28]. The latter method was supplemented with verbal information to incorporate spacing between phosphenes.

The Brussels group of Veraart published several papers in which they mapped phosphenes by letting a subject point to the evoked phosphenes. Phosphenes were evoked with a four-electrode optic-nerve prosthesis that was chronically implanted in a subject suffering from retinitis pigmentosa without useful light perception [8, 29, 30]. The four electrodes were positioned around the right optic nerve.

Absolute phosphene mapping was performed using a method very similar to that described by Brindley and Lewin 30 years earlier. The chronically implanted subject was instructed to point to the location of the perceived phosphene in a hemisphere with a radius of 0.45 m. While the task was performed, the volunteer's head was steadied in front of the hemisphere using a frame that provided support for the forehead, chin and parietal skull. The subject's index finger was placed on the fixation point (a disc in the center of the hemispheric surface) as a proprioceptive reference. The subject was instructed to fix her gaze at the (unseen) fixation point and eye movements were recorded with a camera. Furthermore, electro-oculograms were assessed to monitor eye movements. To help the subject identify phosphenes, electric stimuli were preceded and followed by a tone. The fingers of the right hand were used to indicate the perceived phosphene as a shape on the hemisphere. Various phosphene characteristics were recorded such as position, dimensions and motion.

Interestingly, dependent on the exact stimulation parameters, 64 different phosphenes varying in shape and size could be elicited. More importantly, the phosphenes covered a visual angle of about $+35°$ to $-50°$ vertically and $-30°$ to $+30°$ horizontally, despite the fact that the implant contained just four electrodes.

Subsequent studies on this subject were performed using phosphene mapping methods very similar to the first study. The subject's gaze was steadied and monitored in the same way and phosphenes were localized similarly using the pointing hemisphere. Mapping was performed by an observer who copied the azimuth and elevation coordinates from the hemisphere with the aid of meridians and parallels traced on the hemisphere. These data were then transferred to a digital database in which phosphenes were described as pixels with $1°$ resolution. Phosphene area was defined as the number of pixels within the phosphene.

Again, like in the previous study, phosphenes covered a large area of the visual field (from $-30°$ to $+30°$ horizontally and $+20°$ to $-50°$ vertically), despite the presence of only a very limited number of electrodes. Exact location depended mainly on the electrode position and current level. Each quadrant of the visual field was mostly accommodated by one electrode and higher current levels evoked phosphenes closer to the fixation point. Position was also influenced by duration, number of pulses and pulse rate of the applied pulse trains. Phosphene size and luminosity did not clearly depend on any parameter, but tended to increase at higher stimulus levels. No relation was found between stimulating condition and phosphene color or "texture" [8, 30].

In a later publication they increased the total number of individually addressable phosphenes in the phosphene map to 109, excluding the additional "ghost" phosphenes that appear at higher current [2]. By mathematically fitting model equations relating phosphene location to the afore-mentioned parameters, stimulus conditions could be calculated and assigned to an electrode to elicit a specific phosphene along the visual field. Fitting a camera and processing strategies that translated the perceived image into a phosphene map actually enabled the subject to identify simple patterns. Even though little is known about the long-term stability of phosphenes in such a crude prosthesis, these findings illustrate the importance of accurate phosphene mapping strategies: Phosphene maps proved critical in translating the subjective percepts into discrete maps that could be used for clinical fitting of the visual prosthesis into a functional device.

19.4   Verbal Mapping

Though mapping of phosphenes by retinal and optic nerve stimulation was reported long after the first cortically evoked phosphenes were characterized, the methods used were not very different. Several papers by Humayun et al. mention mapping of retinally evoked phosphenes [18–20], but they do not provide detailed information about mapping conditions, such as gaze-control of the subject or tactile references. In acute experiments Humayun et al. constructed absolute maps by asking the subject to verbally inform the experimenter of the quadrant (1–4), or clock hour (e.g., "9 o'clock") in which the phosphene was perceived [18]. As expected, the results confirmed that subjective phosphene location corresponded well with the electrically stimulated area on the retina. On the basis of these findings, they extended their observations to relative measures by providing simple patterns through multi-electrode probes on the retinal surface and asking the subject about their percepts in acute experiments [19].

Verbal mapping was also applied in a study using TMS to evoke phosphenes in sighted subjects [27]. Subjects were placed in a darkened room with eyes closed to facilitate phosphene percepts elicited by stimulation of the occipital lobes. Subjects reported if the phosphenes appeared in the upper or lower visual field, and whether the percepts were centrally or peripherally located. It appeared that peripheral phosphenes were encountered more often than central ones. Furthermore, phosphenes were more often observed in the lower-field than in the upper field. The first finding is unexpected, since the central visual field at the occipital pole (i.e. the foveal projection) is more accessible for TMS than the peripheral retinal projections located at more rostral aspects from the calcarine fissure. The authors speculated that TMS in this study may have activated peristriate cortical areas.

Fernandez et al. [14] proposed an alternative method of phosphene imaging by verbal communication for sighted people. This method incorporates several training phases by using a clock-face division of visual space. Each of the 12 sectors is labeled accordingly and divided into annuli to produce an inner, middle, and outer portion, representing displacement between the fovea and visual periphery. In the training phase subjects are provided with a computer screen and learn to specify the projection of a light spot over the clock-face frame. Initially, spots of light are presented onto a full outline of the frame and subjects are asked to indicate in which of the 36 sections the spot appears (hour and eccentricity). The second training phase is performed without sector labeling and subjects receive feedback about their performance. The last phase consists of phosphene localization without the frame and subjects again receive feedback on their performance. Each of the training phases are repeated until the subject achieves a required percentage correct performance level, before proceeding to the next phase. After training, testing consists of identification of the location of phosphenes in an imaginary frame. The authors predict that this method should be faster and should yield a higher spatial resolution and more discrete responses from the subjects than other absolute mapping methods. In addition, any effects of visuo-motor transformations required for drawing are eliminated by this method.

The authors mention that blind people could learn this method (though training with a visible frame would be impossible) with the help of a dartboard divided into 12 sectors and three annuli, much like the method of Mladejovsky et al. [26] discussed above.

19.5 Mapping Studies Using Subject Drawings

After their first studies on phosphene mapping using acute retinal stimulation (Sect. 19.3), Humayun et al. chronically implanted a retinitis pigmentosa patient without functional vision with a 16-electrode retinal implant [20]. They mapped the phosphenes by letting the subject draw the percepts on a drawing board positioned on the subject's lap. Similar to their preceding studies using verbal information discussed above, the constructed phosphene maps indicated that percepts correspond well with the retinal layout; i.e., electrodes temporally located evoked nasally perceived phosphenes (and vice versa), and superiorly located electrodes evoke inferiorly perceived phosphenes (and vice versa). Resolution appeared to be 1.5° of visual angle.

In another study a very similar drawing method was used to map phosphenes of sighted subjects with intractable epilepsy who were chronically implanted with subdural electrodes in the extrastriate visual cortex [21]. Subjects were asked to look at the center of a white board positioned 2 m away. The white board was divided into sections by horizontal and vertical median lines. The subjects were instructed to make drawings on a white paper, regarding the outline and location of the phosphenes. The paper was one tenth the size of the white board and had similar dividing lines. Phosphene shape, color and motion were also recorded. These drawings were then used to extract polar angle and eccentricity of the phosphene for mapping purposes. The results of this study showed that retinotopic maps could also be found on the lateral occipital cortex.

Several TMS studies made use of drawings of phosphenes made by subjects, starting with an early study by Marg and Rudiak using normally sighted people [24]. For optimal phosphene perception, subjects were seated in a darkened room with their eyes closed. Subjects made drawings of the phosphenes and reported on characteristics such as the shape, color, brightness/vividness and position and distance in the visual field relative to the fixation point. Besides detailed morphology of TMS-evoked phosphenes, the authors reported that phosphenes in the peripheral field of vision occur more frequently than central ones, in agreement with the findings of Ray et al. [27].

Subject drawings of perceived phosphenes were also applied in a TMS study with sighted subjects and two visually impaired subjects. One of the visually impaired subjects lost all functional vision at the age of 53 (subject was 61 years of age at the time of testing), while the other had a partial vision loss due to severe damage to the left striate cortex at the age of 8 (subject was in his early 40s when testing took place) [5]. Phosphenes were mapped by letting the subjects draw their percepts on a tilted drawing board 57 cm in front of them, while fixing their gaze on the center of the board. In a later phase of the study the subject was seated 50–200 cm from a white wall and was asked to indicate with a laser pointer to the location of the phosphene relative to a reference point on the wall. Subjects also traced the outline of the phosphenes with the laser pointer. The experimenter redrew the phosphene with a pencil. Interestingly, while retinotopy was clearly present in the sighted subjects and the subject with partial vision loss, the retinally blind subject did not show clear retinotopy and had a degraded spatial representation. In contrast, Brindley and Lewin [3] found a clear retinotopy in their blinded patient when using subdural electrodes. The authors speculate that the diffuse cortical excitation inherent in TMS makes precise stimulation impossible when cortical organization is interrupted due to total vision loss [5].

In a similar TMS study the authors made use of a digitizing tablet connected to a personal computer to directly convert the drawings to digital data [14]. After each TMS pulse the subject drew the image on the tablet, which was provided with a central pin for tactile reference to the center of the visual field. This phosphene mapping method showed that TMS was capable of evoking phosphenes in 17 out of 18 sighted people, and that phosphenes could be evoked along the entire visual field by stimulating the occipital cortex with single pulses. Blinded subjects often needed TMS pulse trains instead of single pulses, and in only 54% of these subjects phosphenes could be evoked. TMS is generally used to disrupt cortical function and although many phosphenes appeared as dots of light, spots of darkness ("scotomas") were also reported.

19.6   Recent Simulation Studies Using Phosphene Mapping

Simulation studies on the functionality and capabilities of visual prostheses are becoming more important (Chap. 16). Regarding phosphene mapping, simulation studies can be used to carefully control the test environment and allow a comparison of different mapping strategies.

19.6.1   Tactile Simulations at Shanghai Jiao Tong University

Ren and associates published two papers about novel ways to construct absolute phosphene maps based on simulation studies. Both studies used normally-sighted subjects who were presented with simulated phosphenes using a head-mounted display (HMD). The first study [4] made use of a touch screen (39 cm in width, 30 cm in height) that was placed at eye level. The screen was provided with a tactile reference point in the center of the screen. Subjects were seated in the dark and fixated at the center of the screen by means of a chin rest. Their left index finger was placed on the reference point on the touch screen for tactile feedback, while the right index finger was used to point at the simulated phosphene on the touch screen, much like the pointing hemisphere deployed some 40 years earlier by Brindley and Lewin [3]. The experiment compared two test conditions by presenting phosphenes with and without a reference grid projected in the HMD, which divided the visual field in 6×8 cells. Under both conditions, the experiment was preceded by a training phase in which the subjects could see their own response on the touch screen. In the first phase of training, phosphenes were presented in a predictable way, allowing the subjects to familiarize themselves with the equipment. In the second training phase, phosphenes were randomly presented. After that, the actual test was performed during which subjects could not see their response. Phosphenes were presented at 3°, 11° and 15° eccentricity [4].

The investigated parameters included dispersion of the responses (standard error of the response in mm), accuracy (the distance between phosphene and mean response in mm) and response time. Their results showed that in the presence of the reference grid dispersion, accuracy and response time tended to be lowest. In addition, dispersion and response times increased when phosphenes were presented at larger eccentricity. The authors also showed that dispersion was larger in the left two quadrants of the visual field compared to the right two quadrants. They attributed this result to the fact that the left hand was always used for tactile reference which interfered with pointing to a phosphene in the left half of the visual field.

In a follow-up study, Ren and colleagues [31] used a very similar setup, but adapted their method to improve tactile feedback to the subject by overlaying a 19 touch screen monitor with a 31×31 push-button array (41 cm in width, 35 cm in height). Tactile references were improved by (a) an elevated center button representing the origin and (b) slightly elevated buttons along the horizontal, vertical and diagonals of the push-button array. Subjects could use both hands to localize the phosphene on the push-button array. In contrast to their earlier method, the screen with the push-button array was placed horizontally on a table in front of the subject. Training consisted of three phases. The first training phase permitted the subjects to familiarize themselves with the array by letting them feel the origin and the elevated buttons indicating the dividing lines. The second training phase provided the subjects with phosphenes localized in a restricted portion of the visual field. The third and final training phase provided the subject with 24 random phosphenes and the subject could observe the response in the HMD. The test phase consisted of 98 randomly generated phosphenes. Again, dispersion, accuracy and response time were recorded.

Compared with the unaided touch screen method dispersion was lower and accuracy more constant when using the push-button array. Furthermore, the systematic left hemifield error observed in the first study was absent. However, response times were much higher (25 vs. 3 s in the earlier study) and the authors speculate that subjects spent most of this additional time on finding the tactile references and appropriate button on the array. Another possibility is the fact that the push-button array was placed horizontally in front of the subject, instead of vertically at eye level as in their first study. This setup likely demanded more of the subjects, since they had to translate visual field coordinates to a horizontal surface.

Though testing times may become long, the push-button array may prove a valuable method for testing subjects with little or no residual vision, because of the tactile references, better resolution and reduced localization errors.

19.6.2 Simulations in Our Laboratory

Dagnelie and Vogelstein developed and compared three different phosphene mapping methods based on phosphene localization simulation studies in four normally-sighted volunteers [6, 7]. An HMD with a $40° \times 50°$ binocular display with 5 arcmin resolution was used to present phosphenes (see Chap. 16). The HMD provided subjects with a central fixation point and was equipped with a pupil tracker to monitor eye movement. Trials were aborted if the gaze deviated by more than $0.5°$. All three methods were designed to mimic a prosthesis positioned over the primary visual cortex of one cerebral hemisphere by presenting 32 randomly located phosphenes binocularly in one visual hemifield at eccentricities up to $20°$. Phosphenes were round dots with a diameter of 20 arcmin at the fovea and increasing in diameter to 40 arcmin near $20°$ eccentricity in order to mimic cortical magnification. For each subject, one set of phosphenes was generated for use in all three tests to facilitate comparison between methods within a subject. Additional random sets were used in the touch screen and eye movement procedures, discussed below. Subjects came in for four or five 1-h sessions.

The first method was much like the method of Chai et al. discussed earlier, deploying a touch screen ($18'' \times 12''$, height $\times$ width). Figure 19.1 shows a subject performing this procedure. Subjects sat in front of the screen and held their left index finger on a tactile marker located halfway down the left edge of the screen. Subjects were told to place their right index finger immediately beside the left index finger on the screen and to align their fixation point in the HMD with this (unseen) finger as best they could. A phosphene was then presented, accompanied by a tone and subjects had to slide their right index finger across the screen to the location of the phosphene and lift their finger off the screen. A second tone signaled that the computer had registered the lift-off coordinates. This process was repeated until all 32 phosphene positions were mapped; this process was repeated, for a total of three estimates per position. All data were obtained in a single session.

The second approach recorded a saccade to the remembered phosphene location, using a calibrated pupil tracker in the HMD. The subject fixated on the central dot displayed in the HMD. After a warning tone, a phosphene momentarily appeared (400 ms) after which a saccade was made to the former phosphene location. The subject was required to briefly maintain gaze while the pupil-tracking software recorded the coordinates of the final eye position. This procedure was repeated for all 32 positions and repeated three times in a single session.

In contrast to the absolute phosphene coordinates estimated with the first two techniques, the third method constructed a relative phosphene map. This so-called triadic distance comparison method compares distances among point triads, with map reconstruction through multidimensional scaling (MDS). During the test, subjects

Fig. 19.1 Subject performing the touch screen task employed by Dagnelie and co-workers. The left index finger is placed on a tactile marker representing the center of fixation, placed midway along the left side of the touch screen. The head-mounted display shows an internal fixation point, and the subject attempts to keep the line of sight in the HMD aligned with the tactile fixation marker. A pupil tracking camera inside the HMD is used to monitor steady central fixation while a phosphene is present. The subject's right index finger has just completed tracing towards the perceived phosphene location and is briefly held steady before being taken off the screen, marking the position. The scene camera, visible on the front of the HMD, was not used in this test

again maintained fixation on the central dot on the HMD screen. Three phosphenes appeared sequentially and remained visible for 500 ms. Subjects numbered the phosphenes according to their appearance and reported which two dots were closest and which two were farthest apart. The experimenter keyed in the reply and started a new trial. Testing all possible triads including the 32 phosphene locations would have required 4,960 trials. To reduce this number, the 32 phosphenes were divided into four overlapping groups of 16 from which pseudo-random triads were presented. Each pair in a group of 16 was presented four times, rather than the maximum of 14 times. In this way, 160 triadic comparisons per group were made, for a total of 640 trials to complete all four groups; maps from the groups of 16 could be combined by virtue of the eight common points among "adjacent" groups of 16. Performing the 640 comparisons required two or three sessions.

The MDS procedure consisted of building a similarity matrix for all pairs, in which ternary values were assigned for each response ("2" for closest, "1" for intermediate and "0" for the farthest pair), similar to the method from [26] described in Sect. 19.3. A Kruskal MDS procedure was then performed to reconstruct the two-dimensional dot distribution (mean=0, SD=1) [23]. Note that the resulting map not only needs to be translated and scaled to allow comparison of the reconstructed coordinates with those obtained by the touch screen and eye movement methods but, due to the relative nature of triadic comparisons, it will also require rotation, and possibly a mirror imaging operation.

Figure 19.2 shows results of a complete set of tests for one representative subject. Touch screen results are shown in panel a, eye movement results in panel b, triadic comparison results in panel c, and combined results in panel d. The square symbols and connecting (dashed) lines in each figure represent the stimulus locations and the order in which they were presented; the connecting lines are presented only to facilitate comparison with the corresponding mean responses (drawn black lines; no data points, since each break point in the line is the center of gravity of the responses for that stimulus in multiple trials). Panels a and b show raw responses, while those in panel c have been transformed to optimally fit the stimulus coordinates, as explained above.

A comparison between the three methods suggests that the touch screen and eye movement tests had better relative accuracy with increasing eccentricity, but overall had poor reproducibility (up to 25% test-retest variability). Furthermore, as judged by comparing the line sets in each panel, the touch screen data show a relative expansion along the horizontal axis, and a downward trend, while the eye movement test resulted in a horizontal compression; the break points in the triadic comparison reconstruction map appear much closer to the phosphene coordinates, but this may in part be due to the optimized scaling. For the combination and comparison in panel d we have therefore optimized the fits of the touch screen and eye movement data through translation and isotropic expansion. Averaging all three methods in panel d yields a map in which the lines representing the "grand mean" response bears a good, albeit somewhat distorted, resemblance to the lines connecting the phosphene coordinates.

While these results are those for a single subject, the findings are typical of what was found in a half dozen others: touch screen responses overestimate horizontal eccentricity, while eye movement responses underestimate them. Triadic distance comparison test performance was better than the other two tests. This does not mean, however, that one can rely on this test alone: one or more absolute mapping methods are required to obtain an overall map of phosphene locations.

Also, given the time-consuming nature of the triadic comparison tests, it may be best to concentrate efforts using that test on clusters of closely spaced phosphenes while using the absolute techniques to establish relationships between such clusters. By applying translation, rotation and scaling to achieve maximum correspondence among data from different tests, one can hope to attain the most accurate maps.

One should bear in mind that in an actual prosthesis wearer there will be no stimulus map to which the results can be fitted. Nonetheless the results obtained in our lab inspire some confidence. Dagnelie and coworkers [7] computed a distortion metric from the distance estimation errors for all possible phosphene pairs across the three tests for all five subjects tested with a uniform phosphene set. Three subjects had distortion scores under 15% for all tests. Combining maps by averaging the data of all three tests reduced the errors below 10%, which should enable adequate image recognition. The authors conclude that the three procedures, especially in combination, permit the construction of distortion maps with sufficient fidelity to enable shape recognition by future prosthesis wearers.

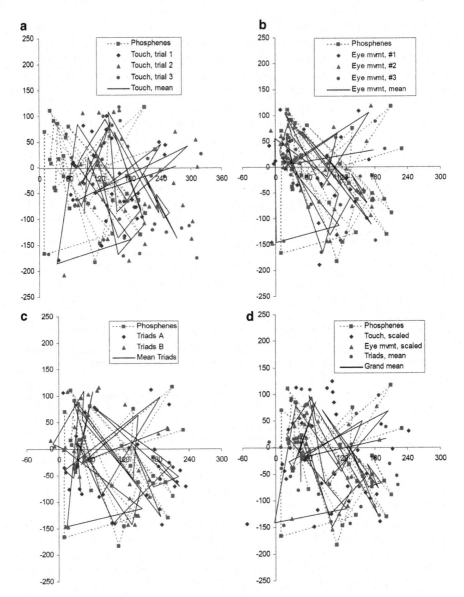

Fig. 19.2 Results of the three phosphene mapping techniques used in our laboratory, and combined results, for one representative subject. Each panel shows the results for one test; phosphene locations are identical in all four panels. Connecting *lines* represent the order of presentation in the touch screen and eye movement tests, and are shown to allow a comparison between the stimuli and corresponding means of the response in multiple trials (**a–c**) or methods (**d**). Coordinates are in screen pixels in the HMD; each pixel subtends an angle of approximately 5 arcmin. Results in (**c**) have been transformed to obtain optimal correspondence with the coordinates of the 32 phosphenes

19.7   Concluding Remarks on Phosphene Mapping Techniques

Various absolute and relative mapping procedures were discussed in this chapter. Absolute mapping provides estimated phosphene coordinates, while relative mapping provides phosphene positions only with respect to each other. Due to eye movements, mapping by sequential electrode activation may still yield unreliable relative phosphene coordinates, but absolute mapping is inherently subject to position errors if gaze is not monitored. Relative mapping of closely spaced phosphenes yields more reliable information about phosphene positions with respect to each other, which will be important when trying to present arbitrary shapes to a prosthesis wearer.

We reviewed more than a dozen absolute mapping techniques using a variety of pointing, drawing, verbal, and eye movement methods. Advantages of most of these absolute mapping procedures are their technical simplicity and the short time required to obtain a phosphene map. Especially when performing acute experiments in the operating room with time and equipment restrictions, absolute mapping by verbal communication may be the most convenient method. Data can be digitized on the spot by a drawing tablet, or recorded by the experimenter in the form of crude coordinates. With chronic implant wearers or visually impaired subjects in a laboratory setting, detailed information can be obtained using a touch screen or a dart board or clock face with tactile markings. Tactile markers and training improve accuracy. Drawings can be advantageous when phosphene shape is of interest.

Disadvantages of these techniques are their inaccuracy and the difficulty resolving phosphenes located closely together, especially by subjects with long-standing vision loss. Visuo-motor translation may affect the results, especially when phosphene location or shape is indicated by drawing. Disadvantages of verbal descriptions, paper drawings and pointing to a surface other than a touch screen include the need to re-draw the data in a visual field map, or to digitize them into a computer. Finally, some of these methods can only be successfully employed by individuals with functional residual vision (e.g., using a laser pointer).

Relative mapping methods require subjects to provide details about the relationship between different phosphenes. The techniques we reviewed varied in phosphene presentation, using timing or other attributes to distinguish two or more phosphenes, but also in response modalities and analysis methods. All these techniques tended to be more complex and time consuming than the absolute techniques. This may not be a serious problem in subjects with long-term implants, as the benefits of careful mapping in increased ability to convey visual information to the prosthesis wearer will far outweigh the cost in time.

Finally we learned that a combination of well-chosen absolute and relative mapping methods may yield accurate maps with acceptably small distortions. There is still a need to further elaborate some of the techniques beyond what has been described in the literature thus far, but many of the elements for reliable and efficient phosphene mapping procedures appear to be available. The principal remaining task is to perform comparisons of promising techniques, and choose optimal combinations.

Acknowledgment Supported in part by PHS grant # EY012843.

References

1. Bak M, Girvin JP, Hambrecht FT, et al. (1990), *Visual sensations produced by intracortical microstimulation of the human occipital cortex.* Med Biol Eng Comput, **28**(3): p. 257–9.
2. Brelen ME, Duret F, Gerard B, et al. (2005), *Creating a meaningful visual perception in blind volunteers by optic nerve stimulation.* J Neural Eng, **2**(1): p. S22–8.
3. Brindley G, Lewin W (1968), *The sensations produced by electrical stimulation of the visual cortex.* J Physiol (Lond), **196**: p. 479–93.
4. Chai XY, Zhang LL, Shao F, et al. (2007), *Tactile based phosphene positioning system for visual prosthesis.* Invest Ophthalmol Vis Sci, **48**: p. ARVO E-Abstr. 662.
5. Cowey A, Walsh V (2000), *Magnetically induced phosphenes in sighted, blind and blind-sighted observers.* Neuroreport, **11**(14): p. 3269–73.
6. Dagnelie G, Vogelstein JV (1999), *Phosphene mapping procedures for prosthetic vision.* In *Vision Science and its Applications,* Optical Society of America, Washington, DC.
7. Dagnelie G, Yin VT, Hess D, Yang L (2003), *Phosphene mapping strategies for cortical visual prosthesis recipients.* J Vis, **3**(12): p. 222.
8. Delbeke J, Oozeer M, Veraart C (2003), *Position, size and luminosity of phosphenes generated by direct optic nerve stimulation.* Vision Res, **43**(9): p. 1091–102.
9. Dobelle WH (2000), *Artificial vision for the blind by connecting a television camera to the visual cortex.* ASAIO J, **46**(1): p. 3–9.
10. Dobelle WH, Mladejovsky MG (1974), *Phosphenes produced by electrical stimulation of human occipital cortex, and their application to the development of a prosthesis for the blind.* J Physiol, **243**(2): p. 553–76.
11. Dobelle WH, Turkel J, Henderson DC, Evans JR (1979), *Mapping the representation of the visual field by electrical stimulation of human visual cortex.* Am J Ophthalmol, **88**(4): p. 727–35.
12. Drasdo N, Fowler CW (1974), *Non-linear projection of the retinal image in a wide-angle schematic eye.* Br J Ophthalmol, **58**: p. 709–14.
13. Everitt BS, Rushton DN (1978), *A method for plotting the optimum positions of an array of cortical electrical phosphenes.* Biometrics, **34**(3): p. 399–410.
14. Fernandez E, Alfaro A, Tormos JM, et al. (2002), *Mapping of the human visual cortex using image-guided transcranial magnetic stimulation.* Brain Res Brain Res Protoc, **10**(2): p. 115–24.
15. Gargini C, Terzibasi E, Mazzoni F, Strettoi E (2007), *Retinal organization in the retinal degeneration 10 (rd10) mutant mouse: a morphological and ERG study.* J Comp Neurol, **500**(2): p. 222–38.
16. Gothe J, Brandt SA, Irlbacher K, et al. (2002), *Changes in visual cortex excitability in blind subjects as demonstrated by transcranial magnetic stimulation.* Brain, **125**(Pt 3): p. 479–90.
17. Holmes G (1918), *Disturbances of vision by cerebral lesions.* Br J Ophthalmol, **2**(7): p. 353–84.
18. Humayun MS, de Juan E, Jr., Dagnelie G, et al. (1996), *Visual perception elicited by electrical stimulation of retina in blind humans.* Arch Ophthalmol, **114**(1): p. 40–6.
19. Humayun MS, de Juan E, Jr., Weiland JD, et al. (1999), *Pattern electrical stimulation of the human retina.* Vision Res, **39**(15): p. 2569–76.
20. Humayun MS, Weiland JD, Fujii GY, et al. (2003), *Visual perception in a blind subject with a chronic microelectronic retinal prosthesis.* Vision Res, **43**(24): p. 2573–81.
21. Kaido T, Hoshida T, Taoka T, Sakaki T (2004), *Retinotopy with coordinates of lateral occipital cortex in humans.* J Neurosurg, **101**(1): p. 114–8.
22. Kandel E, Schwartz J, Jesell T (2000), *Principles of Neural Science,* 4 ed., McGraw-Hill, New York, NY.
23. Kruskal JB (1964), *Nonmetric multidimensional-scaling – a numerical-method.* Psychometrika, **29**(2): p. 115–29.

24. Marg E, Rudiak D (1994), *Phosphenes induced by magnetic stimulation over the occipital brain – description and probable site of stimulation.* Optom Vis Sci, **71**(5): p. 301–11.
25. Milam AH, Li ZY, Fariss RN (1998), *Histopathology of the human retina in retinitis pigmentosa.* Prog Retin Eye Res, **17**(2): p. 175–205.
26. Mladejovsky MG, Eddington DK, Evans JR, Dobelle WH (1976), *A computer-based brain stimulation system to investigate sensory prostheses for the blind and deaf.* IEEE Trans Biomed Eng, **23**(4): p. 286–96.
27. Ray PG, Meador KJ, Epstein CM, et al. (1998), *Magnetic stimulation of visual cortex: factors influencing the perception of phosphenes.* J Clin Neurophysiol, **15**(4): p. 351–7.
28. Schmidt EM, Bak MJ, Hambrecht FT, et al. (1996), *Feasibility of a visual prosthesis for the blind based on intracortical microstimulation of the visual cortex.* Brain, **119**: p. 507–22.
29. Veraart C, Raftopoulos C, Mortimer JT, et al. (1998), *Visual sensations produced by optic nerve stimulation using an implanted self-sizing spiral cuff electrode.* Brain Res, **813**: p. 181–6.
30. Veraart C, Wanet-Defalque MC, Gerard B, et al. (2003), *Pattern recognition with the optic nerve visual prosthesis.* Artif Organs, **27**(11): p. 996–1004.
31. Zhang L, Chai X, Ling S, et al. (2009), *Dispersion and accuracy of simulated phosphene positioning using tactile board.* Artif Organs, **33**(12): p. 1109–16.

Chapter 20
Prosthetic Vision Assessment

Marilyn E. Schneck and Gislin Dagnelie

Abstract As visual prostheses continue to evolve, assessing their efficacy assumes paramount importance. This chapter identifies some of the key questions and issues that arise when planning and designing such assessments, in order to help point the way forward.

High quality evaluations will naturally follow basic scientific principles such as including pre-operative as well as post-operative testing. Evaluations should include both visual function and visual task performance. Improved visual function tests may need to be developed or adapted that are suitable for the levels of vision afforded by current and near-term prosthetics. In assessing task performance, the choice of tasks to be assessed is critical, and can greatly influence the results.

Longer-term follow-up testing after periods of acclimatization and training are also necessary, with control groups receiving alternative training such as more conventional rehabilitation or interventions.

Self-assessment of difficulty in performing daily living tasks is also important, as are the more subjective assessments of user satisfaction.

As the technologies continue to evolve, there will be a changing dynamic involving the steadily improving capabilities of the technology and the unique needs of a growing number and more diverse target population.

Abbreviations

ADL	Activities of daily living
ALS	Activities of life satisfaction
BaLM	Basic light and movement test

M.E. Schneck(✉)
Rehabilitation Engineering and Research Center, The Smith-Kettlewell Eye Research Institute, 2318 Fillmore Street, San Francisco, CA 94115, USA
and
Vision Sciences Program School of Optometry-2020, University of California at Berkeley, Berkeley, CA 94720-2020, USA
e-mail: mes@ski.org

G. Dagnelie (ed.), *Visual Prosthetics: Physiology, Bioengineering, Rehabilitation*,
DOI 10.1007/978-1-4419-0754-7_20, © Springer Science+Business Media, LLC 2011

FrACT   Freiburg acuity test
HR-QOL   Health-related quality of life
IADL   Instrumental activities of daily living
O&M   Orientation and mobility
VFQ   Vision function questionnaire
VEP   Visually evoked cortical potentials

20.1   Introduction

In order to assess the effectiveness of prosthetic vision in context, we must ask the question "Effectiveness for what"? Answering this question requires an understanding of the realistic goals for this new technology. The prospect of "restoring vision to the blind" (e.g., [40]) has, of course, been received with great enthusiasm. However, the richness of our visual experience belies the complexity of the neural system delivering it, thus making it unrealistic for a prosthetic device to provide vision in its full and richly complex form. For instance, the number of electrodes in current devices (16–1,500) is many, many orders of magnitude fewer than required to carry the wealth of the information from the 120 million photoreceptors in each eye along over a million fibers of each optic nerve to the 140 million highly organized neurons in each hemisphere of primary visual cortex, which in turn send them on to the many other regions of the cortex devoted to specific aspects of visual processing.

The multitude of neurons form a complex network with feed-forward, lateral and feed-back signaling giving the visual system its complex imaging power. Subsequent to loss of photoreceptors in outer retinal diseases such as retinitis pigmentosa, there are significant losses in both the inner nuclear layer (e.g., bipolar cells) and ganglion cells [50, 81, 91]. Furthermore, there is significant remodeling of neural retina and thus local neural networks following photoreceptor loss [64, 65]. Thus, whether implanted in the retina or the cortex, the implants will not have the full analytic power of the intact visual system. Currently, prostheses developers are working to increase the number of electrodes and their density to improve resolution (e.g., 1,024 by Troyk's group [98]). Performance has been shown to improve with increasing number of electrodes [30, 42].[1] However, no matter how many electrodes in use, multiple phosphenes (discrete sensations of light), of various sizes shapes and hues, depending on the state of the post-receptoral retina and visual system, will be generated. How will these phosphenes be used to represent the environment? Past visual experience will certainly influence interpretation, but the recipient will have to undergo prolonged training to learn to use these signals.

---

[1] The success of cochlear implants is often cited as a hopeful indication of what can be accomplished by a sensory prosthesis. Only six electrodes stimulating the cells of the auditory nerve enable the wearer to understand speech at near-normal levels. If one assumes the same ratio of electrodes to nerves, hundreds of electrodes are projected to be required [108].

Thus, prosthesis recipients may have to make sense of a "blooming, buzzing confusion" that William James (1842–1910) said faced a newborn.

Bearing these considerations in mind, it becomes clear that "restoring vision to the blind", is an amorphous goal that is unreachable in the foreseeable future. In fact, the goal of prostheses is not to recreate normal vision, but to provide visual perception that however limited in scope is *useful to the individual* [69, 70, 78, 108].

How can we determine whether prosthesis aided or provided vision is useful to the individual? This requires well-defined, measurable goals for evaluation of progress and demonstration of successful outcomes. Widely accepted outcome measures and means of measuring them, and criteria for "success" have not been developed for prosthesis implantation. Standards have not been set. The fact that clinical trials are already in progress (e.g., SecondSight, Intelligent Medical Implants are conducting Phase 2 Clinical Trial) and others are planned makes more pressing the establishment of relevant outcome measures as criteria for success. Once such outcome measures are selected and developed, the means to assess performance with respect to these outcomes are specified, and testing has been carried out, progress can be assessed. The lack of well-specified relevant outcomes and means to assess them is a major hurdle in the future of prostheses.[2]

For most of us, activities of daily life rely on vision. Visual performance (actions incorporating and guided by visual input) is a very complex phenomenon involving other senses, motor skills, memory, prior knowledge, feedback, experience, practice, etc. The basic building blocks for vision performance are sensory visual functions such as motion, color, luminance, contrast, and orientation. These contribute to object identification and localization. These in turn feed into higher order visual areas that integrate the visual information with other senses, motor systems, cognitive systems, and memory the ensemble of which guides the actions that form the tasks of daily living. In assessing prosthetic vision, we should ideally find ways of measuring these functions and task performance.

For simplicity, we restrict the discussion that follows to consideration of individuals with particular characteristics. Most prosthesis recipients to date have lost vision as a result of retinitis pigmentosa (e.g., [21, 22, 30, 51, 105, 118]). Therefore, we assume a target population of individuals with long-standing retinitis pigmentosa (RP), a progressive disease of the photoreceptors. Nearly all RP patients have some residual vision, typically one or more small peripheral islands in an otherwise non-functional retina [22].[3] Since these individuals have had some vision well into

---

[2] Lengthy discussions of these issues took place at a special interest group meeting at ARVO 2007 (organized by author MES and contributed to by author GD) [82] and a symposium hosted by the Smith-Kettlewell Eye Research Institute in San Francisco (October 2007) [83]. These meetings highlighted the complexity of the problems and demonstrated that more work needs to be done before specific recommendations of protocols and tests will be established. Some of the content of this chapter has been gleaned from those meetings.

[3] The degree of residual vision function is highly light-level dependent in RP patients; individuals with retinal disease often require unusually high light levels to attain their best vision function [85, 92].

adulthood, they have not learned many of the adaptive tools of long-term blind individuals. For example, they are not Braille readers, but may use a cane.[4]

Rather than attempt to prescribe specific tests procedures, the remainder of this chapter is intended to raise and discuss the various issues relevant to visual assessment in patients with prostheses.

20.2 Principles for Assessment of Prosthetic Vision

20.2.1 Experimental Design

Studies of the effectiveness of an intervention or technology, such as clinical trials of prostheses will incorporate a repeated measure design in which the individual's performance at one time point (before the intervention) is compared to that at other time points (after the intervention). The comparison of pre- and post-intervention results (the difference) is used as an index of the effect of the intervention. One useful way to express results is to report whether individuals cross some important criterion level of vision for example from "visually impaired" to "normally sighted" following treatment. In many clinical trials, the intervention's success or failure is judged with respect to a criterion level of visual function, usually visual acuity. In the context of prosthesis implantation, many "interventions" will occur (learning, prosthesis implantation, training and rehabilitation). To assess the effect of each intervention requires careful and broad assessment before and after its occurrence using scientific principles.

An appropriate control group is an essential element in clinical testing. Control groups are comprised of individuals who differ from the treatment group, to the extent that is possible, only in that they do not receive the intervention. Comparison of the test results of the two groups is an important component of assessing the effectiveness of the intervention.

---

[4] Another set of issues and considerations arise in the case of long-term blind individuals who have adapted an array of methods that enable them to accomplish most tasks. For simplicity, we do not consider this population, though acknowledge that prostheses may be relevant to this population in the future. There are reports of recovery of vision following prolonged blindness (e.g., [35, 41, 74]). In some cases, sensory vision recovered fairly quickly, though learning to recognize objects and people and to use vision to guide daily activities took considerable time and effort. One such patient never became comfortable relying on vision, preferring instead to close his eyes in tough situations [41]. This indicates that in some conditions leading to blindness, the visual system remains intact, but that things are not immediately recognized by sight; images must become associated with the information from other senses used before the onset of sight.

20.2.2 *The Importance of Pre-operative Testing*

Thorough, scientifically valid pre-implant sensory vision assessment is essential for many reasons. Pre-treatment visual function is typically an inclusion/exclusion criterion for participation, though too rarely are these assessments well conducted or described. Prosthesis developers have suggested pre-operative assessment to select implant recipients [117]. Psychophysical and electrophysiological testing has shown that poorer residual vision is associated with reduced sensitivity to stimulation [117]. In order for prostheses to effectively provide useful vision, the visual apparatuses proximal to the prosthesis (e.g., all post-receptoral elements from eye to visual cortex for sub-retinal implants) must retain functionality, enabling them to receive and transmit the prostheses signal. In cases of retinal disease, such assessment may be accomplished by eliciting phosphenes by electrical stimulation or pressure to the eye, for example to determine whether this elicits a percept. In the case of post-orbital prostheses, the solution is more complex, and may involve magnetic stimulation, for example.

As described in the context of experimental design, only by careful measurement of pre-operative vision can the benefits (or losses) of the implant be determined. This is particularly important because many or most patients/potential recipients will have some rudimentary residual vision. For example, patients with retinitis pigmentosa (RP) form a large portion of the prosthetic implant candidates having undergone and currently undergoing clinical trials [21, 30, 51, 60, 105]. Only very rarely does RP result in total blindness. As mentioned earlier pre-operative testing will allow researchers to determine the degree and characteristics of the residual vision. This is essential for judging whether and to what degree the prosthesis improves vision.

At least as important, but considerably more difficult, is the pre-operative evaluation of the vision-related skills and abilities of the individual. The range of tasks we carry out each day is enormous. Which are most "important" and should be assessed? There is no generally accepted answer. Individual recipients will have different priorities. Skills often assessed in low vision population include reading, face recognition, orientation and mobility and other simple activities of daily living (ADL; e.g., eating, dressing, washing) and more complex instrumental activities of daily living (IADL; more complex ADLs such as shopping, cooking). Creating appropriate tests with quantifiable measures for ADLs and IADLS poses considerable challenges. This issue is discussed in a later section.

Finally, assessment of the degree of difficulty an individual experiences when performing activities (ADL, social, recreational and vocational activities) is of value. A number of questionnaires have been developed and validated for assessment of low vision patients, though none exist for this population of individuals with very little residual vision who become prosthesis recipients. This issue is discussed in a later section.

Discussion of pre-operative (and post-operative) assessment strategies have been also been discussed elsewhere [25].

20.2.3   Post-operative Assessment

The reasons for post-operative assessment are self-evident. Post-operative evaluation will include all of the aspects of vision assessed pre-operatively. Initially, however, the emphasis will be on establishing that the prosthesis is functional and stable (delivers phosphenes in a repeatable manner). Post-operative assessment of the prosthesis recipient will occur over an extended time period, beginning soon after surgery and undoubtedly lasting for years. During this time both learning to use the prosthesis and evolution of the prosthesis will occur.

Learning how to best interpret and use the signals from the prosthesis will be an ongoing process for the recipient (e.g., [25, 51]), most likely carried out interactively with the prosthesis team. Dobelle (2000) reported that a long-time wearer of a cortical implant initially was unable to recognize letters, underwent long-term training (10 days) and after prolonged continued practice had acuity of 20/1,200 [31]. Learning how to interpret signals in order to form basic visual images will likely precede that of more complex visual perceptions and the use of the prosthesis-driven vision to perform tasks. The learning curve for prosthesis use is not known, and will certainly vary among individuals. Meanwhile prosthesis development will be continuously carried out, leading to alterations of the signal processing and stimulation routines. Measurement time points need to be set and all recipients should be seen at each time point using a pre-determined, set protocol, if possible.

Hippocrates put forth the important "First do no Harm" principle of medical intervention. The possibility exists that in some cases prosthesis implantation may negatively affect the recipient's vision. Damaging surrounding functional tissue may reduce or destroy any residual vision. Kiser et al. reported that at least one prosthesis recipient's vision deteriorated due to surgical complications (and three developed cataract leading to a loss of vision) [53]. The use of the prosthesis may reduce visual performance in other ways. Any visual signal, even non-structured, "noise" resulting from the prosthesis may interfere with learned strategies by drawing attention away from reliable information from other senses (e.g., tactile and auditory feedback from a cane). Such losses can only be documented by comparison to pre-operative testing of visual function and performance. One may switch the device on and off to determine its utility and or its interference with performance, but not its potential effect on anatomy.

20.2.4   Methodological Issues in Pre- and Post-operative Vision Assessment

20.2.4.1   Potential Approaches

Two approaches offer themselves for assessing vision: psychophysics and electrophysiology (recording of electrical responses from the visual system).

Psychophysics is a discipline that determines the relationship between dimensions of a physical stimulus and perception. Psychophysical techniques can be used to measure threshold (minimal detectable or minimally discriminable) and supra-threshold vision. Virtually all standard clinical measures of vision are psychophysical threshold measurements (for example, visual acuity, contrast sensitivity, visual fields, color discrimination). Supra-threshold measures are more often made in the laboratory and include for example, contrast matching, figure-ground segregation, brightness matching, visual search etc. Psychophysics is an elaborate, well-defined discipline with its own rules of operation. Psychophysical procedures are designed to minimize observer and tester bias (see e.g., [23]).

Visual electrophysiology involves the recording of electrical signals (voltage changes) generated in response to the stimuli of interest. The visually evoked (cortical) response (VEP), is the most commonly used method for assessing basic visual function. The VEP can be used to assess both threshold and supra-threshold vision. It is recorded by placing electrodes on the scalp over the primary visual cortex (at the back of the head) and other associated visual areas of interest. In recent years, increased signal to noise ratio (SNR), new methods of signal processing and data extraction have greatly extended the range of visual functions that can be measured and the sensitivity of these tools. Currently sweep VEP measures [103] are used to determine visual acuity, contrast thresholds and Vernier thresholds in clinical settings (e.g., [1, 73, 107]). Many other characteristics of vision, including higher order visual processes, can also be measured by the VEP. The VEP is an objective (bias-free) measure.

Though electrophysiology (and other imaging techniques not described) certainly have a role in prosthesis assessment (particularly the VEP in pre-operative and early post-operative settings for non-cortical implants), psychophysical testing will undoubtedly dominate vision assessment. VEPs measure activity of the primary visual cortex. It is possible that VEPs are recordable, but that the individual has no access to this information, i.e., cannot use it to see.

If the fellow eye retains some vision, then testing of the *prosthesis* is most easily carried out with the fellow eye patched (i.e., monocularly). However, assessment of the *individual's* vision and functional ability should always be made binocularly, as that is the way one goes about his/her business.

20.2.4.2 Avoidance of Bias

Two sources of unwanted bias may affect experimental results. Experimenter bias occurs when a researcher unconsciously manipulates procedures to achieve an expected or desired outcome, potentially skewing the results. Patient bias arises from either a "placebo effect" (effect based on the power of suggestion and/or induced by having been seen by a trusted expert and receiving a "treatment"), or the desire to please the investigator.

To avoid these biases a double blind approach is often used for testing. In this case, neither the examiner nor the patient is aware of whether the subject has received the treatment/intervention or not. To assure that the masking is effective, placebo treatments and/or sham procedures that closely resembles the intervention may be performed. Comparison of the sham or control subjects' and patients' performance reveals the true (unbiased) treatment effect.

Change over time in the sham and treated groups may also be compared. If the change across time in the sham group is as large as and in the same direction as that of the treatment group, it is concluded that the treatment (implant) has not affected performance.

Appropriate psychophysical techniques for vision measures within the double blind (or masked) protocol (in which neither the experimenter nor subject is aware of the subject's group (treatment vs. control)) further reduce both sources of bias. Methods for development, administration and analysis of questionnaires using a masked protocol reduce bias for these instruments

20.2.4.3 Criteria for Sound Testing

To be scientifically sound and produce credible results, test procedures must be carried out in such a way as to (1) specify stimulus characteristics such as size, intensity, spectral characteristics, and duration in sufficient detail to be replicated; (2) use an objective, criterion free response (such as offered by a multiple alternative forced choice (see below)); (3) assure that the outcome cannot be attributable to extraneous (non-visual) cues (including bias); (4) allow specification of the likelihood that the reported outcome occurred by chance; (5) have predefined methods of analysis and predefined criteria for success/failure. These principles apply to both sensory vision and vision performance testing. We address these principles more fully below.

20.2.4.4 Forced Choice Procedures

Forced choice psychophysical procedures minimize observer and tester bias. As the name implies, in this procedure the stimulus is present in one of a number of spatial or temporal intervals, and the observer's task is to select the correct interval. Forced choice procedures are equally appropriate and simple to use with supra-threshold stimuli. In this context, the observer's task is to choose from among several stimuli to pick the one with the attribute of interest (brightest, highest contrast, fastest, etc). Insertion of "blank" or "catch" trials further allows the experimenter to assess and remove biases. To determine whether the observer did in fact detect/identify the target correctly, performance is corrected for guessing (expected percent of the trials the patient would get correct by guessing) as follows:

$$\text{True percent correct} = \frac{(\text{observed percent correct} - \text{expected percent correct by chance})}{(1 - \text{expected percent correct by chance})}.$$

There is a lower limit on the number of trials necessary for assessing whether performance is beyond chance. In effect, increasing the number of alternatives on a trial increases the information from that trial and reduces the number of trials required to assure that the observer's performance is not based on chance. Minimizing the number of trials per test is crucial, particularly when many measures are to be made. It is thus important to use an efficient rule for varying the parameter of interest to arrive at "threshold" with the minimum number of stimulus presentations (and minimize errors based on assumptions of the slopes of the psychometric functions). There are many in use (e.g., Pest, bestPest, Quest, Psi, adaptive staircase, Method of Adjustment, the Method of Limits, and Method of Constant Stimuli) [23]. Threshold is the minimum level of some stimulus parameter that can be detected, resolved or discriminated reliably. In practice, threshold is typically defined as the stimulus level that produces a criterion percent correct after correction for guessing.

To more thoroughly consider response biases, and make better use of the information to be gained from each trial one may analyze data within the context of signal detection theory [39, 59].

20.2.4.5   Response Time

In psychophysical testing, the time between stimulus presentation and the response (often called "reaction time") provides additional important information. It is an indication of the individual's confidence in his/her response and thus the difficulty of the task. For example, responses are faster to stimuli that are well above threshold than to near-threshold stimuli. In the present context, response time has other significance. To be useful in the "real world", performance must be at must be at least as efficient as that obtainable by other (non-visual) methods. An individual may be able to locate a banana on a table using prosthetic vision after several minutes of searching, but be able to accomplish the task by searching using his/her hands in seconds. In other situations, such as crossing the street, fast responses are more critical. If an extended time is required for image interpretation and arrival at correct decisions, the information may be of limited value.

A criterion response time for passing/failing can be developed for each measure based on (multiples of) the response time of normally sighted observers, the time required for a blind individual to achieve the task using other strategies, or with respect to the window over which the response would be useful. Another important consideration is the degree to which the individual desires independence. An individual who does not want to be helped may tolerate much longer performance times.

20.2.4.6   Task (Perceptual) Learning

An individual's performance on a task improves with repeated performance of the task and finally asymptotes at peak performance when learning is "complete" (i.e., asymptotic performance along the dimension of interest is reached). At each measurement time point (both pre- and post-operatively), it is crucial that testing continue until this asymptote is reached. This is important for pre-operative testing because we cannot otherwise distinguish whether any observed benefit at a later time point is due simply to continued learning or from the advantage conferred by the implanted device. Conversely, we may underestimate benefit by not measuring maximum performance at the later time point. Repeated testing of a task after task performance has reached its asymptote also provides an estimate of test-re-test repeatability. The 95% confidence limit of this variability sets the criterion for true change imparted by the implant (and subsequent interventions including training and rehabilitation). This is crucial for establishing whether any change is statistically significant at a criterion probability.

20.2.4.7 Establishing Criteria for Meaningful Change

After defining significance in the statistical sense, there remains the question as to whether the measured benefit or loss has any clinical or practical significance. For visual acuity, this criterion is generally 0.3 log units [24], which is equivalent to a halving of the required target size, for example going from 20/2,000 to 20/1,000 or 20/40 to 20/20. Individuals with acuity improvements of this magnitude also demonstrate large gains on the vision specific subscales of the VFQ-25 questionnaire [18]. However, these criteria are based on statistical considerations (test-re-test repeatability). A similar criterion (0.3 log units) is also useful for contrast sensitivity. Criteria of meaningful change have not been established for other aspects of visual function, or visual performance. In the case of visual performance (of tasks or activities of daily life), defining a criterion for meaningful improvement may be more difficult.

20.2.4.8 Light Level

There are standards for light level for acuity measures (80–120 cd/m^2) [14]. In normally-sighted people, acuity declines with luminance below the standard values and is constant across higher light levels. In those with disease, however, both sensory vision and function are highly dependent on light level over a broader range (e.g., [85, 92]). Many affected individuals require higher-than-normal light levels to see and to function. For example, Kuyk et al. showed that reducing light level had an adverse effect on mobility in patients with age-related macular disease [54]. In RP patients, including those with Usher's syndrome, field size is critically dependent on light level, shrinking with diminishing light levels. This is evident in six patients with Usher's syndrome illustrated in Fig. 20.1 (Haegerstrom-Portnoy, personal communication). The field diameter of Subject TL increases by a factor of 10, from a mere 2.5 degrees to 25 degrees, over the light range tested (1 to 1,000 cd/m^2) and would most likely continue to increase with further increases in luminance. *Therefore, pre- and post-implant testing should be carried out over a range of light levels,* including those

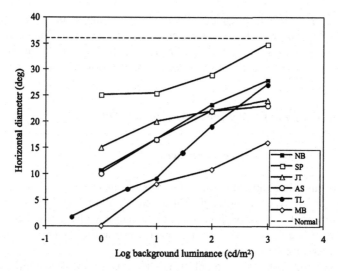

Fig. 20.1 Horizontal diameters of visual fields are plotted across luminance. Each symbol identifies an individual subject

well above what is normally required, say at daylight levels (10,000+ cd/m²) in order to maximize the contribution from the remaining functional retina beyond the prosthesis. This may be done in log unit steps of 10 cd/m² (white paper under living room level), 100 cd/m² (white paper under office light levels), 1,000 cd/m² (white paper outdoors in overcast) and 10,000 cd/m² (white paper outdoors in sun).

20.3 Vision Assessment in Prosthesis Recipients: Overview

We consider vision assessment as two components. They are, in our terminology, visual function and visual performance. Visual performance is itself comprised of two components: objective (measured) and subjective (self-reported) ability to accomplish tasks. In this regard, the point of view of this chapter is similar to the three-pronged approach suggested by Wilke et al. [112].

20.3.1 Visual Function Assessment: Overview

Visual function forms the basis of evidence for evaluating benefit/risk to relevant parties (e.g., other researchers, funding agencies, the FDA). Acceptable outcome measures of clinical trials have classically been visual function, most notably visual acuity, as described above. Four aspects of vision accepted by the FDA are, with constraints, visual acuity, visual fields, contrast sensitivity, and color vision [24]. Visual function may be more reliable, objective, and sensitive than either aspect of visual performance; however, it may be the least relevant to real-life conditions to be faced by the prosthetic recipient. Detailed description of prosthesis-generated visual function will guide future prostheses work. As noted by Chader et al. [20], "what is needed is an accurate and reproducible method to link visual testing with real-world functional capacity in individuals with very low vision".

20.3.2   Visual Performance Assessment: Overview

20.3.2.1   Measured Visual Performance

Both sighted and blind individuals integrate input from a variety of sense modalities as well the motor system, and cognition to learn skills to accomplish tasks. The accomplishment of tasks through vision in combination with other senses and skills is the most complete, and, arguably the most important index of the value of the vision afforded prosthetic device.

Numerous studies have shown only weak (though statistically significant) associations between visual function and task performance, including performance of ADL (e.g., [47, 54, 93]). This is not surprising since with training and practice (i.e., rehabilitation), most tasks can be accomplished without any vision. An individual's visual performance cannot be inferred from even an extensive array of well-selected sensory vision measures. In practice, a very limited number of such measures will be made. Thus, it is necessary to meet the challenge of developing tests of components of tasks of interest, or means of measuring directly performance of the actual tasks. Development of valid, relevant performance tasks is a hurdle that must be overcome to advance the field of visual prostheses.

20.3.2.2   Self-Reported Visual Performance

Self-report, using properly developed questionnaires and appropriate analysis tools, serves many functions. These instruments enable us to: assess extraneous variables that could affect intervention success (e.g., depression and other co-morbidities); assess a wider spectrum of functional and performance aspects of vision than are practical via direct measurement; assess the impact of interventions on global indices of well-being such as quality of life and self-perceived disability. Instruments also efficiently gather large amounts of information [61].

20.4   Visual Function Assessment

Given that function is only loosely related to sensory vision (except in the extremes), the choice of visual attributes to be measured could be made on the basis of a number other criteria such as acceptability to relevant agencies as an outcome measure, value in guiding further prosthesis refinement, scientific "curiosity", or determining the individual's legal eligibility for benefits.

The selection and design of vision assessment tests must take into account the expected level of function but at the same time cover a wide level of function to be suitable for pre-implant and post-implant assessment. As noted, prosthetic vision in the near future is likely to be fairly crude, but it will exceed the visual function with which the subject presents. For preoperative assessment, tests should be designed for those with extremely limited vision. However, the prostheses may result in large improvements, so measures must be able to assess the higher function as well. There is no consensus on which aspects to measure, but the battery must be sufficiently brief to assure patient comfort. Tests that are inexpensive and easy to administer are most likely to gain wide acceptance.

20.4.1   Candidate Measures

Light perception is the ability to tell whether one is in light or darkness, or more specifically, whether the visual field is light or dark. This level of vision is well below the target for outcomes in implant trials. However, measurement of light perception, is invaluable in pre-operative testing, particularly for determining (1) whether the visual pathway proximal to the implant is functional and (2) and to define minimal light levels for further testing. Dagnelie [25] has suggested measuring light perception (detection) as a threshold task; at what level can the individual first detect light (threshold) and out to what range? The idea behind this test is that, considering the vast range of light levels over which the visually normal person can function, determining the operating range may have a greater ability to classify severe vision loss than almost any other measure. Knowledge whether an individual has light perception prior to implantation is necessary for comparison to post-implant visual function gain. However, vision at the level of light perception is of very limited value in terms of the recipient's daily function.

Light projection (or localization) is the ability to indicate from which direction a light originates. Assuming clear ocular media, light projection in the eye follows the laws of geometric optics, so that the retinal location of the illuminated area (nasal vs. temporal, up vs. down) is predictable. However, in implant recipients who have no vision outside the areas driven by the implant, perception of location is likely to be driven by prosthesis location with respect to the fovea or fixation location, at least initially. The recipient will, presumably, learn to remap visual space based on the implant's location and function and to "fit" visual space into this area.[5]

---

[5] A difficult aspect of any vision task using a target of limited spatial extent will be locating the target in visual space. This is most difficult for devices with which field of view does not follow eye movements, which are currently the most common. Individuals can learn to suppress eye movements in favor of head movements, but this is difficult and perhaps inadequate. Some prosthesis developers have addressed this problem by yoking the external camera or its image to eye movements or implanting the photo-detector/camera in the eye (e.g., [22, 37]). Though prostheses are placed to tap into foveal processes, the (retinal) prosthesis may be displaced from the fovea. Prosthesis wearers must unlearn the tendency to move the eyes to foveate. For prostheses in which the receiving and stimulating elements are co-located in the parafovea, the recipient may need eccentric viewing training, currently used for patients with age-related macular degeneration with absolute scotomas that involve the fovea [72].

Light projection enables the individual to locate light sources (such as windows or doors) with respect to his or her position and can thus aid navigation.

The *visual field* is the area of space within which an individual can detect the presence of a visual stimulus. The visual field of a normal human eye measures (from the point of fixation) 100 degrees temporally, 60 nasally, 75 superiorly and 60 inferiorly [8]. Binocular (using both eyes) visual fields are approximately 200 degrees wide and 135 degrees tall, with a region of binocular overlap that is 120 degrees wide.

In visually normal observers, sensitivity varies considerably within these limits [97], so that the size of the measured visual field is strongly dependent on target size and luminance. This dependence is more dramatic in those with RP. Very small fields impair mobility.

The integrity of the field is also an important measure. If a bigger or brighter target than normal is required for detection in a region, that region has a *relative* field loss or relative scotoma in that region. An absolute field loss, such as an absolute scotoma, is a region in which the patients cannot detect *any* target.

Visual field testing (perimetry) merely require target detection, rather than localization and is carried out two basic ways: with stationary (static perimetry) or moving (kinetic perimetry) targets. In kinetic testing, targets are slowly brought from random locations outside the far peripheral field toward the point of fixation. In contrast, in static perimetry stationary targets appear briefly at any random location irrespective of distance from the fovea. Large differences in results between fields measured with static and kinetic perimetry are often seen in patients with larger fields measured for moving targets.

Commercially available field devices such as the Humphrey field analyzer (HFA) are not typically appropriate for assessing fields in prosthesis recipients for a number of reasons including limitations on target size, intensity, and the relatively limited (40 degrees) central region tested. However, a few means of field assessment for those with low vision have been developed and are considered in a later section.

One may question the value of visual field measures in the presence of a visual prosthesis. Surely field results can readily be predicted based on the dimensions of the electrode array (in degrees), the magnification or minification of the image processing unit, the density of electrodes, and the array location (for retinal implants particularly but cortical implants as well). Whether this holds true within the area "covered" by the prosthesis remains to be seen. One report indicated that some individuals implanted with the Artificial Silicon Retina had larger fields post-op, but that others showed shrinkage due to complications [53]. Certainly, field measurements are essential for describing (residual) vision outside of this region that will contribute not only to field dimensions but also to task performance. Measurement of visual fields requires stable fixation at a known (pre-determined) location. For individuals with poor vision, placement of a finger at the fixation location is extremely helpful.

Visual acuity is an index of the finest discernable detail.[6] Visual acuity is typically measured using targets approaching 100% contrast (black and white) because

---

[6] See [14, 48] for a detailed discussion of acuity measurement.

resolution improves with contrast. As noted earlier, visual acuity has for some time been the predominant outcome measure in intervention studies, and is the primary visual descriptor of participants in vision studies, and of populations [24]. At the coarse end, visual acuity is clinically described in terms of whether or not the individual can detect hand motion (HM) or count fingers (CF) at a specified test distance.

It is preferable, of course, to use targets with more precisely controlled and defined specifications. Gratings and optotypes are the most common types of acuity targets. Grating targets measure the minimum separable resolution whereas optotype acuity is a form of recognition acuity. The relative value of measuring grating and optotype acuity is a matter of some debate, and also a matter of circumstances. Grating acuity can be quantitatively related to optotype acuity in visually normal individuals, but this association breaks down when disease is present (e.g., [32, 36, 110]). When using grating stimuli, aliasing associated with under-sampling or other distortions associated with abnormal retina or the prosthesis a concern and can lead to an over-estimate of resolution. Aliasing is the situation in which a high spatial frequency target is miss-perceived as a stimulus of lower spatial frequency or a distorted grating [17, 94, 95, 113, 114].

Optotype acuity has won out in clinical settings. Common optotype targets are simple shapes, letters, numbers, the tumbling E (formerly "illiterate E"), and Landolt rings (also called Landolt C's). The smallest optotype target size (in terms of visual angle) that the patient can identify is determined [98]. For the tumbling E targets, the observer's task is to indicate in which of the four cardinal directions direction the "tines" are pointing. The Landolt C target is a circle with a gap in it. The gap is presented in four or eight locations (the four cardinal plus the four obliques) and the observer's task is to indicate the location of the gap in each ring.

Provided that the subject is required to continue to attempt to identify or guess until some criterion is reached (e.g., three out of five optotypes are identified incorrectly), acuity measures are criterion and bias free. It is recommended that acuity be scored letter-by-letter rather than line by line [15].

Though standard, commercially available letter charts such as the ETDRS acuity chart [34] or the Bailey-Lovie Chart [16] were not designed to measure extremely poor acuity, the lower end of their range can be extended into the range of interest by simply decreasing the test distance. At the standard (20 ft. or 6 m) test distance, the largest letters on the Bailey-Lovie Chart correspond to an acuity of 20/125; at a 10 ft. test distance 20/250, and down to 20/2,500 at 1 ft.

Tests specifically designed to measure acuity for low vision are discussed in a later section. An important gain from using optotypes to measure acuity is that the presence of measurable optotype acuity provides evidence of form vision capability.

20.4.1.1   Contrast Sensitivity (Contrast Detection)

Contrast (of a grating; Peak-to-peak contrast or Michelson contrast) is defined as

$$C_m = (L_{max} - L_{min}) / (L_{max} + L_{min}).$$

Contrast can vary from 0 to 1, and is more often specified as a percentage (0–100%). Contrast sensitivity is the inverse of contrast at threshold. Within the linear systems approach to vision, a description of an individual's contrast sensitivity function (CSF, i.e., the minimum contrast required to see a grating, measured as a function of spatial frequency, or bar width) provides a means of knowing the visual system's response to any stimulus defined by luminance contrast. For practical purposes, though, contrast sensitivity testing is typically limited to a single large (relative to acuity) target size, specifying one point on the CSF near the peak of the CSF.

In clinical settings, optotype measures are more commonly used than grating targets [9]. Unlike grating stimuli, optotype targets, such as those on the Pelli-Robson Chart [76] are specified in Weber contrast, which is defined as

$$C_w = (L_{max} - L_{min}) / L_{max}.$$

Note that the two measures are the same if the mean grating luminance is $L_{max}/2$. Contrast sensitivity deficits are present in RP patients, even in those with normal or near-normal acuity [4, 7].

Despite a strong correlation between contrast sensitivity and visual acuity, one cannot predict one from another on an individual basis [44], and therefore, both should be measured. An individual with very poor acuity, but fairly good contrast sensitivity and fields of reasonable size will have no trouble navigating and moving through the environment but probably will not be able to read well or at all. The converse is also true. An individual with a very small visual field, good contrast sensitivity and good acuity will have great difficulty moving about the world or finding targets; however, once the targets are "found" (are placed within the functional field) they will have no trouble identifying the target or reading print.

Reports suggest that contrast sensitivity better predicts performance than other measures (e.g., acuity). Associations have been reported between contrast sensitivity and reading performance [57, 111], ambulation mobility [38, 45, 55, 66, 99], driving [115, 116], face recognition [75, 109], and tasks of daily living [79, 80, 109].

20.4.1.2 Contrast Discrimination

Most natural images contain both high and low contrast. In scenes, features to be detected are frequently observed in the presence of other supra-threshold (visible) background structures. Detection of such features requires contrast discrimination, which is necessary for the subsequent process of object recognition. Contrast discrimination is impaired in RP patients, even those who have good acuity to moderately reduced contrast sensitivity [6].

No simple chart or other test of contrast discrimination is available though, in principle, one could be developed fairly easily. Such a chart might consist of sets of stimuli each with at least two elements ranging in contrast, with the patient's task being, for example to identify the stimulus with the highest contrast, with the difference between the contrasts among stimuli in stimulus sets decreasing down the chart. PC-based tests on the same principle would be more flexible.

20.4.1.3 Motion Perception

The ability to detect a luminance-defined, or color-defined target as moving and to judge motion speed and direction requires detection and localization over an extended retinal area, as well as interactions between neighboring areas and, intact temporal processing. Human visual motion occurs beyond the retina [12], so that intact temporal processing at earlier stages is crucial. The temporal sequence of retinal stimulation in a degenerated retina may become highly distorted, so that the signals reaching the cortex become ambiguous [25]. Temporal processing and displacement discrimination are abnormal in RP patients [3, 5]. Prosthesis processing delays and those from degenerating retina may favor slow stimuli of coarse spatial grain. Movement perception is likely to be greatly impaired in retinal degenerations [25].

In fact, perceived motion in the presence of a prosthesis is most like apparent motion, or sampled motion, which is the perception of smooth motion from sequential presentation of discrete stationary targets, in this case, electrode-generated phosphenes. The appropriate combinations of temporal and spatial interval characteristics for apparent motion have been worked out for normally sighted individuals [19]. However, the same relationships are unlikely to hold for prosthetic vision.

Bearing these considerations in mind, motion processing may thus be severely impaired in (potential) prosthesis recipients. The loss and rewiring of post-receptoral elements is an additional factor in the case of sub-retinal implants.

In patients with retinal degenerations who retain relatively good form vision, aspects of motion perception that have been assessed include judgments of motion displacement thresholds [5] and heading direction [101]. Minimum displacement thresholds are increased and maximum displacement thresholds decreased, greatly restricting the range of detectable motion in patients with RP, even when visual acuity showed only minor reductions (acuity of 20/40 or better) [5]. Patients with retinal degenerations with form vision have elevated thresholds, reduced maximum velocity and/or direction discrimination for two-dimensional (2D) motion. Yanai et al. [118], testing three RP patients implanted with a 16-electrode prosthesis, found that performance on a motion discrimination task was above chance only so long as the subjects were allowed to move their heads to scan.

2D motion perception imparts important information unrelated to perception of motion per sé. For example, motion parallax is an important depth cue, particularly for those lacking binocular vision, among whom are prosthesis recipients. Object (or person) motion facilitates detection, particularly in complex environments.

In terms of survival skills, the ability to judge motion toward or away from oneself may be more important. Our movement through an environment is guided by optic flow, perceived visual direction, and judgment of focus of expansion, among other things. Turano et al. (2005) reported a change in the ability of individuals with field changes to utilize optic flow for judging heading direction, important for orientation and mobility [101]. It is equally important to judge a moving object's path with respect to oneself (to avoid collision or catch a ball). This is based on, for example, looming and zooming cues, the detection and interpretation of which may be greatly impaired in prosthesis recipients.

Assessment of the many aspects of motion perception is unpractical. Further, many aspects of motion perception require some level of spatial vision that may not be met by the prosthesis patient. A practical test of motion direction discrimination might be to use a bright large dot or line moving on a dark screen in a dark room along one of four (cardinal directions) or eight (with diagonals) and requiring that the subject identify the direction of motion. A similar approach using a penlight instead of a dot can be used in cases of high light thresholds an has been shown to have good reproducibility within intact visual field areas [49]. However, it is uncertain that such a test will tell us much about many aspects of motion. Tests of seemingly unrelated aspects, such as the presence of long range interactions and the ability to judge the relative timing of flashes may be much more informative, as both are prerequisites for perception of motion.

Long-range spatial interactions are key to integrating information within a scene and detecting motion. The most striking demonstrations of long-range spatial interactions are illusory or subjective contours [52], examples of which are shown in Fig. 20.2. As can be seen on the right, a central inverted triangle appears though there are no lines to demark it. Its presence is induced by the corner elements, and despite its lack of true form, it is seen as occluding the upright "triangle" inferred from its corners. The simpler form of the induced contour is shown in the left half of Fig. 20.2. Perception of these contours demonstrates the presence of the capacity for long-range interactions necessary for motion perception as well as judgments such as figure ground. These functions serve many important purposes in making sense of visual scenes. Long-range interactions are also necessary for recognizing partially occluded objects in a cluttered environment.

Fig. 20.2 Examples of illusory contours produced in the presence of elements that appear partially occluded. *Left*: One illusory square. *Right*: Two illusory triangles

20.4.1.4 Depth Perception

Stereopsis is the perception of depth produced by binocular retinal disparity. Stereopsis subserves fine depth discriminations for near objects. Prosthesis recipients will not have stereopsis. These individuals will necessarily rely on monocular depth cues including those mentioned previously (occlusion and motion parallax), as well as lighting and shadows, linear perspective, texture gradient, height in field, to name but a few.

Color is one of the most compelling aspects of vision, adding beauty to our visual experience as well as aiding in detection and identification of objects. Some reports indicate that induced phosphenes are of many colors [51, but cf. 31]. Nonetheless, due to the complexity of color-conferring circuitry and crudeness of available stimulation methods, veridical color vision will not be afforded by prostheses in the foreseeable future.

20.4.2 Tests Used in Prosthesis Trials

In the absence of a well-described, validated set of tests, measuring visual function has been and still is up to the ingenuity of those working on the projects. Dobelle (2000) measured visual acuity in three recipients of their 64-element cortical implant and reported acuity and visual field size for one [31]. For this subject acuity measured using tumbling E's and Landolt C's show excellent agreement (20/1,200). Second Sight Medical Products (Sylmar, CA) has published a report on visual function in three recipients implanted with their 16-electrode prosthesis [117] based on three to four alternative tests (locate and count objects such as common household items, discriminate among those items, determine the orientation of a large L, and identify the direction of object motion). A grating orientation discrimination task was used to measure acuity. Kiser et al. used standard static (HFA to test central 30 degrees) and kinetic perimetry (Goldmann to test far periphery) methods to test vision of eight Artificial Silicon Retina implant (Optobionics Corp.) [53]. Yanai [118] found that, with repeated testing, three implanted subjects could discriminate between plate, knife or cup against a dark background.

20.4.3 Tests that Have Been Designed for Use with Prostheses

A few laboratories have developed new tests specifically for assessing prosthesis recipients, but have not yet tested recipients with them. In many instances the devices have been used in the context of simulated prosthetic (pixelized) vision (e.g., [26, 27, 87, 96]), but have not been used to assess prosthesis candidates or recipients. In addition, simulations in subjects with better vision have been used to estimate how well a person may perform using a prosthetic, with the idea that the pixels correspond to electrodes [25, 106].

20.4.4   Vision Tests for Very Low Vision

PC-based tests to assess basic aspects of visual function in prosthesis recipients have recently been developed.

The BaLM test (Basic Light, Localization and Motion; Zrenner and Wrobel, Retina Implant AG) [112], measures several visual functions, including light detection, light location, temporal resolution and motion direction discrimination. All tests involve forced-choice responses, provide optional auditory feedback and allow the number of trials to be varied. Output includes percent correct and response times.

The *Berkeley Rudimentary Vision Test* was created by Ian Bailey and co-workers to assess individuals with very low vision, within the range typically described as hand motion or count fingers (i.e., worse than 20/800 (LogMar 1.6)). It contains a light and basic form perception test (BFPT) and an acuity screening and measurement test, the Single Tumbling E's Test (STET). Visual fields can also be measured.

FrACT, the Freiburg Acuity and Contrast Test [11] is available online at http://www.michaelbach.de/fract/index.html. It uses Landolt C's to assess very coarse acuity in the range of hand motion to count fingers, and contrast sensitivity. It also assesses vernier acuity. By combining the results of FrACT, clinical acuity measures, and ETDRS acuity, Schulze-Bonsel et al. found that CF and HM acuity can be reproducibly assessed and correspond to acuities of 20/1,400 and 20/4,000 [84].

Dagnelie and co-workers have developed a number of PC-based tests appropriate in this context. These include a visual field measure [13, 28], yet to be validated, that can bridge the gap between crude localization and standard field measures.

20.5   Visual Performance Assessment

As has been noted [25], it is not what the prosthesis recipient can see, but what they can do that is critical. The aim of prostheses is to improve the ability of the recipient to perform activities of daily living (ADL), instrumental tasks of daily living (IADL) and what we may call activities of life satisfaction (ALS). Visual performance affects independence, quality of life and "visual disability" [67]. Assessing this aspect of vision remains a major hurdle in the path of prosthesis evaluation and progress. Assessment of visual performance (pre- and post-implantation) addresses questions such as: How has everyday task performance changed since implantation? Is the increment in performance sufficient to be of real value to the recipient? Does the prosthesis increase efficiency, reduce risk and increase independence?

This aspect of visual assessment has two branches: measured performance and self-reported performance. Direct measurement has been argued to be superior to self-report, at least in aging [43]. Theoretical advantages of measuring function include better reliability and validity, greater sensitivity to change and less influence by confounding factors such as culture and language [45]. However, questionnaires enable us to assess important issues beyond the realm of performance, such as whether the implant was of any benefit (e.g., all of the tasks can now do) to the recipient, the difficulty an individual has performing particular IADLs, and the impact that difficulty or inability has on quality of life. Questionnaires provide valuable information that may guide the choice of performance measurement. Both approaches are important for demonstration to patients, care-givers, government, and funding organizations the effectiveness of the prosthesis.

20.5.1 Measured Performance

Observation of blind individuals who have undergone rehabilitation reminds us that there are fewer tasks than we realize that cannot be accomplished without vision. However, the return of the ability to perform tasks visually is desirable to those with little or no vision – particularly those who, like the subjects in the current implant studies, have lost their vision later in life.

Among tasks that cannot be performed at all without vision, one of the most important is reading printed text with all its images, information provided by formatting (e.g., headings, emphasis), potential for scanning to find information, etc. (Braille comes in only one "font" and letter size). Other desirable tasks that require vision are driving and the ability to identify and locate at a distance (Josh Miele, personal communication).

There is no generally accepted test battery of even rudimentary task performance measures for use in low vision patients (including prosthesis recipients). Many of the performance batteries that have been developed are not relevant to or have not been assessed in those with very low vision (e.g., [46, 54, 93, 102]). In choosing items of ADL to assess for their index, Haymes et al. (2001) considered how common the ADLs were on existing instruments (questionnaires) and whether they were consistent with daily living problems reported by a very large number of people with vision impairment [46]. Unfortunately, the tasks chosen which include reading print in various contexts (e.g., newspaper, medication label), using a telephone, recognizing faces, threading a needle, require vision superior to that currently afforded by prostheses.

Optimally, performance would be measured *in situ*. However, it is more practical to measure tasks under controlled (laboratory) conditions. In either case careful consideration of task relevance is crucial.

The most commonly used tasks are reading, face recognition, and mobility and orientation (wayfinding). As an alternative or in conjunction with measurements of ADLs and IADLs, which are complex, one may assess "component" tasks such as eye-hand coordination, visual search, figure ground discrimination, and finding embedded objects [2, 54, 58].

Wilke et al. (2007) developed two sets of test which we shall refer to as "at table" tasks and orientation and mobility tasks for use in visual prosthesis recipients [112].

The at table tasks incorporate an important component of many activities, eye hand coordination. The orientation and mobility test uses projected images of a street scene as viewed from different distances and notes the "viewing distance" at which particular scene items are first seen. Importantly, both tasks incorporate a measure of the time required to complete tasks.

Turano and coworkers have developed both real [99] and virtual [100] environments in which to assess mobility performance. Velikay-Parel et al. (2007) have developed a mobility task for individuals with very low vision [104].

One of the authors (GD) has developed a set of task performance measures beginning with a search task, locating and counting white checkerboard squares [27, 106]. This activity is followed by a measure of eye-hand coordination: placing black checkers on the white checkerboard squares they had previously counted. Scoring is in terms of time to complete the task (speed) and the number of checks that are not or are incompletely covered (accuracy). Another test measuring eye hand coordination is a maze tracing task, scored in terms of speed and accuracy (or rather errors: cumulative area spanned by tracing outside the borders) [71]. In addition, a complete record of the performance is recorded for more detailed analysis. To date, these tests have only been used to assess simulated prosthetic vision (coarsely pixelized vision).

Wayfinding by individuals with poor vision is of great interest to low vision researchers. Wayfinding includes mobility ability, orientation skills, the ability to form mental maps or learn a route. Because of safety concerns focus has been on two scenarios: visually guided travel in the laboratory and cane-assisted travel in everyday environments. A difficulty facing this area of research is that individuals learn test routes quickly, limiting the number of "trials" that can be used. Real-world wayfinding is subject to numerous uncontrollable variables and may require the presence of an O & M instructor, further limiting its practicality. There is a continuing effort to overcome these problems (e.g., [56, 104]). Velikay-Parel et al. (2007) addresses the difficult issue of repeatability of the measures [104].

Evaluation of task performance is based on speed and accuracy. These are easily quantified for simple tasks. However, for more complex tasks, such as ADLs and IADLs, an occupational therapist and orientation and mobility trainer, masked as to whether the individual has received a prosthesis, may better evaluate task performance.

In summary, measured task performance brings us closest to knowing the benefit gained by the individual with respect to everyday activities. Establishment of a battery that is both relevant and appropriate for those with very low vision, and validating and standardizing such a battery would be of enormous value to the field of prostheses.

20.5.2 Self-Reported Performance (Questionnaires)

Funding organizations such as the NEI require the inclusion of a patient-reported outcome for clinical trials of any disease intervention or treatment or assistive device, and the FDA also considers information from visual function questionnaires

(VFQs) [24, 77]. Instruments have long been the means of assessing the success or failure of rehabilitation programs (e.g., [10, 29, 88, 90]).

The rising use and interest in questionnaires has been driven by both need and, largely, by improvements in the methods of development and data analysis. The latter charge was lead in large part by Robert Massof. He applies techniques including Item Response Theory and Rasch analysis that have greatly increased the value and "interpretability" of the information gathered. Massof (2007) has quite clearly laid out the merits of and need for these techniques [68].

Visual function questionnaires contains activities to which difficulty ratings (related to vision) are assigned by study subjects/patients. There is an abundance of VFQs, including the NEI VFQ-25 [62], VAQ (visual activities questionnaire) [86], and Veterans Affairs questionnaires [89] for low vision (VA LV VFQ-48 and LV VFQ-20).

There are also many instruments addressing Activities of Daily Living (e.g., [46, 63]).

More recently, instruments assessing Health-Related Quality of Life (HR-QOL) or Quality of Life (QOL) have come into use. The content of HR-QOL questionnaires typically includes assessment of the ability to perform tasks of daily living, interactions with other people, emotional well-being and independence [33].

The broad range of instruments available, with varying content, and some developed before and some after the changes in design and scoring alluded to above, makes choosing the appropriate instrument tricky. The means to assess the quality of an instrument and to decide whether it is appropriate for use in a particular context as well as key issues for questionnaire development have recently been nicely laid out [77].

20.6 Summary

Assessing the efficacy of visual prostheses is a complex undertaking. There are many issues to consider in the assessment of visual prostheses. As the technologies continue to evolve, there will be a changing dynamic involving the steadily improving capabilities of the technology and the unique needs of the growing number of target populations. What is common in all circumstances is the need to assess visual function, visual performance, and self-perceived visual ability using tasks that are appropriate to the target population in terms of functional level and needs, with methodologies that meet generally adopted criteria scientific rigor, including adequate pre- and post-implant evaluation and appropriate control procedures.

Acknowledgments The authors gratefully acknowledge the helpful discussions with and suggestions of Dr. Gunilla Haegerstrom-Portnoy, Dr. Lori Lott and Dr. John Brabyn. The authors also thank Prof. Michael Bach and Dr. Walter G. Wrobel for providing us with the BaLM test for examination and Prof. Ian Bailey for providing us with the Berkeley Rudimentary Vision Test.

Supported by The Smith-Kettlewell Eye Research Institute and NEI Grant EY09588 to John Brabyn.

References

1. Ahmadi H, Bradfield YS (2007) *Chorioretinopathy and microcephaly with normal development.* Ophthalmic Genet, **28**(4): p. 210–15.
2. Aki E, Atasavun S, Turan A, Kayihan H (2007) *Training motor skills of children with low vision.* Percept Mot Skills, **104**(3): p. 1328–36.
3. Alexander KR, Barnes CS, Fishman GA (2003) *Deficits in temporal integration for contrast processing in retinitis pigmentosa.* Invest Ophthalmol Vis Sci, **44**(7): p. 3163–9.
4. Alexander KR, Derlacki DJ, Fishman GA (1992) Contrast thresholds for letter identification in retinitis pigmentosa. Invest Ophthalmol Vis Sci, **33**(6): p. 1846–52.
5. Alexander KR, Derlacki DJ, Xie W, Fishman GA, et al. (1998) *Discrimination of spatial displacements by patients with retinitis pigmentosa.* Vision Res, **38**: p. 1171–81.
6. Alexander KR, Pokorny J, Smith VC, Fishman GA, et al. (2001) *Contrast discrimination deficits in retinitis pigmentosa are greater for stimuli that favor the magnocellular pathway.* Vision Res, **41**(5): p. 671–83.
7. Alexander KR, Rajagopalan AS, Seiple W, Zemon VM, et al. (2005) *Contrast response properties of magnocellular and parvocellular pathways in retinitis pigmentosa assessed by the visual evoked potential.* Invest Ophthalmol Vis Sci, **46**(8): p. 2967–73.
8. Anderson, DR (1987*) Perimetry With and Without Automation*: Mosby, St. Louis.
9. Arditi A (2005*) Improving the design of the letter contrast sensitivity tests.* Invest Ophthalmol Vis Sci, **46**(6): p. 2225–9.
10. Babcock JL, Goodrich GL, Head DN, Boyless JA (2000) *Developing geriatric training outcome assessments in vision rehabilitation.* J Vis Impair Blind, **94**(5): p. 307–21.
11. Bach M (1996) *The "Freiburg Visual Acuity Test" – Automatic measurement of visual acuity.* Optom Vis Sci, **73**(1): p. 49–53.
12. Bach M, Hoffmann MB (2000) *Visual motion detection in man is governed by non-retinal mechanisms.* Vision Res, **40**(18): p. 2379–85.
13. Bahrami H, Melia M, Dagnelie G (2006). *Lutein supplementation in retinitis pigmentosa: PC-based vision assessment in a randomized double-masked placebo-controlled clinical trial.* BMC Ophthalmol, **6**:23.
14. Bailey IL (1998) *Visual Acuity.* Borisch Benjamin WJ, Editor. *Borisch's Clinical Refraction*: WB Saunders Co., Philadelphia.
15. Bailey IL, Bullimore MA, Raasch TW, Taylor HR (1991) *Clinical grading and the effects of scaling.* Invest Ophthalmol Vis Sci, **32**(2): p. 422–32.
16. Bailey IL, Lovie JE (1976) *New design principles for visual acuity letter charts.* Am J Optom Physiol Opt, **53**: p. 740–45.
17. Brendan BT, Pacey IE, Bradley A, Thibos LN, et al. (2003) *Nonveridical visual perception in human amblyopia.* Invest Ophthalmol Vis Sci, **44**(4): p. 1555–67.
18. Brown DM, Kaiser PK, Michels M, et al., ANCHOR Study Group (2006) *Ranibizumab versus verteporfin for neovascular age-related macular degeneration.* N Engl J Med, **355**(14): p. 1432–44.
19. Bundesen C (1989) *Spatio-temporal conditions for apparent movement.* Phys Scr, **39**: p. 128–32.
20. Chader GJ, Weiland J, Humayun MS (2009) *Artificial vision: needs, functioning, and testing of a retinal electronic prosthesis.* Prog Brain Res, **175**: p. 317–32.
21. Chow AY, Chow VY, Packo KH, Pollack JS, et al. (2004) *The artificial silicon retina microchip for the treatment of vision loss from retinitis pigmentosa.* Arch Ophthalmol, **122**(4): p. 460–9.
22. Cohen ED (2007) *Prosthetic interfaces with the visual system: Biological issues.* J Neural Eng, **4**(2): p. R14–31.
23. Corliss DA, Norton TT (2002) *Principles of psychological measurement.* Norton TT, Corliss DA, Bailey IL, Editors. *The Psychophysical Measurement of Visual Function*: Richmond Products, Inc., Albuquerque.

24. Csaky KG, Richman EA, Ferris FL, 3rd (2008) *Report from the NEI/FDA Ophthalmic Clinical Trial Design and Endpoints Symposium.* Invest Ophthalmol Vis Sci, **49**(2): p. 479–89.
25. Dagnelie G (2008) *Psychophysical evaluation for visual prosthesis.* Annu Rev Biomed Eng. **10**: p. 339–68.
26. Dagnelie G, Keane P, Narla V, et al. (2007) *Real and virtual mobility performance in simulated prosthetic vision.* J Neural Eng, **4**: p. S92–101.
27. Dagnelie G, Walter M, Yang L (2006) *Playing checkers: Detection and eye-hand coordination in simulated prosthetic vision.* J Modern Optics, **53**: p. 1325–42.
28. Dagnelie G, Yang L, Eshraghi F, Lewis NL, et al. (2002) *Validated vision test battery for the home PC.* Invest Ophthalmol Vis Sci, **43**: ARVO E-abstract 3811.
29. De l'Aune WR, Welsh RL, Williams MD (2000) *A national outcomes assessment of the rehabilitation of adults with visual impairments.* J Vis Impair Blind, **94**(5): p. 281–91.
30. Delbeke J, Wanet-Defalque MC, Gérard B, Troosters M, et al. (2002) *The microsystems based visual prosthesis for optic nerve stimulation.* Artif Organs, **26**(3): p. 232–4.
31. Dobelle WH (2000) *Artificial vision for the blind by connecting a television camera to the visual cortex.* ASAIO J, **46**(1): p. 3–9.
32. Dobson V, Quinn GE, Tung B, Palmer EA, et al. (1995) *Comparison of recognition and grating acuities in very-low-birth-weight children with and without retinal residua of retinopathy of prematurity. Cryotherapy for Retinopathy of Prematurity Cooperative Group.* Invest Ophthalmol Vis Sci, **36**(3): p. 692–702.
33. Elliot DB, Pesudovs, K, Mallinson T (2007*) Vision related quality of life.* Optom Vis Sci, **84**(8): p. 656–8.
34. Ferris FL 3rd, Kassoff A, Bresnick GH, Bailey I (1982) *New visual acuity charts for clinical research.* Am J Ophthalmol, **94**(1): p. 91–6.
35. Fine I, Wade AR, Brewer AA, May MG, et al. (2003) *Long-term deprivation affects visual perception and cortex.* Nat Neurosci, **6**(9): p. 915–6.
36. Fosse P, Valberg A, Arnliot HM (2001) *Retinal illuminance and the dissociation of letter and grating acuity in age-related macular degeneration.* Optom Vis Sci, **78**(3): p. 162–8.
37. Gekeler F, Zrenner E (2005) *Status of the subretinal implant project. An overview.* Ophthalmologe, **102**(10): p. 941–9.
38. Geruschat DR, Turano KA, Stahl JW (1998) *Traditional measures of mobility performance and retinitis pigmentosa.* Optom Vis Sci, **75**(7): p. 525–37.
39. Green DM, Swets JA (1966) *Signal Detection Theory and Psychophysics*: Wiley, NY.
40. Greenberg RJ (2008), *Visual prosthesis: A review.* Neuromodulation, **3**(3): p. 161–5.
41. Gregory RL (1977) *Eye and Brain. The Psychology of Seeing. 3rd Ed.* Weidenfeld and Nicolson, London.
42. Grove N (2007) *Retinal implant offers spatial vision to blind patients.* Ophthalmology Times. Nov. 1 2007: p. 15.
43. Guralnik JM, Branch LG, Cummings SR, Curb JD (1989) *Physical performance measures in aging research.* J Gerontol, **44**(5): M1410146.
44. Haegerstrom-Portnoy G, Schneck ME, Lott LA, Brabyn JA (2000) *The relation between visual acuity and other spatial vision measures.* Optom Vis Sci, **77**(12): p. 653–62.
45. Haymes S, Guest D, Heyes A, Johnston A (1996) *Mobility of people with retinitis pigmentosa as a function of vision and psychological variables.* Optom Vis Sci, **73**(10): p. 621–37.
46. Haymes SA, Johnston AW, Heyes AD (2001) *The development of the Melbourne Low-Vision ADL Index: A measure of disability.* Invest Ophthalmol Vis Sci, **42**(6): p. 1215–25.
47. Haymes SA, Johnston AW, Heyes AD (2002) *Relationship between vision impairment and ability to perform activities of daily living.* Ophthalmic Physiol Opt, **22**(2): p.79–91.
48. Holladay JT (2004) *Visual acuity measurements.* J Cataract Refract Surg, **30**(2): p. 287–90.
49. Humayun MS, de Juan E, Jr., del Cerro M, Dagnelie G, et al. (2000) *Human neural retinal transplantation.* Invest Ophthalmol Vis Sci, **41**(10): 3100–6.
50. Humayun MS, Prince M, de Juan E Jr., Barron Y, et al. (1999) *Morphometric analysis of the extramacular retina from postmortem eyes with retinitis pigmentosa.* Invest Ophthalmol Vis Sci, **40**(1): p. 143–48.

51. Humayun MS, Weiland JD, Fujii GY, Greenberg R, et al. (2003) *Visual perception in a blind subject with a chronic microelectronic retinal prosthesis.* Vis Res, **43**(24): p. 2573–81.

52. Kanizsa G (1976) *Subjective contours.* Sci Am, **234**(4): p. 48–52.

53. Kiser AK, Dagnelie G, Schuchard RA, Pollack JS, et al. (2006) *Changes in visual field among subjects implanted with the Optobionics' ASR^{TM} Device.* Invest Ophthalmol Vis Sci, **47**: ARVO E-Abstract 3213.

54. Kuyk T, Elliott JL (1999) *Visual factors and mobility in persons with age-related macular degeneration.* J Rehabil Res Dev, **36**(4): p. 303–12.

55. Kuyk T, Elliott JL, Fuhr PS (1998) *Visual correlates of mobility in real world settings in older adults with low vision.* Optom Vis Sci, **75**(7): p. 538–47.

56. Leat SJ, Lovie-Kitchin JE (2006) *Measuring mobility performance: Experience gained in designing a mobility course.* Clin Exp Optom, **89**(4): p. 215–28.

57. Legge GE (1993) *The role of contrast in reading: Normal and low vision.* Shapley R, Lam DM-K, Editors. *Contrast Sensitivity*: MIT Press, Cambridge.

58. Liu L, Kuyk T, Fuhr P (2007) *Visual search training in subjects with severe to profound low vision.* Vision Res, **47**(20): p. 2627–36.

59. MacMillan NA, Creelman CD (2004) *Detection Theory: A User's Guide. 2nd Ed.* Psychology Press, NY.

60. Mahadevappa M, Weiland JD, Yanai D, Fine I, et al. (2005) *Perceptual thresholds and electrode impedance in three retinal prosthesis subjects.* IEEE Trans Neural Syst Rehabil Eng, **13**(2): p. 201–6.

61. Mallinson T (2007) *Why measurement matters for measuring patient vision outcomes.* Optom Vis Sci, **84**(8): p. 675–82.

62. Mangione CM, Lee PP, Gutierrez PR, Spritzer K, et al. (2001) *Development of the 25-item National Eye Institute Visual Function Questionnaire.* Arch Ophthalmol, **119**(7): 1050–8.

63. Mangione C, Phillips R, Seddon J, et al. (1992) *Development of the "Activities of Daily Vision Scale."* Med Care, **30**: p. 1111–26.

64. Marc RE, Jones BW, Watt CB, Strettoi E (2003) *Neural remodeling in retinal degeneration.* Prog Retin Eye Res, **22**(5): p. 607–55.

65. Margalit E, Maia M, Weiland JD, Greenberg RJ, et al. (2002) *Retinal prosthesis for the blind.* Surv Ophthalmol, **47**(4): p. 335–56.

66. Marron JA, Bailey IL (1982) *Visual factors and orientation: Mobility performance.* Am J Optom Physiol Opt, **59**(5): p. 413–26.

67. Massof RW (2002) *The measurement of visual disability.* Optom Vis Sci, **79**(8): p. 516–52.

68. Massof RW (2007) *An interval-scaled scoring algorithm for visual function questionnaires.* Optom Vis Sci, **84**(8): p. 689–704.

69. Maynard EM (2001) *Visual prostheses.* Annu Rev Biomed Eng, **3**: p. 145–68.

70. Merabet LB, Rizzo JF 3rd, Pascual-Leone A, Fernandez E (2007) *'Who is the ideal candidate?': Decisions and issues relating to visual neuroprosthesis development, patient testing and neuroplasticity.* J Neural Eng, **4**(1): p. S130–5.

71. Mueller V, Wang L, Ostrin LA, Barnett GD, et al. (2007) *Meander mazes: Eye-hand coordination in simulated prosthetic vision.* Invest Ophthalmol Vis Sci, **48**: ARVO E-abstract #2548.

72. Nilsson UL, Frennesson C, Nilsson SE (2003) *Patients with AMD and a large absolute central scotoma can be trained successfully to use eccentric viewing, as demonstrated in a scanning laser ophthalmoscope.* Vision Res, **43**(16): p. 1777–87.

73. Orel-Bixler D, Haegerstrom-Portnoy G, Hall A (1989) *Visual assessment of the multiply X handicapped patient.* Optom Vis Sci, **66**(8): p. 530–6.

74. Ostrovsky Y, Andalman A, Sinha P (2006) *Vision following extended congenital blindness.* Psychol Sci, **17**(12): p. 1009–14.

75. Owsley C, Sloane ME (1987) *Contrast sensitivity, acuity, and the perception of 'real-world' targets.* Br J Ophthalmol, **71**(10): p. 791–96.

76. Pelli DG, Robson JG (1991) *Are letters better than gratings?* Clin Vis Sci, **6**: p. 409–11.

77. Pesudovs K, Burr JM, Harley C, Elliott DB (2007) *The development, assessment, and selection of questionnaires.* Optom Vis Sci, **84**(8): p. 663–74.

78. Rizzo JF III, Wyatt J (1997) *Prospects for a visual prosthesis.* Neuroscientist, **3**(4): p. 251–62.
79. Rubin GS (2001) *Prevalence of visual disabilities and their relationship to visual impairments.* Massof RW, Lidoff L, Editors. *Issues in Low Vision Rehabilitation: Service Delivery, Policy, and Funding*: American Foundation for the Blind Press, New York, p. 27–38.
80. Rubin GS, Roch KB, Prasada-Rao P, Fried LP (1994) *Visual impairment and disability in older adults.* Optom Vis Sci, **71**(12): p. 750–60.
81. Santos A, Humayun MS, de Juan E Jr, Greenburg RJ, et al. (1997) *Preservation of the inner retina in retinitis pigmentosa. A morphometric analysis.* Arch Ophthalmol, **115**(4): p. 511–15.
82. Schneck ME (2007) *Visual Prostheses: Towards Standards for Outcomes and Their Evaluation.* Special Interest Group (SIG) Session at the Association for Research in Vision and Ophthalmology (ARVO) Annual Meeting: May 9, 2007. Ft. Lauderdale, FL. Organized by M.E. Schneck.
83. Schneck ME (Organizer) (2007) *Visual Prostheses: Towards Standards for Outcomes and Evaluation II.* The Smith-Kettlewell Eye Research Institute Symposium: October 12–14, 2007.
84. Schulze-Bonsel K, Feltgen N, Burau H, Hansen L, et al. (2006) *Visual acuities "hand motion" and "counting fingers" can be quantified with the Freiburg Visual Acuity Test.* Invest Ophthalmol Vis Sci, **47**(3): p. 1236–40.
85. Sloan LL (1969) *Variation of acuity with luminance in ocular diseases and anomalies.* Doc Ophthalmol, **26**: p. 384–93.
86. Sloane ME, Ball K, Owsley C, Bruni JR, et al. (1992) *The Visual Activities Questionnaire: Developing an instrument for assessing problems in everyday visual tasks.* Noninvasive Assessment of the Visual System: Technical Digest Series. Optical Society of America, **1**: 26–9.
87. Sommerhalder J, Rappaz B, de Haller R, et al.(2004) *Simulation of artificial vision: II. Eccentric reading of full-page text and the learning of this task.* Vision Res, **44**: p. 1693–706.
88. Stelmack JA, Massof RW (2007) *Using the VA LV VRQ-48 and LV VFQ-20 in low vision rehabilitation.* Optom Vis Sci, **84**(8): p. 705–9.
89. Stelmack J, Szlyk JP, Stelmack T, Demers-Turco P, et al. (2006) *Measuring outcomes of vision rehabilitation with the Veterans Affairs Low Vision Visual Functioning Questionnaire.* Invest Ophthalmol Vis Sci, **47**: p. 3253–61.
90. Stelmack JA, Tang XC, Reda DJ, Moran D, et al. (2007*) The Veterans Affairs Low Vision Intervention Trial (LOVIT): Design and methodology.* Clinical Trials, **4**(6): p. 650–60.
91. Stone JL, Barlow WE, Humayun MS, de Juan E Jr, Milam AH (1992) *Morphometric analysis of macular photoreceptors and ganglion cells in retinas with retinitis pigmentosa.* Arch Ophthalmol, **110**(11): p. 1634–9.
92. Sunness JS, Rubin GS, Broman A, Applegate CA, Bressler NM, Hawkins BS (2008) *Low luminance visual dysfunction as a predictor of subsequent visual acuity loss from geographic atrophy in age-related macular degeneration.* Ophthalmology, **115**(9): p. 1480–8.
93. Szlyk JP, Seiple W, Fishman GA, Alexander KR, et al. (2001) *Perceived and actual performance of daily tasks: Relationship to visual function tests in individuals with retinitis pigmentosa.* Ophthalmology, **108**(1): p. 65–75.
94. Thibos LN, Cheney RE, Walsh DJ (1987) *Retinal limits to the detection and resolution of gratings.* J Opt Sci Am A, **4**(8): p. 1524–29.
95. Thibos LN, Walsh DJ (1985) *Detection of high frequency gratings in the periphery.* J Opt Soc Am A, **2**: p. 64.
96. Thompson RJ Jr, Barnett GD, Humayun MS, Dagnelie G (2003) *Facial recognition using simulated prosthetic pixelized vision.* Invest Ophthalmol Vis Sci, **44**(11): p. 5035–42.
97. Traquair HM (1946) *Introduction to Clinical Perimetry*: Mosby, St. Louis.
98. Troyk PR, Bradley D, Bak M, Cogan S, et al. (2005) *Intracortical visual prosthesis research: Approach and progress.* Conf Proc IEEE Eng Med Biol Soc. Shanghai, China, Sept. 1–4, p. 7376–79.
99. Turano KA, Geruschat DR, Stahl JW, Massof RW (1999) *Perceived visual ability for independent mobility in persons with retinitis pigmentosa.* Invest Ophthalmol Vis Sci, **40**(5): p. 865–77.

100. Turano KA, Rubin GS, Quigley HA (1999) *Mobility performance in glaucoma.* Invest Ophthalmol Vis Sci, **40**(12): p. 2803–9.
101. Turano KA, Yu D, Hao L, Hicks JC (2005) *Optic-flow and egocentric-direction strategies in walking: Central vs. peripheral visual field.* Vision Res, **45**(25–26): p. 3117–32.
102. Turco PD, Connolly J, McCabe P, Glynn RJ (1994) *Assessment of function vision performance: A new test for low vision patients.* Ophthalmic Epidemiol, **1**(1): p. 15–25.
103. Tyler CW, Apkarian P, Levi DM, Nakayama K (1979) *Rapid assessment of visual function: An electronic sweep technique for the pattern visual evoked potential.* Invest Ophthalmol Vis Sci, **18**(7): p. 703–13.
104. Velikay-Parel M, Ivastinovic D, Koch M, Hornig R, et al. (2007) *Repeated mobility testing for later artificial visual function evaluation.* J Neural Eng, **4**(1): p. S102–7.
105. Veraart C, Raftopoulos C, Mortimer JT, Delbeke J, et al. (1998) *Visual sensations produced by optic nerve stimulation using an implanted self-sizing spiral cuff electrode.* Brain Res, **813**(1): p. 181–6.
106. Walter M, Yang L, Dagnelie G (2007) *Prosthetic vision simulation in fully and partially sighted individuals.* Humayun MS, Weiland JD, Chader G, Greenbaum E, Editors. *Artificial Sight: Basic Research, Biomedical Engineering, and Clinical Advances*: Springer, New York.
107. Watson T, Orel-Bixler D, Haegerstrom-Portnoy G (2007) *Longitudinal quantitative assessment of vision function in children with cortical visual impairment.* Optom Vis Sci, **84**(6): p. 471–80.
108. Weiland JD, Liu W, Humayun MS (2005) *Retinal prosthesis.* Annu Rev Biomed Eng, **7**: p. 361–401.
109. West SK, Rubin GS, Broman AT, Munoz B, et al. (2002) *How does visual impairment affect performance on tasks of everyday life? The SEE Project. Salisbury Eye Evaluation.* Arch Ophthalmol, **120**(6): p. 774–80.
110. White JM, Loshin DS (1989) *Grating acuity overestimates Snellen acuity in patients with age-related maculopathy.* Optom Vis Sci, **66**(11): p. 751–5.
111. Whittaker SG, Lovie-Kitchin J (1993) *Visual requirements for reading.* Optom Vis Sci. **70**(1): p. 54–65.
112. Wilke R, Bach M, Wilhelm B, Durst W, et al. (2007) *Testing visual functions in patients with visual prostheses.* Humayun MS, Weiland JD, Chader G, Greenbaum E, Editors, *Artificial Sight*: Springer.
113. Williams DR (1985) *Aliasing in human foveal vision.* Vis Res, **25**(2): p. 195–205.
114. Williams DR (1985) *Visibility of interference fringes near the resolution limit.* J Opt Soc Am A, **2**(7): p. 1087–93.
115. Wood JM, Dique T, Troutbeck R (1993) *The effect of artificial visual impairment on functional visual fields and driving performance.* Clin Vis Sci, **8**: p. 563–75.
116. Wood JM, Troutbeck R (1995) *Elderly drivers and simulated visual impairment.* Optom Vis Sci, **72**(2): p. 115–24.
117. Yanai D, Lakhanpal RR, Weiland JD, Mahadevappa M, et al. (2003) *The value of preoperative tests in the selection of blind patients for a permanent microelectronic implant.* Trans Am Ophthalmol Soc, **101**: p. 223–8.
118. Yanai D, Weiland JD, Mahadevappa M, Greenberg RJ, et al. (2007). *Visual performance using a retinal prosthesis in three subjects with retinitis pigmentosa.* Am J Ophthalmol, **143**(5): p. 820–7.

Chapter 21
Activities of Daily Living and Rehabilitation with Prosthetic Vision

Duane R. Geruschat and James Deremeik

Abstract Now that technology has the capability to provide ultra-low vision to individuals who are functionally blind, there is a recognized need for vision rehabilitation to become part of the process of adaptation. This chapter will present concepts of rehabilitation as they relate to prosthetic vision, describe approaches to evaluation and instruction, address issues related to measuring outcomes, and offer thoughts on the future of rehabilitation for individuals with prosthetic vision.

The purpose of this chapter is to describe the challenges and opportunities of prosthetic vision in the context of using such vision for activities of daily living and to propose rehabilitation techniques that could assist patients as they adapt and integrate prosthetic vision into their lives. The chapter will be divided into four sections: Concepts of Functional Vision and Rehabilitation, Evaluation and Intervention with Prosthetic Vision, Measuring Functional Outcomes, and The Future.

Abbreviations

ADL	Activities of daily living
CCTV	Closed circuit television
ETA	Electronic travel aid
O&M	Orientation and mobility

D.R. Geruschat (✉)
Lions Vision Research & Rehabilitation Center, Wilmer Eye Institute, Johns Hopkins
University School of Medicine, 550 N. Broadway, 6th Floor, Baltimore, MD 21205, USA
e-mail: dgeruschat@jhmi.edu

G. Dagnelie (ed.), *Visual Prosthetics: Physiology, Bioengineering, Rehabilitation,*
DOI 10.1007/978-1-4419-0754-7_21, © Springer Science+Business Media, LLC 2011

21.1 Concepts of Functional Vision and Rehabilitation

21.1.1 Application to Orientation and Mobility

Ultra-low vision is at the lower end of the clinical visual acuity continuum, which includes light perception, light projection, and form perception; it can have a functional impact on individuals with visual impairments by enhancing or improving their orientation and mobility (O&M) skills. For example, when walking the halls of a residential school for students who are blind it is not uncommon to see two to three totally blind students holding the arm of and following behind the one student who has form perception. The lead student, using form perception, can see the lights in the ceiling and visually trails the lights to maintain a straight line of travel down the corridor. Another utility of ultra-low vision is demonstrated by the fully sighted person who wakes in a hotel room in the night and uses the moonlight shining through a gap in the curtain or the ambient light from the alarm clock to orient himself and locate the entry to the bathroom without turning on a light that might disturb his spouse.

The ability of a person with ultra-low vision to visually detect contrast can enhance her awareness of her location in a room. The left panel of Fig. 21.1 shows a white door in a white room (low contrast); the right panel shows the same door but with a dark-colored robe on the door's hook (high contrast), which makes the door easier to identify visually. In this example, a simple environmental feature (the placement of a robe) can enhance movement through the room for a person with ultra-low vision. The benefits afforded by the ability to perceive light or see contrasting colors illustrates why we believe that prosthetic vision can be useful for orientation and mobility (O&M).

Fig. 21.1 Effect of contrast on visibility: a dark robe on a light door

Although there are a variety of technological approaches to providing an individual with prosthetic vision, when the technology allows the individual to reverse contrast, this feature may enhance the individual's ability to detect objects and the like. Many patients who utilize a closed circuit television (CCTV), for example, prefer to do so by making the letters white and the background black (that is, by reversing the contrast of the monitor). This technique could be applied to mobility, for finding a doorway out of a well-lit room, if the individual using prosthetic vision reversed contrast to show a bright door opening in a dark wall.

The foregoing examples described potential enhancements to orientation with no descriptions of benefits to mobility, because the current prosthetic vision technology is not sufficient to replace or eliminate the need for a long cane or a guide dog for independent travel in unfamiliar environments. The point is that ultra-low vision can have a positive impact on functional orientation but not on mobility in novel environments.

Today's prosthetic vision may be potentially safe enough for an individual to use it as their primary source of mobility information (no cane or guide dog) indoors in a controlled and familiar space or when locating furniture or objects with high contrast. A "controlled space" is an indoor environment in which changes in elevation (stairs) are not present or their location is known, and in which furniture and other room elements maintain the same location over time. Travel in unknown and/or complex environments (crossing the street or walking in a shopping mall) requires the use of a long cane or guide dog. In such a situation, prosthetic vision can be used as a supplementary source of information to enhance the individual's orientation while other sensory information (audition, tactual) is combined with primary sources of information for mobility: the long cane or guide dog.

21.1.2 Application for Activities of Daily Living

Because prosthetic vision provides very low levels of visual acuity, activities of daily living (ADLs) that require detailed vision (sewing, reading, or the recognition of facial features) are not envisioned as being amenable to prosthetic visual rehabilitation until the level of resolution the technology provides has been substantially improved. The opportunity presented by the current technology, which allows users to perceive high contrast can be of benefit with a variety of ADLs, including personal care and personal management. For example, in the area of personal care, the ability to identify toothpaste on a toothbrush might be accomplished with the use of high contrast (green toothpaste on a white-bristled toothbrush). Visually locating soap or a shampoo bottle in a bathtub may be possible with high contrast. The ability to apply lipstick may be enhanced with ultra-low vision. The ability to visually sort dark- and light-colored socks, to identify a white shirt from a dark-colored shirt may be possible with prosthetic vision. The use of contrasting colors in the kitchen could prove to be beneficial for people with prosthetic vision, who may be able to

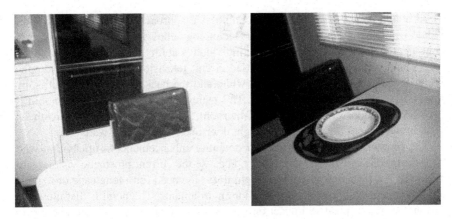

Fig. 21.2 Effect of contrast on visibility: a dark placemat and a light dish

differentiate milk from juice or mayonnaise from ketchup in the refrigerator. The use of high contrast between a dark placemat and a light-colored dish can enhance a client's ability to locate the dish, as illustrated in Fig. 21.2.

21.1.3 Patient Lifestyle and Expectations

Early chapters of this book concentrated on the visual system to the exclusion of personal history. As rehabilitation specialists, we think of vision in the context of the person, their history, lifestyle, expectations, and acknowledge that these personal elements influence the way vision is used. Consider two patients who have the same clinical vision status (visual acuity, contrast sensitivity, visual field); one patient uses a cane and minimizes the use of their remaining vision, and the other travels without a cane, and utilizes optical equipment to read street signs and view traffic lights. Individuals who maximize the use of their remaining vision (light perception) prior to implant tend to have the best prognosis for integrating prosthetic vision into their lifestyle and to experience more benefits after implantation.

The management of patient expectations is a key element to successful functional outcomes with prosthetic vision and must be considered part of the rehabilitation process. Patients want to know how their life will change or improve with prosthetic vision. Will prosthetic vision resolve their functional problems? Research on the most common functional problems in mobility clearly shows that managing illumination (light adaptation, low-light environments), detecting changes in elevation such as drop-offs (curbs, stairs), and crossing the street are three of the most common problems for patients with low vision [4]. Our experience with prosthetic vision suggests that these leading low vision mobility problems may not be addressed by the current technologies. The implant wearers we have seen at the time this chapter is being written are, in the context of O&M, quite similar. They are all independent travelers who use some combination of long cane, guide dog, and/or remaining vision. They travel in familiar and unfamiliar areas, ride public transportation, and do not report a serious limitation to their independent travel because of the loss of vision. Because the prosthetic vision systems we have worked with provide ultra-low vision, we have not identified anyone for whom prosthetic vision has been sufficient to replace the long cane or the guide dog when walking in unfamiliar areas. Until the technology improves visual acuity and visual field, prosthetic vision is viewed as an additional travel aid, an enhancement to travel, specifically orientation, rather than a substitution system that would supplant the primary travel aid. Therefore, specialized instruction from properly trained rehabilitation professionals will benefit a prosthetic vision program.

21.1.4 Congenital and Adventitious Vision Loss

We assume there is an interest in offering prosthetic vision to those with congenital blindness. There is a significant difference between the visual abilities and the psychological adjustment process of someone with adventitious vision loss who has had his sight restored (cataract extraction, corneal transplant) and an adult with congenital vision loss who has been provided sight for the first time as an adult. Personal accounts such as the experience of Mike May, described in the book *Crashing Through* [7], show that clearing the optical pathway does not result in an immediate improvement in functional vision for someone who is congenitally blind. In fact, the more common experience involves a sense of being overwhelmed and confused [6]. It is important to recognize that an individual who has lived as a blind person does not suddenly benefit from visual input. If the patient has a congenital vision loss, a significant period of adaptation, learning to interpret this novel sensory input and time to develop a visual memory, will be required. If the patient lost vision later in life, the age when it was lost, their ability to use low levels of vision as their vision gradually decreased, the primary learning modality for gathering information from the environment (visual, tactual, auditory), and the amount of remaining visual memory are a few of the issues to consider before implantation. These issues may also become a part of the screening and selection process for those who will participate in any type of prosthetic vision rehabilitation program because different strategies may be needed when providing rehabilitation training to these individuals.

Another often repeated description from those who gained sight as adults involves the amount of effort required to process visual input. Mike May describes the need to close his eyes to process information and to feel calm [7]. As revealed in May's experiences, there are unknown challenges that await the patient who gains sight after leading the life of someone who is congenitally blind.

21.2 Evaluation and Intervention with Prosthetic Vision

21.2.1 Evaluation

The process of rehabilitation always begins with an evaluation. The evaluation of prosthetic vision should begin with an assessment of functional vision [2, 8, 9]. The typical functional vision assessment is hierarchical and begins with evaluating the ability to respond to a light source, determining if the patient can localize, fixate, track, and scan the light. This is followed by an assessment of visual motor skills then higher order perceptual skills (color identification; three- and two-dimensional shape recognition; and symbol, letter, and word recognition).

The standardized functional vision assessment is followed by a comprehensive evaluation of the use or non-use of vision for mobility [4], personal care, and/or personal management. A few examples of this assessment include the ability to detect changes in contrast of open doorways, the ability to locate windows, to visually trace the lights in a hallway, and the ability to identify the location of a white napkin on a dark table.

Information gained from such assessments can be useful for developing an intervention program. For example, let's assume one patient with no residual vision travels independently without a long cane or guide dog, but only in her home and within a fenced back yard. The patient has small children and frequently steps on or kicks toys and bicycles that are left on the floor or grass. During the evaluation it is determined that the goal for this patient is to improve her ease of travel in the home and around the yard through improving her ability to locate high-contrast objects. Prosthetic vision (form perception) could enhance this patient's life by reducing the frequency with which she kicks her children's toys. The intervention may involve teaching the patient scanning techniques to locate obstacles. It may also be necessary to teach the patient how to interpret prosthetic vision, to essentially determine the identity of the low-resolution images her prosthetic vision provides. This patient's vision rehabilitation program will emphasize integrating prosthetic with other sensory information to reduce the mental effort that occurs when she experiences a new type of sensory input.

Another patient with minimal residual vision travels independently on public transportation using a long cane supplemented by light projection. Prosthetic vision for this patient may provide form perception and the ability to differentiate areas of high contrast such as the sidewalk (light) from the pavement of the street (dark). This patient would be introduced to the same concepts that are currently taught to someone with severe low vision. Specifically, instruction would address the issue of when it is safe to use vision only and when vision is only safe to supplement the use of the long cane. This is one of the more challenging skills to acquire. It is difficult to always know when vision alone can be the primary modality and provides sufficient information for making decisions about safety.

21.2.2 Intervention

As the technology improves, allowing for sharper visual acuity, contrast sensitivity, and an increase of peripheral visual fields, we expect that some type of instruction to enhance the use of prosthetic vision will be beneficial. We assume there would be two distinct approaches to instruction, one approach for patients who present with congenital blindness (no visual memory) and an approach for patients who lost vision later in life.

The descriptions from prior decades on the effect of adult onset of vision could be instructive for anticipating and understanding the challenges to be faced by an adult with late onset of vision [3]. We would expect to observe a patient with limited ability to comprehend what they were seeing. The work of Mary Anne Frostig and Natalie Barraga during the 1960s [1], specifically their instructional procedures that follow the process of visual perceptual development, would be a useful place to begin instruction.

The large body of literature on instructional strategies for children with low vision, as well as the literature on visual perceptual instruction that has evolved during the past 40 years, provides useful concepts and instructional sequences that could be adapted for an adult population. Since the adult patient with congenital blindness and adult onset of vision is functioning visually at an earlier developmental level, materials written for children may prove to be useful. The American Printing House for the Blind has a collection of materials such as Bright Sights that are designed primarily for children with low vision. The materials provide lesson plans of sequential lessons as well as assessment tools to monitor progress. For use with adults, the materials and lessons would need to be modified to be age appropriate. Isolating the visual system in the early developmental stages before integrating visual information through a multisensory approach has been demonstrated to be effective with young children. Experience will be required to determine if this same approach would be useful for adults with congenital blindness and adult onset of vision.

For adults who have visual memories, early visual developmental skills should still be present or could be re-acquired fairly quickly. Intervention strategies for these adults could include the introduction of specific visual skills such as establishing a consistent response to a given visual stimulus (type of light source), systematic scanning to localize objects, fixate on the object, tracking and shifting gaze between objects, as well as some perceptual training to learn to (re)interpret the visual world. Instruction may also involve teaching low vision skills such as scanning to define the borders of an area, practicing the important ability of being systematic, and scanning for objects in the direction perpendicular to their primary orientation [4].

Localizing and fixating are specific techniques that can be impaired by ocular pathology. We do not know the effects of prosthetic vision on these skills. There may be a need to introduce eccentric viewing to improve visual clarity, as well as the concept of turning the head to an eccentric position to improve visual ability.

Tracking and shifting gaze are important skills for personal care and mobility. For example, applying make-up involves eye-hand coordination and the ability to shift gaze from a cosmetic to the reflection of one's face in the mirror. In mobility the ability to track a moving vehicle to assess time-to-contact (impact) and shifting gaze to acquire information at a two-way street crossing are common mobility tasks.

21.3 Measuring Functional Outcomes

It is important to determine the effects of prosthetic vision on functional performance. Did the treatment make a difference? If so, what kind and how much of a difference? Is there a difference in the patient's posture, body position, and/or head position? Did the treatment result in greater safety or was the level of independence improved? Does the patient experience greater visual independence when preparing dinner? Have additional responsibilities in the home been absorbed by the patient such as sorting laundry or selecting the proper placemat and plate while setting the dinner table?

It is critical to understand what each patient's goal is for prosthetic vision. If the goal is for prosthetic vision to replace or supplement the individual's need to travel with a long cane, then the resulting prosthetic vision will need to be of sufficient resolution to allow the person to reliably and safely detect changes in elevation (curbs, stairs), confidently travel in a variety of light levels, accurately detect objects in the travel path, and consistently identify changes in the travel surface (gravel, grass, pavement). If the goal is simply to enhance the currently existing travel skills, then the resolution provided by the prosthetic vision can be more modest.

Another prosthetic vision recipient may have sufficient functional vision to travel safely in familiar indoor and outdoor areas, but use a long cane when traveling in unfamiliar areas. In this example, the question is what type and how much prosthetic vision will be required to improve travel skills (light perception to light projection or hand motion)?

Assuming the subject has some amount of independent mobility prior to the treatment, the effect will either be to change the mode of travel (non-visual to visual) or to enhance the current approach to travel. Herein is the challenge. If the patient is an independent traveler prior to the introduction of prosthetic vision, prosthetic vision may not increase their level of independence. The best possible outcome would be an enhancement of independent travel, that is, improved orientation in unfamiliar areas. It is quite difficult in general to measure the enhancement of travel. If the expected outcome of prosthetic vision is to change the mode of travel (cane or guide dog to vision only), the technology is not yet capable of providing that level of visual input. Both examples present challenges to measuring outcomes. For example, if the patient is an independent cane traveler or guide dog user, taking the cane or dog away may result in a degradation of performance until the new visual approach has been mastered, assuming the future of prosthetic vision provides for better visual acuity and visual field. In this example, a degradation of performance would be expected for the short term with the hope that performance will improve as mastery of prosthetic vision occurs. However, even in this example, the best we can hope for is for the patient to transition from being an independent blind traveler to an independent visual traveler. When vision is considered as an enhancement to the current mode of travel the measurement challenge is more extreme. Attributes such as ease, anticipation, or previewing are difficult to quantify.

One approach to measuring mobility performance would be to isolate the visual aspects of mobility and to concentrate testing on these visual tasks. For example, walking down a hallway while following the ceiling lights could be compared to walking down the hallway without the ceiling lights, identifying the location of windows in a room, or visually identifying the fourth intersecting sidewalk are all discrete elements of O&M that can only be done visually.

Another approach to measuring mobility performance with prosthetic vision is to measure mental effort. The underlying assumptions of this approach are:

1. Patients with low vision rarely bump into obstacles
2. It requires more cognitive attention to the environment to travel with no vision than low vision
3. This attention can be measured as mental effort

Experiments have shown that measures of mental effort through the use of a secondary task are responsive to variation in environmental complexity [10] and to varying extent of visual field [5]. Assessing mental effort may be an approach with potential for measuring outcomes with prosthetic vision. We assume, however, that the introduction of prosthetic vision would itself impose a secondary task and at least initially result in an increase of mental effort until the patient adapted to the new visual input. We have observed patients with recent implants whose performance is initially degraded as they adjust to the prosthetic vision. At times we have also observed that patients ignore other sensory information as they strive to utilize their new vision. Time is required for patients to complete their adjustment to the prosthesis and reintegrate all the sensory information it provides.

In the context of ADLs, the issue of independence of people who are blind is also present. Since many patients can perform ADLs without sight, it is challenging to measure the effects of a prosthetic implant on living skills. One approach is to isolate the visual elements of a task and to concentrate the performance measure on those specific elements. For example, the identification of a white shirt may be done tactually via the feel of the cloth, the location of the shirt in a closest, or a tactual marking on the inside of the collar. However, it may also be possible to sort shirts based upon visual input, separating white shirts from dark shirts, and this is what should be evaluated in people with prosthetic vision. Another example is the height of flame on a stove. The common approach is to attach tactual markers to the stove dials that indicate the relative height of the flame. Prosthetic vision may be used to visually detect the flame.

21.4   The Future

Although prosthetic vision does not currently offer those who receive it much assistance with tasks that require good functional visual acuity, we anticipate that the next generation of this technology, if it offers a modest improvement in visual acuity, may be of benefit for tasks that require high contrast and do not require good visual acuity. For example, a black hairbrush on a white countertop in the bathroom might be located through the use of artificial sight. In the kitchen, counting and placing strawberries on a white cutting board, working with peeled white potatoes on a dark cutting board, locating food such as beef on a white plate, determining how much liquid is in a glass (see Fig. 21.3), and sorting laundry into dark and light piles are activities that could be aided by the use of artificial vision.

Anticipating that the technology will improve over the next decade, we would expect that tasks requiring higher levels of visual acuity will become possible through the use of prosthetic vision.

Prosthetic vision has been described as being analogous to cochlear implants. Although this analogy may prove to be accurate, we provide a different analogy that offers a word of caution. Electronic travel aids (ETAs) have been designed to provide information about the environment with the goal of offering an improved preview of the environment. Common examples of ETAs include the laser cane and the sonic guide. These two devices provide detailed information about the environment. Users of such devices find that they compete with the naturally occurring sensory information and tend to only use them in specific and isolated situations. For example, the sonic guide is effective for following the barrier that separates paid customers from the general public in a subway station, allowing the user of the sonic guide to follow the barrier without touching it. In other situations such as a grocery store, users describe the overwhelming amount of information it provides

Fig. 21.3 Effect of contrast on visibility: dark liquid in a clear glass

as competing with the naturally occurring sensory input. Mike May's descriptions of his experience with regaining vision are similar to the reactions of users of the sonic guide in sensory-rich environments. We wonder if improvements in the level of resolution of prosthetic vision will result in clients receiving too much visual information to process, inhibiting their functional performance.

As the quality and quantity of prosthetic vision improves, we assume that patients who are congenitally blind will receive implants in greater numbers. The history of adult-onset vision, as previously mentioned, suggests there will be challenges. A few of these may include

- Learning to interpret the image
- Adapting to the new sensory input
- Integrating vision into a lifestyle of non-visual independence

We believe it will be necessary to educate low vision rehabilitation service providers in how to work with individuals who have prosthetic vision. We do not know if the standard low vision rehabilitation techniques will apply to this population or if entirely new strategies will need to be developed or if the modifications to the existing approaches is all that will be required. Low vision rehabilitation professionals working with recipients of prosthetic vision will need to be familiar and competent with non-visual techniques and strategies in the performance of ADL tasks as these skills may continue to be essential for the individual with prosthetic vision. We have emphasized addressing mobility skills with this population, but as the technology improves applications for near-point activities (reading and writing) may also involve changes to the current regime of rehabilitation strategies. Assuming the technology will ultimately be funded by third-party payers, if the rehabilitation strategies are highly specialized, it will be necessary to add new certification requirements and a greater body of knowledge including the non-visual strategies and techniques for rehabilitation specialists serving this population.

In conclusion, the use of prosthetic vision must be understood in the context of the client and his goals, lifestyle, and ability to adapt to change. Realistic expectations and a high level of independence will enhance the chances of a positive outcome with prosthetic vision. Patients who are properly selected to participate in prosthetic vision rehabilitation need extensive education in regard to what prosthetic vision intervention can and cannot do to assure they have realistic expectations, which are vital to the success of rehabilitation. A realistic understanding of the potential of the technology, in combination with rehabilitation instruction, is one key to a successful outcome.

References

1. Barraga, N. C. (Ed.). (1970). *Teacher's guide for development of visual learning abilities and utilization of low vision*. Louisville, KY: American Printing House for the Blind.
2. Corn, A. L., & Erin, J. N. (2010). *Foundations of low vision: Clinical and functional perspectives* (2nd ed.). New York: AFB Press.

3. Fine, I., Wade, A. R., Brewer, A. A., May, M. G., Goodman, D. F., Boynton, G. M., Wandell, B. A., & MacLeod, D. I. A. (2003). The effects of long-term deprivation on visual perception and visual cortex. *Nature Neuroscience, 6*(9), 915–916.
4. Geruschat, D. R., & Smith, A. J. (2010). Low vision for orientation and mobility. In W. R. Wiener, R. L. Welsh, & B. B. Blasch (Eds.), *Foundations of orientation and mobility, vol I. History and theory* (3rd ed., pp. 63–83). New York: AFB Press.
5. Geruschat, D. R., & Turano, K. A. (2007). Estimating the amount of mental effort required for independent mobility: Persons with glaucoma. *Investigative Ophthalmology and Vision Science, 48*(9), 3988–3994.
6. Gregory, R. L. (1997). *Eye and brain: The psychology of seeing* (5th ed.). Princeton, NJ: Princeton University Press.
7. Kurson, R. (2007). *Crashing through: A true story of risk, adventure, and the man who dared to see.* New York: Random House.
8. Lueck, A. H. (2004). *Functional vision: A practitioner's guide to evaluation and intervention.* New York: AFB Press.
9. Silverstone, B., Lang, M. A., Rosenthal, B., & Faye, E. E. (Eds.). (2000). *The lighthouse handbook on vision impairment and vision rehabilitation vol I part VII.* New York: Oxford University Press.
10. Turano, K. A., Geruschat, D. R., & Stahl, J. W. (1998). Mental effort required for walking: Effects of retinitis pigmentosa. *Optometry and Vision Science, 75*(12), 879–886.

Author Index

G. Dagnelie (ed.), *Visual Prosthetics: Physiology, Bioengineering, Rehabilitation*,
DOI 10.1007/978-1-4419-0754-7, © Springer Science+Business Media, LLC 2011

425

Subject Index